DIFFERENTIAL GEOMETRY OF COMPLEX VECTOR BUNDLES

PUBLICATIONS OF THE MATHEMATICAL SOCIETY OF JAPAN

1. The Construction and Study of Certain Important Algebras. By Claude Chevalley.
2. Lie Groups and Differential Geometry. By Katsumi Nomizu.
3. Lectures on Ergodic Theory. By Paul R. Halmos.
4. Introduction to the Problem of Minimal Models in the Theory of Algebraic Surfaces. By Oscar Zariski.
5. Zur Reduktionstheorie Quadratischer Formen. Von Carl Ludwig Siegel.
6. Complex Multiplication of Abelian Varieties and its Applications to Number Theory. By Goro Shimura and Yutaka Taniyama.
7. Équations Différentielles Ordinaires du Premier Ordre dans le Champ Complexe. Par Masuo Hukuhara, Toshihusa Kimura et Mme Tizuko Matuda.
8. Theory of Q-varieties. By Teruhisa Matsusaka.
9. Stability Theory by Liapunov's Second Method. By Taro Yoshizawa.
10. Fonctions Entières et Transformées de Fourier. Application. Par Szolem Mandelbrojt.
11. Introduction to the Arithmetic Theory of Automorphic Functions. By Goro Shimura. (Kanô Memorial Lectures 1)
12. Introductory Lectures on Automorphic Forms. By Walter L. Baily, Jr. (Kanô Memorial Lectures 2)
13. Two Applications of Logic to Mathematics. By Gaisi Takeuti. (Kanô Memorial Lectures 3)
14. Algebraic Structures of Symmetric Domains. By Ichiro Satake. (Kanô Memorial Lectures 4)
15. Differential Geometry of Complex Vector Bundles. By Shoshichi Kobayashi. (Kanô Memorial Lectures 5)

PUBLICATIONS OF THE MATHEMATICAL SOCIETY OF JAPAN
15

DIFFERENTIAL GEOMETRY OF COMPLEX VECTOR BUNDLES

by

Shoshichi Kobayashi

KANÔ MEMORIAL LECTURES 5

Iwanami Shoten, Publishers
and
Princeton University Press
1987

Kanô Memorial Lectures

In 1969, the Mathematical Society of Japan received an anonymous donation to encourage the publication of lectures in mathematics of distinguished quality in commemoration of the late Kôkichi Kanô (1865–1942).

K. Kanô was a remarkable scholar who lived through an era when Western mathematics and philosophy were first introduced to Japan. He began his career as a scholar by studying mathematics and remained a rationalist for his entire life, but enormously enlarged the domain of his interest to include philosophy and history.

In appreciating the sincere intentions of the donor, our Society has decided to publish a series of "Kanô Memorial Lectures" as a part of our Publications. This is the fifth volume in the series.

Clothbound editions of Princeton University Press books are printed on acid-free paper, and binding materials are chosen for strength and durability

Co-published for
The Mathematical Society of Japan
by
Iwanami Shoten, Publishers
and
Princeton University Press

Printed in the United States of America by
Princeton University Press, Princeton, New Jersey

Dedicated to
Professor Kentaro Yano

It was some 35 years ago that I learned from him Bochner's method of proving vanishing theorems, which plays a central role in this book.

Preface

In order to construct good moduli spaces for vector bundles over algebraic curves, Mumford introduced the concept of a stable vector bundle. This concept has been generalized to vector bundles and, more generally, coherent sheaves over algebraic manifolds by Takemoto, Bogomolov and Gieseker. As the differential geometric counterpart to the stability, I introduced the concept of an Einstein-Hermitian vector bundle. The main purpose of this book is to lay a foundation for the theory of Einstein-Hermitian vector bundles. We shall not give a detailed introduction here in this preface since the table of contents is fairly self-explanatory and, furthermore, each chapter is headed by a brief introduction.

My first serious encounter with stable vector bundles was in the summer of 1978 in Bonn, when F. Sakai and M. Reid explained to me the work of Bogomolov on stable vector bundles. This has led me to the concept of an Einstein-Hermitian vector bundle. In the summer of 1981 when I met M. Lübke at DMV Seminar in Düsseldorf, he was completing the proof of his inequality for Einstein-Hermitian vector bundles, which rekindled my interest in the subject.

With this renewed interest, I lectured on vanishing theorems and Einstein-Hermitian vector bundles at the University of Tokyo in the fall of 1981. The notes taken by I. Enoki and published as Seminar Notes 41 of the Department of Mathematics of the University of Tokyo contained good part of Chapters I, III, IV and V of this book. Without his notes which filled in many details of my lectures, this writing project would not have started. In those lectures I placed much emphasis——perhaps too much emphasis——on vanishing theorems. In retrospect, we need mostly vanishing theorems for holomorphic sections for the purpose of this book, but I decided to include cohomology vanishing theorems as well.

During the academic year 1982/83 in Berkeley and in the summer of 1984 in Tsukuba, I gave a course on holomorphic vector bundles. The notes of these lectures ("Stable Vector Bundles and Curvature" in the "Survey in Geometry" series) distributed to the audience, consisted of the better part of Chapters I through V. My lectures at the Tsukuba workshop were supplemented by talks by T. Mabuchi (on Donaldson's work) and by M. Itoh (on Yang-Mills theory). In writing Chapter VI, which is mainly on the work of Donaldson on stable bundles over algebraic surfaces, I made good use of Mabuchi's notes.

During the fall of 1985 in Berkeley, H.-J. Kim gave several seminar talks on moduli of Einstein-Hermitian vector bundles. Large part of Chapter VII is based on his Berkeley thesis as well as Itoh's work on moduli of anti-self-dual connections on Kähler surfaces. While I was revising the manuscript in this final form, I had occasions to talk with Professor C. Okonek on the subject of stable bundles, and I found discussions with him particularly enlightening.

In addition to the individuals mentioned above, I would like to express my gratitude to the National Science Foundation for many years of financial support, to Professor F. Hirzebruch and Sonderforschungsbereich in Bonn, where I started my work on Einstein-Hermitian vector bundles, and to Professors A. Hattori and T. Ochiai of the University of Tokyo and the Japan Society for Promotion of Sciences for giving me an opportunity to lecture on holomorphic vector bundles. I would like to thank also Professor S. Iyanaga for inviting me to publish this work in Publications of the Mathematical Society of Japan and Mr. H. Arai of Iwanami Shoten for his efficient co-operation in the production of this book.

February, 1986

S. Kobayashi

Contents

Chapter I

Connections in vector bundles

Although our primary interest lies in holomorphic vector bundles, we begin this chapter with the study of connections in differentiable complex vector bundles. In order to discuss moduli of holomorphic vector bundles, it is essential to start with differentiable complex vector bundles. In discussing Chern classes it is also necessary to consider the category of differentiable complex vector bundles rather than the category of holomorphic vector bundles which is too small and too rigid.

Most of the results in this chapter are fairly standard and should be well known to geometers. They form a basis for the subsequent chapters. As general references on connections, we mention Kobayashi-Nomizu [1] and Chern [1].

§ 1 Connections in complex vector bundles (over real manifolds)

Let M be an n-dimensional real C^∞ manifold and E a C^∞ complex vector bundle of rank (=fibre dimension) r over M. We make use of the following notations:

A^p = the space of C^∞ complex p-forms over M,

$A^p(E)$ = the space of C^∞ complex p-forms over M with values in E.

A *connection* D in E is a homomorphism

$$D\colon A^0(E) \longrightarrow A^1(E)$$

over $\mathbf{C}$ such that

$$(1.1) \qquad D(f\sigma) = \sigma df + f \cdot D\sigma \qquad \text{for} \quad f \in A^0, \quad \sigma \in A^0(E),$$

Let $s = (s_1, \cdots, s_r)$ be a local frame field of E over an open set $U \subset M$, i.e.,

i) $s_i \in A^0(E|_U) \qquad i = 1, \cdots, r,$

ii) $(s_1(x), \cdots, s_r(x))$ is a basis of E_x for each $x \in U$.

Then given a connection D, we can write

$$(1.2) \qquad Ds_i = \sum s_j \omega_i^j, \qquad \text{where} \quad \omega_i^j \in A^1|_U$$

We call the matrix 1-form $\omega = (\omega_i^j)$ the *connection form* of D with respect to the frame field s. Considering $s = (s_1, \cdots, s_r)$ as a row vector, we can rewrite (1.2) in matrix notations as follows:

$$(1.2)' \qquad Ds = s \cdot \omega .$$

If $\xi = \sum \xi^i s_i$, $\xi^i \in A^0|_U$, is an arbitrary section of E over U, then (1.1) and (1.2) imply

$$(1.3) \qquad D\xi = \sum s_i(d\xi^i + \sum \omega^i_j \xi^j) .$$

Considering $\xi = {}^t(\xi^1, \cdots, \xi^r)$ as a column vector, we may rewrite (1.3) as follows:

$$(1.3)' \qquad D\xi = d\xi + \omega\xi .$$

We call $D\xi$ the *covariant derivative of* ξ.

Evaluating D on a tangent vector X of M at x, we obtain an element of the fibre E_x denoted by

$$(1.4) \qquad D_X\xi = (D\xi)(X) \in E_x .$$

We call $D_X\xi$ the *covariant derivative of* ξ *in the direction of* X.

A section ξ is said to be *parallel* if $D\xi = 0$. If $c = c(t)$, $0 \leqq t \leqq a$, is a curve in M, a section ξ defined along c is said to be *parallel along* c if

$$(1.5) \qquad D_{c'(t)}\xi = 0 \qquad \text{for} \quad 0 \leqq t \leqq a ,$$

where $c'(t)$ denotes the velocity vector of c at $c(t)$. In terms of the local frame field s, (1.5) can be written as a system of ordinary differential equations

$$(1.5)' \qquad \frac{d\xi^i}{dt} + \sum \omega^i_j(c'(t))\xi^j = 0 .$$

If ξ_0 is an element of the initial fibre $E_{c(0)}$, it extends uniquely to a parallel section ξ along c, called the *parallel displacement of* ξ_0 *along* c. This is a matter of solving the system of ordinary differential equations (1.5)′ with initial condition ξ_0. If the initial point and the end point of c coincide so that $x_0 = c(0) = c(a)$, then the parallel displacement along c induces a linear transformation of the fibre E_{x_0}. The set of endomorphisms of E_{x_0} thus obtained from all closed curves c starting at x_0 forms a group, called the *holonomy group* of the connection D with reference point x_0.

We shall now study how the connection form ω changes when we change the local frame field s. Let $s' = (s'_1, \cdots, s'_r)$ be another local frame field over U. It is related to s by

$$(1.6) \qquad s = s' \cdot a ,$$

where $a\colon U \to GL(r; \boldsymbol{C})$ is a matrix-valued function on U. Let $\omega' = (\omega'^i_j)$ be the connection form of D with respect to s'. Then

$$\omega = a^{-1}\omega' a + a^{-1} da \,. \tag{1.7}$$

In fact,

$$s\omega = Ds = D(s'a) = (Ds')a + s'da = s'\omega'a + s'da = s(a^{-1}\omega'a + a^{-1}da)\,.$$

We extend a connection $D\colon A^0(E) \to A^1(E)$ to a $\boldsymbol{C}$-linear map

$$D\colon A^p(E) \longrightarrow A^{p+1}(E)\,, \qquad p \geqq 0\,, \tag{1.8}$$

by setting

$$D(\sigma \cdot \varphi) = (D\sigma) \wedge \varphi + \sigma \cdot d\varphi \qquad \text{for} \quad \sigma \in A^0(E)\,, \quad \varphi \in A^p\,. \tag{1.9}$$

Using this extended D, we define the *curvature* R of D to be

$$R = D \circ D\colon A^0(E) \longrightarrow A^2(E)\,. \tag{1.10}$$

Then R is A^0-linear. In fact, if $f \in A^0$ and $\sigma \in A^0(E)$, then

$$D^2(f\sigma) = D(\sigma df + f \cdot D\sigma) = D\sigma \wedge df + df \wedge D\sigma + fD^2\sigma = fD^2\sigma\,.$$

Hence, R is a 2-form on M with values in $\mathrm{End}(E)$. Using matrix notations of (1.2)′, the *curvature form* Ω of D with respect to the frame field s is defined by

$$s\Omega = D^2 s\,. \tag{1.11}$$

Then

$$\Omega = d\omega + \omega \wedge \omega\,. \tag{1.12}$$

In fact,

$$s\Omega = D(s\omega) = Ds \wedge \omega + sd\omega = s(\omega \wedge \omega + d\omega)\,.$$

Exterior differentiation of (1.12) gives the *Bianchi identity*:

$$d\Omega = \Omega \wedge \omega - \omega \wedge \Omega\,. \tag{1.13}$$

If ω' is the connection form of D relative to another frame field $s' = sa^{-1}$ as in (1.6) and (1.7), the corresponding curvature form Ω' is related to Ω by

$$\Omega = a^{-1}\Omega' a\,. \tag{1.14}$$

In fact,

$$s\Omega = D^2 s = D^2(s'a) = D(Ds'a + s'da) = D^2 s'a - Ds' \wedge da + Ds' \wedge da$$
$$= s'\Omega' a = s a^{-1}\Omega' a\,.$$

Let $\{U, V, \cdots\}$ be an open cover of M with a local frame field s_U on each U. If $U \cap V \neq \varnothing$, then

$$s_U = s_V g_{VU} \qquad \text{on} \quad U \cap V\,, \tag{1.15}$$

where $g_{VU}: U \cap V \to GL(r; \boldsymbol{C})$ is a C^∞ mapping, called a transition function. Given a connection D in E, let ω_U be the connection form on U with respect to s_U. Then (1.7) means

$$\omega_U = g_{VU}^{-1}\omega_V g_{VU} + g_{VU}^{-1} dg_{VU} \qquad \text{on} \quad U \cap V\,. \tag{1.16}$$

Conversely, given a system of $\mathfrak{gl}(r; \boldsymbol{C})$-valued 1-forms ω_U on U satisfying (1.16), we obtain a connection D in E having $\{\omega_U\}$ as connection forms.

If Ω_U is the curvature form of D relative to s_U, then (1.14) means

$$\Omega_U = g_{VU}^{-1}\Omega_V g_{VU} \qquad \text{on} \quad U \cap V\,. \tag{1.17}$$

It is sometimes more convenient to consider a connection in E as a connection in the associated principal $GL(r; \boldsymbol{C})$-bundle P. In general, let G be a Lie group and $\mathfrak{g}$ its Lie algebra identified with the tangent space T_eG at the identity element e of G. Let P be a principal G-bundle over M. Let $\{U\}$ be an open cover of M with local sections $\{s_U\}$ of P. Let $\{g_{VU}\}$ be the family of transition functions defined by $\{(U, s_U)\}$; $g_{VU}: U \cap V \to G$ is defined by

$$s_U(x) = s_V(x) g_{VU}(x) \qquad x \in U \cap V\,.$$

A *connection* in P is given by a family of $\mathfrak{g}$-valued 1-forms ω_U on U satisfying (1.16).

A connection in P induces a connection in every bundle associated to P. In particular, a connection in a principal $GL(r; \boldsymbol{C})$-bundle P induces a connection in the vector bundle $E^\rho = P \times_\rho \boldsymbol{C}^N$ associated to P by any representation ρ: $GL(r; \boldsymbol{C}) \to GL(N; \boldsymbol{C})$.

For further details on connections in principal bundles, see Kobayashi-Nomizu [1].

§ 2 Flat bundles and flat connections

Let E be a C^∞ complex vector bundle over a real manifold M as in the preceding section. A *flat structure* in E is given by an open cover $\{U, s_U\}$ with local frame fields such that the transition functions $\{g_{VU}\}$ (see (1.15)) are all constant matrices in $GL(r; \boldsymbol{C})$. A vector bundle with a flat structure is said to be *flat*. On the other hand, a connection D in a vector bundle E is said to be *flat* if its curvature R vanishes.

A flat vector bundle E admits a natural flat connection D; namely, if the flat structure is given by $\{U, s_U\}$, then D is defined by

$$Ds_U = 0 . \tag{2.1}$$

Since the transition functions $\{g_{VU}\}$ are all constants, the condition $Ds_U = 0$ and $Ds_V = 0$ are compatible on $U \cap V$ and the connection D is well defined. We note that if ω_U is the connection form of D relative to s_U, then (2.1) is equivalent to

$$\omega_U = 0 . \tag{2.1$'$}$$

From (2.1) (or (2.1)′), it follows that the curvature of D vanishes.

Conversely, a vector bundle E with a flat connection D admits a natural flat structure $\{U, s_U\}$. To construct a local frame field s_U satisfying (2.1), we start with an arbitrary local frame field s' on U and try to find a function $a\colon U \to GL(r; \boldsymbol{C})$ such that $s_U = s'a$ satisfies (2.1)′. Let ω' be the connection form of D relative to s'. By (1.7), the condition $\omega_U = 0$ is equivalent to

$$a^{-1}\omega' a + a^{-1} da = \omega_U = 0 . \tag{2.2}$$

This is a system of differential equations where ω' is given and a is the unknown. Multiplying (2.2) by a and differentiating the resulting equation

$$\omega' a + da = 0 ,$$

we obtain the integrability condition

$$0 = (d\omega')a - \omega' \wedge da = (d\omega')a + (\omega' \wedge \omega')a = \Omega' a .$$

So the integrability condition is precisely vanishing of the curvature $\Omega' = 0$.

In general, if s and s' are two local parallel frame fields, i.e., if $Ds = Ds' = 0$, then $s = s'a$ for some constant matrix $a \in GL(r; \boldsymbol{C})$ since the connection forms ω and ω' vanish in (1.7). Hence, if $Ds_U = Ds_V = 0$, the g_{VU} is a constant matrix. This proves that a flat connection D gives rise to a flat structure $\{U, s_U\}$.

Let E be a vector bundle with a flat connection D. Let x_0 be a point of M and π_1 the fundamental group of M with reference point x_0. Since the connection is flat, the parallel displacement along a closed curve c starting at x_0 depends only on the homotopy class of c. So the parallel displacement gives rise to a representation

(2.3)
$$\rho : \pi_1 \longrightarrow GL(r; \boldsymbol{C}) .$$

The image of ρ is the holonomy group of D defined in the preceding section.

Conversely, given a representation (2.3), we can construct a flat vector bundle E by setting

(2.4)
$$E = \tilde{M} \times_\rho \boldsymbol{C}^r ,$$

where $\tilde{M}$ is the universal covering of M and $\tilde{M} \times_\rho \boldsymbol{C}^r$ denotes the quotient of $\tilde{M} \times \boldsymbol{C}^r$ by the action of π_1 given by

$$\gamma : (x, v) \in \tilde{M} \times \boldsymbol{C}^r \longmapsto (\gamma(x), \rho(\gamma)v) \in \tilde{M} \times \boldsymbol{C}^r , \qquad \gamma \in \pi_1$$

(We are considering π_1 as the covering transformation group acting on $\tilde{M}$). It is easy to see that E carries a natural flat structure coming from the product structure of $\tilde{M} \times \boldsymbol{C}^r$. The vector bundle defined by (2.4) is said to be *defined by the representation* ρ. In summary, we have established the following

(2.5) **Proposition** *For a complex vector bundle E of rank r over M, the following three conditions are equivalent*:

(1) *E is a flat vector bundle,*

(2) *E admits a flat connection D,*

(3) *E is defined by a representation $\rho : \pi_1 \to GL(r; \boldsymbol{C})$.*

A connection in a vector bundle E may be considered as a connection in the associated principal $GL(r; \boldsymbol{C})$-bundle. More generally, let G be a Lie group and P a principal G-bundle over M. Let $\{U\}$ be an open cover of M with local sections $\{s_U\}$ of P. Let $\{g_{VU}\}$ be the family of transition functions defined by $\{(U, s_U)\}$; each g_{VU} is a C^∞ maps from $U \cap V$ into G. A *flat structure* in P is given by $\{(U, s_U)\}$ such that $\{g_{VU}\}$ are all constant maps. A connection in P is said to be *flat* if its curvature vanishes identically.

Let $\tilde{M}$ be the universal covering space of M; it is considered as a principal π_1-bundle over M, where π_1 is the fundamental group of M acting on $\tilde{M}$ as the group of covering transformations. Given a homomorphism ρ: $\pi_1 \to G$, we obtain a principal G-bundle $P = \tilde{M} \times_\rho G$ by "enlarging" the structure group from π_1 to G. Then P inherits a flat structure from the natural flat structure of the product bundle $\tilde{M} \times G$ over $\tilde{M}$. The following proposition is a straightforward generalization of (2.5).

(2.6) **Proposition** *For a principal G-bundle P over M, the following three* *conditions are equivalent:*

(1) *P admits a flat structure,*

(2) *P admits a flat connection,*

(3) *P is defined by a representation $\rho: \pi_1 \to G$.*

Applied to the principal $GL(r; \boldsymbol{C})$-bundle associated to a vector bundle E, (2.6) yields (2.5). We shall now consider the case where G is the projective linear group $PGL(r; \boldsymbol{C}) = GL(r; \boldsymbol{C})/\boldsymbol{C}^* I_r$, (where $\boldsymbol{C}^* I_r$ denotes the center of $GL(r; \boldsymbol{C})$ consisting of scalar multiples of the identity matrix I_r). Given a vector bundle E, let P be the associated principal $GL(r; \boldsymbol{C})$-bundle. Then $\hat{P} = P/\boldsymbol{C}^* I_r$ is a principal $PGL(r; \boldsymbol{C})$-bundle. We say that E is *projectively flat* when $\hat{P}$ is provided with a flat structure. A connection D in E (i.e., a connection in P) is said to be *projectively flat* if the induced connection in $\hat{P}$ is flat. As a special case of (2.6), we have

(2.7) **Corollary** *For a complex vector bundle E of rank r over M with the associated principal $GL(r; \boldsymbol{C})$-bundle P, the following three conditions are equivalent:*

(1) *E is projectively flat,*

(2) *E admits a projectively flat connection,*

(3) *The $PGL(r; \boldsymbol{C})$-bundle $\hat{P} = P/\boldsymbol{C}^* I_r$ is defined by a representation $\rho: \pi_1 \to PGL(r; \boldsymbol{C})$.*

Let $\mu: GL(r; \boldsymbol{C}) \to PGL(r; \boldsymbol{C})$ be the natural homomorphism and $\mu': \mathfrak{gl}(r; \boldsymbol{C}) \to \mathfrak{pgl}(r; \boldsymbol{C})$ the corresponding Lie algebra homomorphism. If R denotes the curvature of a connection D in E, then the curvature of the induced connection in $\hat{P}$ is given by $\mu'(R)$. Hence,

(2.8) **Proposition** *A connection D in a complex vector bundle E over M is projectively flat if and only if its curvature R takes values in scalar multiples of the identity endomorphism of E, i.e., if and only if there exists a complex 2-form α on M such that*

$$R = \alpha I_E .$$

Let $V = \boldsymbol{C}^r$. If $A: V \to V$ is a linear transformation of the form aI_r, $a \in \boldsymbol{C}^*$, then $A \otimes {}^t A^{-1}$ is the identity transformation of $\mathrm{End}(V) = V \otimes V^*$. Hence,

(2.9) **Proposition** *If a complex vector bundle E is projectively flat, then the bundle* $\mathrm{End}(E) = E \otimes E^*$ *is flat in a natural manner.*

§3 Connections in complex vector bundles (over complex manifolds)

Let M be an n-dimensional complex manifold and E a C^∞ complex vector bundle of rank r over M. In addition to the notations A^p and $A^p(E)$ introduced in Section 1, we use the following:

$A^{p,q}$ = the space of (p, q)-forms over M,

$A^{p,q}(E)$ = the space of (p, q)-forms over M with values in E,

so that

$$A^r = \sum_{p+q=r} A^{p,q}, \qquad A^r(E) = \sum_{p+q=r} A^{p,q}(E).$$

$$d = d' + d'', \quad \text{where} \quad d' : A^{p,q} \longrightarrow A^{p+1,q} \quad \text{and} \quad d'' : A^{p,q} \longrightarrow A^{p,q+1}.$$

Let D be a connection in E as defined in Section 1. We can write

$D = D' + D''$, where

$$D' : A^{p,q}(E) \longrightarrow A^{p+1,q}(E) \quad \text{and} \quad D'' : A^{p,q}(E) \longrightarrow A^{p,q+1}(E).$$

Decomposing (1.1) and (1.9) according to the bidegree, we obtain

$$\text{(3.1)} \qquad \begin{aligned} D'(\sigma\varphi) &= D'\sigma \wedge \varphi + \sigma d'\varphi \\ D''(\sigma\varphi) &= D''\sigma \wedge \varphi + \sigma d''\varphi \end{aligned} \qquad \sigma \in A^0(E), \quad \varphi \in A^{p,q}.$$

Let R be the curvature of D, i.e., $R = D \circ D \in A^2(\mathrm{End}(E))$. Then

$$\text{(3.2)} \qquad R = D' \circ D' + (D' \circ D'' + D'' \circ D') + D'' \circ D'',$$

where

$$D' \circ D' \in A^{2,0}(\mathrm{End}(E)), \qquad D'' \circ D'' \in A^{0,2}(\mathrm{End}(E)),$$
$$D' \circ D'' + D'' \circ D' \in A^{1,1}(\mathrm{End}(E)).$$

Let s be a local frame field of E and let ω and Ω be the connection and the curvature forms of D with respect to s. We can write

$$\text{(3.3)} \qquad \omega = \omega^{1,0} + \omega^{0,1},$$

$$\text{(3.4)} \qquad \Omega = \Omega^{2,0} + \Omega^{1,1} + \Omega^{0,2}.$$

We shall now characterize in terms of connections those complex vector bundles which admit holomorphic structures.

(3.5) **Proposition** *Let E be a holomorphic vector bundle over a complex manifold M. Then there is a connection D such that*

(3.6) $$D'' = d''\,.$$

For such a connection, the $(0, 2)$*-component* $D'' \circ D''$ *of the curvature R vanishes.*

Proof Let $\{U\}$ be a locally finite open cover of M and $\{p_U\}$ a partition of 1 subordinate to $\{U\}$. Let s_U be a holomorphic frame field of E on U. Let D_U be the flat connection in $E|_U$ defined by $D_U(s_U)=0$. Then $D=\sum p_U D_U$ is a connection in E with the property that $D''=d''$. The second assertion is obvious. Q.E.D.

Conversely, we have

(3.7) **Proposition** *Let E be a C^∞ complex vector bundle over a complex manifold M. If D is a connection in E such that $D'' \circ D''=0$, then there is a unique holomorphic vector bundle structure in E such that $D''=d''$.*

Proof We define an almost complex structure on E by specifying a splitting of the complex cotangent spaces into their $(1,0)$- and $(0,1)$-components and verify the integrability condition. Intrinsically, such a splitting is obtained by identifying the horizontal subspace of each tangent space of E with the corresponding tangent space of M. But we shall express this construction more explicitly in terms of local coordinates. We fix a local trivialization $E|_U = U \times \boldsymbol{C}^r$. Let $(z^1, \cdots, z^n)$ be a coordinate system in U, and $(w^1, \cdots, w^r)$ the natural coordinate system in $\boldsymbol{C}^r$. Let $\omega = (\omega^i_j)$ be the connection form in this trivialization. Let

$$\omega^i_j = \omega'^i_j + \omega''^i_j$$

be the decomposition into $(1,0)$- and $(0,1)$-components. Now we define an almost complex structure on E by taking

(3.8) $$\{dz^\alpha,\ dw^i + \sum \omega''^i_j w^j\}$$

as a basis for the space of $(1,0)$-forms on E. Since

$$d(dw^i + \sum \omega''^i_j w^j) \equiv \sum (d\omega''^i_j + \sum \omega''^i_k \wedge \omega''^k_j) w^j \equiv \sum d'\omega''^i_j w^j \equiv 0$$

modulo the ideal generated by (3.8), this almost complex structure is integrable. In order to show that this holomorphic structure of E has the desired property, it suffices to verify that if a local section s of E satisfies the equation $D''s=0$, then it pulls back every $(1,0)$-form of E to a $(1,0)$-form of the base manifold M. Let s be given locally by

$$s: U \longrightarrow U \times \boldsymbol{C}^r, \qquad s(z) = (z, \xi(z)).$$

Then the condition $D''s = 0$ is given by

$$d''\xi^i + \sum \omega_j''^i \xi^j = 0.$$

Pulling back the $(1,0)$-forms in (3.8) we obtain $(1,0)$-forms:

$$s^*(dz^\alpha) = dz^\alpha,$$
$$s^*(dw^i + \sum \omega_j''^i w^j) = d\xi^i + \sum \omega_j''^i \xi^j = d'\xi^i$$

The uniqueness is obvious. Q.E.D.

(3.9) **Proposition** *For a connection D in a holomorphic vector bundle E, the following conditions are equivalent*:

(a) $D'' = d''$;

(b) *For every local holomorphic section s, Ds is of degree* $(1,0)$;

(c) *With respect to a local holomorphic frame field, the connection form is of degree* $(1,0)$.

The proof is trivial.

§4 Connections in Hermitian vector bundles

Let E be a C^∞ complex vector bundle over a (real or complex) manifold M. An *Hermitian structure* or *Hermitian metric* h in E is a C^∞ field of Hermitian inner products in the fibers of E. Thus,

$$(4.1) \qquad \begin{aligned} & h(\xi, \eta) \text{ is linear in } \xi, \qquad \text{where } \xi, \eta \in E_x, \\ & h(\xi, \eta) = \overline{h(\eta, \xi)} \\ & h(\xi, \xi) > 0 \qquad \text{for } \xi \neq 0 \\ & h(\xi, \eta) \text{ is a } C^\infty \text{ function if } \xi \text{ and } \eta \text{ are } C^\infty \text{ sections}. \end{aligned}$$

We call (E, h) an *Hermitian vector bundle*.

Given a local frame field $s_U = (s_1, \cdots, s_r)$ of E over U, we set

$$(4.2) \qquad h_{ij} = h(s_i, s_j), \qquad i, j = 1, \cdots, r,$$

and

$$H_U = (h_{i\bar{j}}) . \tag{4.3}$$

Then H_U is a positive definite Hermitian matrix at every point of U. When we are working with a single frame field, we often drop the subscript U. We say that s_U is a *unitary frame field* or *orthonormal frame field* if H_U is the identity matrix. Under a change of frame field (1.15) given by $s_U = s_V g_{VU}$, we have

$$H_U = {}^t g_{VU} H_V \bar{g}_{VU} \quad \text{on} \quad U \cap V . \tag{4.4}$$

A connection D in (E, h) is called an *h-connection* if it preserves h or makes h parallel in the following sense:

$$d(h(\xi, \eta)) = h(D\xi, \eta) + h(\xi, D\eta) \qquad \text{for} \quad \xi, \eta \in A^0(E) . \tag{4.5}$$

Let $\omega = (\omega^i_j)$ be the connection form relative to s_U defined by (1.2). Then setting $\xi = s_i$ and $\eta = s_j$ in (4.5), we obtain

$$dh_{i\bar{j}} = h(Ds_i, s_j) + h(s_i, Ds_j) = \sum \omega^a_i h_{a\bar{j}} + h_{i\bar{b}} \bar{\omega}^b_j . \tag{4.6}$$

In matrix notation,

$$dH = {}^t\omega H + H\bar{\omega} . \tag{4.6$'$}$$

Applying d to (4.6)$'$ we obtain

$${}^t\Omega H + H\bar{\Omega} = 0 . \tag{4.7}$$

Let H be the identity matrix in (4.6)$'$ and (4.7). Thus,

$${}^t\omega + \bar{\omega} = 0 , \qquad {}^t\Omega + \bar{\Omega} = 0 \qquad \text{if} \quad s_U \quad \text{is unitary} , \tag{4.8}$$

that is, ω and Ω are skew-Hermitian with respect to a unitary frame. This means that ω and Ω take values in the Lie algebra $\mathfrak{u}(r)$ of the unitary group $U(r)$.

We shall now study holomorphic vector bundles. If E is a holomorphic vector bundle (over a complex manifold M), then an Hermitian structure h determines a natural h-connection satisfying (3.6). Namely, we have

(4.9) **Proposition** *Given an Hermitian structure h in a holomorphic vector bundle E, there is a unique h-connection D such that $D'' = d''$.*

Proof Let $s_U = (s_1, \cdots, s_r)$ be a local holomorphic frame field on U. Since $Ds_i = D's_i$, the connection form $\omega_U = (\omega^i_j)$ is of degree $(1, 0)$. From (4.6) we obtain

$$d'h_{i\bar{j}} = \sum \omega^a_i h_{a\bar{j}} \qquad \text{or} \quad d'H_U = {}^t\omega_U H_U . \tag{4.10}$$

This determines the connection form ω_U, i.e.,

$$(4.11) \qquad {}^t\omega_U = d'H_U H_U^{-1},$$

proving the uniqueness part. To prove the existence, we compare ${}^t\omega_U$ with ${}^t\omega_V = d'H_V H_V^{-1}$. Using (4.4) we verify by a straightforward calculation that ω_U and ω_V satisfy (1.16). Then the collection $\{\omega_U\}$ defines the desired connection. Q.E.D.

We call the connection given by (4.9) the *Hermitian connection* of a holomorphic Hermitian vector bundle (E, h). Its connection form is given explicitly by (4.11).

Its curvature $R = D \circ D$ has no $(0,2)$-component since $D'' \circ D'' = d'' \circ d'' = 0$. By (4.7) it has no $(2,0)$-component either. So the curvature

$$(4.12) \qquad R = D' \circ D'' + D'' \circ D'$$

is a $(1,1)$-form with values in $\mathrm{End}(E)$.

With respect to a local holomorphic frame field, the connection form $\omega = (\omega^i_j)$ is of degree $(1,0)$, see (3.9). Since the curvature form Ω is equal to the $(1,1)$-component of $d\omega + \omega \wedge \omega$, we obtain

$$(4.13) \qquad \Omega = d''\omega .$$

From (4.11) we obtain

$$(4.14) \qquad {}^t\Omega = d''d'H \cdot H^{-1} + dH' \cdot H^{-1} \wedge d''H \cdot H^{-1} .$$

We write

$$(4.15) \qquad \Omega^i_j = \sum R^i_{j\alpha\bar\beta} dz^\alpha \wedge d\bar z^\beta$$

so that

$$(4.16) \qquad R_{j\bar k\alpha\bar\beta} = \sum h_{i\bar k} R^i_{j\alpha\bar\beta} = -\partial_{\bar\beta}\partial_\alpha h_{j\bar k} + \sum h^{a\bar b}\partial_\alpha h_{j\bar b}\partial_{\bar\beta} h_{a\bar k} ,$$

where $\partial_\alpha = \partial/\partial z^\alpha$ and $\partial_{\bar\beta} = \partial/\partial \bar z^\beta$.

The second part of the following proposition follows from (3.7).

(4.17) **Proposition** *The curvature of the Hermitian connection in a holomorphic Hermitian vector bundle is of degree* $(1,1)$.

If (E, h) *is a* C^∞ *complex vector bundle over a complex manifold with an Hermitian structure* h *and if* D *is an* h*-connection whose curvature is of degree* $(1,1)$, *then there is a unique holomorphic structure in* E *which makes* D *the Hermitian connection of the Hermitian vector bundle* (E, h).

(4.18) **Proposition** *Let (E, h) be a holomorphic Hermitian vector bundle and D the Hermitian connection. Let $E' \subset E$ be a C^∞ complex vector subbundle invariant under D. Let E'' be the orthogonal complement of E' in E. Then both E' and E'' are holomorphic subbundles of E invariant by D, and they give a holomorphic orthogonal decomposition*:

$$E = E' \oplus E'' .$$

Proof. Since E' is invariant under D, so is its orthogonal complement E''. Let s be a local holomorphic section of E and let $s = s' + s''$ be its decomposition according to the decomposition $E = E' \oplus E''$. We have to show that s' and s'' are holomorphic sections of E. Since $D = D' + d''$ by (4.9), we have $Ds = D's$. Comparing $Ds = Ds' + Ds''$ with $D's = D's' + D's''$, we obtain $Ds' = D's'$ and $Ds'' = D's''$. Hence, $d''s' = 0$ and $d''s'' = 0$. Q.E.D.

In Riemannian geometry, normal coordinate systems are useful in simplifying explicit local calculations. We introduce analogous holomorphic local frame fields $s = (s_1, \cdots, s_r)$ for an Hermitian vector bundle (E, h). We say that a holomorphic local frame field s is *normal* at $x_0 \in M$ if

$$\text{(4.19)} \qquad \begin{aligned} &h_{i\bar{j}} = \delta_{ij} && \text{at} \quad x_0 \\ &\omega^i_j = \sum h^{i\bar{k}} d' h_{j\bar{k}} = 0 && \text{at} \quad x_0 . \end{aligned}$$

(4.20) **Proposition** *Given a holomorphic Hermitian vector bundle (E, h) over M and a point x_0 of M, there exists a normal local frame field s at x_0.*

Proof We start with an arbitrary holomorphic local frame field s around x_0. Applying a linear transformation to s, we may assume that the first condition of (4.19) is already fulfilled. Now we apply a holomorphic transformation

$$s \longrightarrow sa$$

such that $a = (a^i_j)$ is a holomorphic matrix which reduces to the identity matrix at x_0. Then the Hermitian matrix H for h changes as follows:

$$H \longrightarrow {}^t\bar{a}Ha .$$

We have to choose a in such a way that $d'({}^t\bar{a}Ha) = 0$ at x_0. Since $d''{}^t\bar{a} = 0$ and $a^i_j = \delta^i_j$ at x_0, we have

$$d'({}^t\bar{a}Ha) = d'H + d'a \qquad \text{at} \quad x_0 .$$

It is clear that there is a solution of the form

$$a^i_j = \delta^i_j + \sum c^i_{jk} z^k , \qquad \text{where} \quad c^i_{jk} \in \boldsymbol{C} . \qquad \text{Q.E.D}$$

We shall now combine flat structures discussed in Section 2 with Hermitian structures. A *flat Hermitian structure* in a C^∞ complex vector bundle E is given by an open cover $\{U\}$ of M and a system of local frame fields $\{s_U\}$ such that the transition functions $\{g_{VU}\}$ are all constant unitary matrices in $U(r)$. The following statement can be verified in the same manner as (2.5).

(4.21) **Proposition** *For a C^∞ complex vector bundle E of rank r over M, the following conditions are equivalent.*

(1) *E admits a flat Hermitian structure;*

(2) *E admits an Hermitian structure h and a flat h-connection D;*

(3) *E is defined by a representation $\rho: \pi_1(M) \to U(r)$ of $\pi_1(M)$.*

Let P be the principal $GL(r; \boldsymbol{C})$-bundle associated to E. Then E admits a flat Hermitian structure if and only if P can be reduced to a principal $U(r)$-bundle admitting a flat structure in the sense of Section 2.

Let $\hat{P}$ be the principal $PGL(r; \boldsymbol{C})$-bundle associated to E as in Section 2, i.e., $\hat{P} = P/\boldsymbol{C}^* I_r$. Let $PU(r)$ be the projective unitary group defined by

$$PU(r) = U(r)/U(1)I_r ,$$

where $U(1)I_r$ denotes the center of $U(r)$ consisting of matrices of the form cI_r, $|c| = 1$. Given an Hermitian structure h in E, we obtained a reduction of P to a principal $U(r)$-bundle P', which in turn gives rise to a principal $PU(r)$-subbundle $\hat{P}' = P'/U(1)I_r$ of $\hat{P}$. Conversely, a reduction $P' \subset P$ (or $\hat{P}' \subset \hat{P}$) gives rise to an Hermitian structure h in E. If $\hat{P}'$ is flat, or equivalently, if $\hat{P}'$ admits a flat connection, we say that the Hermitian structure h is *projectively flat*. From (2.6) we obtain

(4.22) **Proposition** *For a complex vector bundle E of rank r over M, the following conditions are equivalent:*

(1) *E admits a projectively flat Hermitian structure h;*

(2) *The $PGL(r; \boldsymbol{C})$-bundle $\hat{P} = P/\boldsymbol{C}^* I_r$ is defined by a representation $\rho: \pi_1(M) \to PU(r)$ of $\pi_1(M)$.*

Letting $A = aI$, $|a| = 1$, in the proof of (2.9), we obtain

(4.23) **Proposition** *If a complex vector bundle E admits a projectively flat Hermitian structure h, then the bundle* $\mathrm{End}(E)=E\otimes E^*$ *admits a flat Hermitian structure in a natural manner.*

§5 Connections in associated vector bundles

Let E be a C^∞ complex vector bundle over a real manifold M. Let E^* be the dual vector bundle of E. The dual pairing

$$\langle\ ,\ \rangle : E_x^* \times E_x \longrightarrow C$$

induces a dual pairing

$$\langle\ ,\ \rangle : A^0(E^*)\times A^0(E) \longrightarrow A^0 .$$

Given a connection D in E, we define a connection, also denoted by D, in E^* by the following formula:

$$d\langle\xi,\eta\rangle=\langle D\xi,\eta\rangle+\langle\xi,D\eta\rangle \qquad \text{for} \quad \xi\in A^0(E)\,, \quad \eta\in A^0(E^*)\,. \tag{5.1}$$

Given a local frame field $s=(s_1,\cdots,s_r)$ of E over an open set U, let $t=(t^1,\cdots,t^r)$ be the local frame field of E^* dual to s so that

$$\langle t^i, s_j\rangle=\delta^i_j \qquad \text{or} \qquad \langle t,s\rangle=I_r\,,$$

where s is considered as a row vector and t as a column vector. If $\omega=(\omega^i_j)$ denotes the connection form of D with respect to s so that (see (1.2))

$$Ds_i=\sum s_j\omega^j_i \qquad \text{or} \qquad Ds=s\omega\,, \tag{5.2}$$

then

$$Dt^i=-\sum\omega^i_j t^j \qquad \text{or} \qquad Dt=-\omega t\,. \tag{5.3}$$

This follows from

$$0=d\delta^i_j=\langle Dt^i,s_j\rangle+\langle t^i,Ds_j\rangle=\langle Dt^i,s_j\rangle+\omega^i_j\,.$$

If $\eta=\sum\eta_i t^i$ is an arbitrary section of E^* over U, then (1.1) and (5.3) imply

$$D\eta=\sum(d\eta_i-\sum\omega^j_i\eta_j)t^i \tag{5.4}$$

Considering $\eta=(\eta_1,\cdots,\eta_r)$ as a row vector, we may rewrite (5.4) as

$$D\eta=d\eta-\eta\omega\,. \tag{5.4$'$}$$

If Ω is the curvature form of D with respect to the frame field s so that (see

(1.11))

$$D^2 s = s\Omega , \tag{5.5}$$

then with respect to the dual frame field t we have

$$D^2 t = -\Omega t . \tag{5.6}$$

This follows from

$$D^2 t = -D(\omega t) = -(d\omega t - \omega Dt) = -(d\omega + \omega \wedge \omega)t = -\Omega t .$$

Now, let $\bar{E}$ denote the (complex) conjugate bundle of E. There is a natural conjugation map

$$^{-} : E \longrightarrow \bar{E} \tag{5.7}$$

such that $\overline{\lambda\xi} = \bar{\lambda}\bar{\xi}$ for $\xi \in E$ and $\lambda \in \boldsymbol{C}$. The transition functions of $\bar{E}$ are the complex conjugates of those for E in a natural manner.

Given a connection D in E we can define a connection D in $\bar{E}$ by

$$D\bar{\xi} = \overline{D\xi} \qquad \text{for} \quad \xi \in A^0(E) . \tag{5.8}$$

If $s = (s_1, \cdots, s_r)$ is a local frame field for E, then $\bar{s} = (\bar{s}_1, \cdots, \bar{s}_r)$ is a local frame field for $\bar{E}$. If $\omega = (\omega^i_j)$ and Ω denote the connection and curvature forms of D with respect to s, then we have clearly

$$D\bar{s}_i = \sum \bar{s}_j \bar{\omega}^j_i \qquad \text{or} \qquad D\bar{s} = \bar{s}\bar{\omega} , \tag{5.9}$$

$$D^2 \bar{s} = \bar{s}\bar{\Omega} . \tag{5.10}$$

We shall now consider two complex vector bundles E and F over the same base manifold M. Let D_E and D_F be connections in E and F, respectively. Then we can define a connection $D_E \oplus D_F$ in the direct (or Whitney) sum $E \oplus F$ and a connection $D_{E\otimes F}$ in the tensor product $E \otimes F$ in a natural manner. The latter is given by

$$D_{E\otimes F} = D_E \otimes I_F + I_E \otimes D_F , \tag{5.11}$$

where I_E and I_F denote the identity transformations of E and F, respectively. If we denote the curvatures of D_E and D_F by R_E and R_F, then

$$R_E \oplus R_F = \text{the curvature of } D_E \oplus D_F , \tag{5.12}$$

$$R_E \otimes I_F + I_E \otimes R_F = \text{the curvature of } D_{E\otimes F} . \tag{5.13}$$

If $s = (s_1, \cdots, s_r)$ is a local frame field of E and $t = (t_1, \cdots, t_p)$ is a local frame field of F and if ω_E, ω_F, Ω_E, Ω_F are the connection and curvature forms with respect to these frame fields, then in a natural manner the connection and curvature forms of $D_E + D_F$ are given by

$$\begin{pmatrix} \omega_E & 0 \\ 0 & \omega_F \end{pmatrix} \quad \text{and} \quad \begin{pmatrix} \Omega_E & 0 \\ 0 & \Omega_F \end{pmatrix}, \tag{5.14}$$

and those of $D_{E \otimes F}$ are given by

$$\omega_E \otimes I_p + I_r \otimes \omega_F \quad \text{and} \quad \Omega_E \otimes I_p + I_r \otimes \Omega_F , \tag{5.15}$$

where I_p and I_r denote the identity matrices of rank p and r.

All these formulas ((5.11)–(5.15)) extend in an obvious way to the direct sum and the tensor product of any number of vector bundles. Combined with formulas ((5.1)–(5.6)), they give formulas for the connection and curvature in

$$E^{\otimes p} \otimes E^{* \otimes q} = E \otimes \cdots \otimes E^* \otimes \cdots \otimes E^*$$

induced from a connection D in E.

As a special case, we consider $\mathrm{End}(E) = E \otimes E^*$. With respect to a local frame field $s = (s_1, \cdots, s_r)$ of E and the dual frame field $t = (t^1, \cdots, t^r)$, let

$$\xi = \sum \xi^i_j s_i \otimes t^j .$$

be a p-form with values in $\mathrm{End}(E)$, where ξ^i_j are differential p-forms. Then

$$\begin{aligned} D\xi &= \sum (d\xi^i_j s_i \otimes t^j + (-1)^p \xi^i_j Ds_i \otimes t^j + (-1)^p \xi^i_j s_i \otimes Dt^j) \\ &= \sum (d\xi^i_j + (-1)^p \sum \xi^k_j \wedge \omega^i_k - (-1)^p \sum \xi^i_k \wedge \omega^k_j) s_i \otimes t^j . \end{aligned} \tag{5.16}$$

The curvature R of D in E is a 2-form with values in $\mathrm{End}(E)$. Write

$$R = \sum \Omega^i_j s_i \otimes t^j .$$

Then

$$DR = \sum (d\Omega^i_j + \sum \omega^i_k \wedge \Omega^k_j - \sum \Omega^i_k \wedge \omega^k_j) s_i \otimes t^j .$$

Hence, by (1.13)

$$DR = 0 , \tag{5.17}$$

which is nothing but the Bianchi identity.

The p-th exterior power $\bigwedge^p E$ of E is a direct summand of the p-th tensor power $E^{\otimes p}$. It is easy to verify that the connection D in $E^{\otimes p}$ induced by a connection D of E leaves $\bigwedge^p E$ invariant, i.e.,

$$D(A^0(\textstyle\bigwedge^p E)) \subset A^1(\textstyle\bigwedge^p E) .$$

For example,

$$D(\sum \xi^{ijk} s_i \wedge s_j \wedge s_k) = \sum \nabla\xi^{ijk} s_i \wedge s_j \wedge s_k ,$$

where

$$\nabla\xi^{ijk} = d\xi^{ijk} + \sum \omega^i_a \xi^{ajk} + \sum \omega^j_a \xi^{iak} + \sum \omega^k_a \xi^{ija} .$$

In particular, for the line bundle $\det(E) = \bigwedge^r E$, called *the determinant bundle* of E, we have

$$D(s_1 \wedge \cdots \wedge s_r) = (\sum \omega^i_i) s_1 \wedge \cdots \wedge s_r ,$$

i.e., the connection form for D in $\det(E)$ is given by the trace of ω:

(5.18) $$\operatorname{tr}\omega = \sum \omega^i_i .$$

Similarly, its curvature is given by

(5.19) $$\operatorname{tr}\Omega = \sum \Omega^i_i .$$

Let h be an Hermitian structure in E. Since it defines an Hermitian map $E_x \times E_x \to C$ at each point x of M, it may be considered as a section of $E^* \otimes \bar{E}^*$, i.e.,

$$h = \sum h_{i\bar{j}} t^i \otimes \bar{t}^j .$$

If D is any connection in E, then

(5.20) $$Dh = \sum (dh_{i\bar{j}} - \sum h_{k\bar{j}} \omega^k_i - \sum h_{i\bar{k}} \bar{\omega}^k_j) t^i \otimes \bar{t}^j .$$

Let E be a complex vector bundle over M and let N be a another manifold. Given a mapping $f\colon N \to M$, we obtain an induced vector bundle f^*E over N with the commutative diagram:

(5.21) $$\begin{array}{ccc} f^*E & \xrightarrow{f} & E \\ \downarrow & & \downarrow \\ N & \xrightarrow{f} & M \end{array}$$

Then a connection D in E induces a connection in f^*E in a natural manner; this induced connection will be denoted by f^*D. If $s = (s_1, \cdots, s_r)$ is a local frame field defined on an open set U of M, then we obtain a local frame field f^*s of f^*E over $f^{-1}U$ in a natural manner. If ω is the connection form of D with respect to s, then $f^*\omega$ is the connection form of f^*D with respect to f^*s. Similarly, if Ω is the curvature form with respect to s, $f^*\Omega$ is the curvature form with respect to f^*s.

Let h be an Hermitian structure in a complex vector bundle E. Then the dual bundle E^* has a naturally induced Hermitian structure h^*. If $s = (s_1, \cdots, s_r)$ is a local frame field and $t=(t^1, \cdots, t^r)$ is the dual frame field, then h and h^* are related by

$$h=\sum h_{i\bar{j}} t^i \otimes \bar{t}^j, \qquad h^*=\sum h^{i\bar{j}} s_i \otimes \bar{s}_j, \tag{5.22}$$

where $(h^{i\bar{j}})$ is the inverse matrix of $(h_{i\bar{j}})$ so that

$$\sum h^{i\bar{k}} h_{j\bar{k}} = \delta^i_j .$$

This relationship is compatible with that between a connection in E and the corresponding connection in E^*. Namely, if a connection D in E preserves h, then the corresponding connection D in E^* preserves h^*. In particular, if E is holomorphic and D is the Hermitian connection defined by h, then the corresponding connection D in E^* is exactly the Hermitian connection defined by h^*.

Similarly, for the conjugate bundle $\bar{E}$, we have an Hermitian structure $\bar{h}$ given by

$$\bar{h}=\sum \bar{h}_{i\bar{j}} \bar{t}^i \otimes t^j = \sum h_{j\bar{i}} \bar{t}^i \otimes t^j . \tag{5.23}$$

Given Hermitian structures h_E and h_F in vector bundles E and F over M, we can define Hermitian structure $h_E \oplus h_F$ and $h_E \otimes h_F$ in $E \oplus F$ and $E \otimes F$ in a well known manner. In the determinant bundle $\det(E)=\bigwedge^r E$, we have a naturally induced Hermitian structure $\det(h)$. If $H=(h_{i\bar{j}})$ is defined by (4.2), then

$$\det(h)(s_1 \wedge \cdots \wedge s_r, s_1 \wedge \cdots \wedge s_r) = \det H . \tag{5.24}$$

If (E, h) is an Hermitian vector bundle with Hermitian connection D and curvature form $\Omega=(\Omega^i_j)$ with respect to $s=(s_1, \cdots, s_r)$, then $(\det(E), \det(h))$ is an Hermitian line bundle with curvature form $\operatorname{tr}\Omega=\sum \Omega^i_i$, (see (5.19)). By (4.15) we have the *Ricci form*

$$\operatorname{tr}\Omega = \sum R_{\alpha\bar{\beta}} dz^\alpha \wedge d\bar{z}^\beta \tag{5.25}$$

where

$$R_{\alpha\bar{\beta}} = \sum R^i_{i\alpha\bar{\beta}} = -\partial_\alpha \partial_{\bar{\beta}} \det(h_{i\bar{j}}) .$$

Finally, given (E, h) over M, a mapping $f\colon N \to M$ induces an Hermitian structure f^*h in the induced bundle f^*E over N.

All these constructions of Hermitian structures in various vector bundles are compatible with the corresponding constructions of connections.

What we have done in this section may be best described in terms of principal bundles and representations of structure groups. Thus, if P is the principal $GL(r, \boldsymbol{C})$-bundle associated to E and if $\rho: GL(r; \boldsymbol{C}) \to GL(k; \boldsymbol{C})$ is a representation, we obtain a vector bundle $E^\rho = P \otimes_\rho \boldsymbol{C}^k$ of rank k. A connection in E is really a connection in P, and the latter defines a connection in every associated bundle, in particular in E^ρ. An Hermitian structure h in E corresponds to a principal $U(r)$-subbundle P'. Considering E^ρ as a bundle associated to the subbundle P' by restricting ρ to $U(r)$, we obtain an Hermitian structure h^ρ in E^ρ. A connection in E preserving h is a connection in P' and induces a connection in E^ρ preserving h^ρ.

§6 Subbundles and quotient bundles

Let E be a holomorphic vector bundle of rank r over an n-dimensional complex manifold M. Let S be a holomorphic subbundle of rank p of E. Then the quotient bundle $Q=E/S$ is a holomorphic vector bundle of rank $r-p$. We can express this situation as an exact sequence

$$0 \longrightarrow S \longrightarrow E \longrightarrow Q \longrightarrow 0. \tag{6.1}$$

Let h be an Hermitian structure in E. Restricting h to S, we obtain an Hermitian structure h_S in S. Taking the orthogonal complement of S in E with respect to h, we obtain a complex subbundle $S^\perp$ of E. We note that $S^\perp$ may not be a holomorphic subbundle of E in general. Thus

$$E = S \oplus S^\perp \tag{6.2}$$

is merely a C^∞ orthogonal decomposition of E. As a C^∞ complex vector bundle, Q is naturally isomorphic to $S^\perp$. Hence, we obtain also an Hermitian structure h_Q in a natural way.

Let D denote the Hermitian connection in (E, h). We define D_S and A by

$$D\xi = D_S\xi + A\xi \qquad \xi \in A^0(S), \tag{6.3}$$

where $D_S\xi \in A^1(S)$ and $A\xi \in A^1(S^\perp)$. Then

(6.4) **Proposition** (1) *D_S is the Hermitian connection of* (S, h_S);
(2) *A is a* $(1,0)$*-form with values in* $\mathrm{Hom}(S, S^\perp)$, *i.e.*, $A \in A^{1,0}(\mathrm{Hom}(S, S^\perp))$.

Proof Let f be a function on M. Replacing ξ by $f\xi$ in (6.3), we obtain

$$D(f\xi) = D_S(f\xi) + A(f\xi).$$

On the other hand,

$$D(f\xi) = df \cdot \xi + fD\xi = df \cdot \xi + fD_S\xi + fA\xi.$$

Comparing the S- and $S^\perp$-components of the two decompositions of $D(f\xi)$, we conclude

$$D_S(f\xi) = df\cdot\xi + fD_S\xi\,, \qquad A(f\xi) = fA\xi\,.$$

The first equality says that D_S is a connection and the second says that A is a 1-form with values in $\mathrm{Hom}(S, S^\perp)$. If ξ in (6.3) is holomorphic, then $D\xi$ is a $(1,0)$-form with values in E and, hence, $D_S\xi$ is a $(1,0)$-form with values in S while A is a $(1,0)$-form with values in $\mathrm{Hom}(S, S^\perp)$. Finally, if $\xi, \xi' \in A^0(S)$, then

$$\begin{aligned} d(h(\xi, \xi')) &= h(D\xi, \xi') + h(\xi, D\xi') \\ &= h(D_S\xi + A\xi, \xi') + h(\xi, D_S\xi' + A\xi') \\ &= h(D_S\xi, \xi') + h(\xi, D_S\xi')\,, \end{aligned}$$

which proves that D_S preserves h_S. Q.E.D.

We call $A \in A^{1,0}(\mathrm{Hom}(S, S^\perp))$ the *second fundamental form* of S in (E, h). With the identification $Q = S^\perp$, we consider A as an element of $A^{1,0}(\mathrm{Hom}(S, Q))$ also.

Similarly, we define $D_{S^\perp}$ and B by setting

$$D\eta = B\eta + D_{S^\perp}\eta\,, \qquad \eta \in A^0(S^\perp)\,, \tag{6.5}$$

where $B\eta \in A^1(S)$ and $D_{S^\perp}\eta \in A^1(S^\perp)$. Under the identification $Q = S^\perp$, we may consider $D_{S^\perp}$ as a mapping $A^0(Q) \to A^1(Q)$. Then we write D_Q in place of $D_{S^\perp}$.

(6.6) **Proposition** (1) *D_Q is the Hermitian connection of (Q, h_Q);*

(2) *B is a $(0,1)$-form with values in $\mathrm{Hom}(S^\perp, S)$, i.e., $B \in A^{0,1}(\mathrm{Hom}(S^\perp, S))$;*

(3) *B is the adjoint of $-A$, i.e.,*

$$h(A\xi, \eta) + h(\xi, B\eta) = 0 \qquad \text{for} \quad \xi \in A^0(S)\,, \quad \eta \in A^0(S^\perp)\,.$$

Proof As in (6.4) we can see that $D_{S^\perp}$ defines a connection in $S^\perp$ which preserves $h_{S^\perp}$ and that B is an element of $A^{0,1}(\mathrm{Hom}(S^\perp, S))$.

Let $\tilde\eta$ be a local holomorphic section of Q, η the corresponding C^∞ section of $S^\perp$ under the identification $Q = S^\perp$, and ζ a local holomorphic section of E representing $\tilde\eta$. Let

$$\zeta = \xi + \eta \qquad \text{with} \quad \xi \in A^0(S)\,. \tag{6.6}$$

Applying D to (6.6) and making use of (6.3) and (6.5), we obtain

$$\begin{aligned} D\zeta &= D\xi + D\eta = D_S\xi + A\xi + B\eta + D_{S^\perp}\eta \\ &= (D_S\xi + B\eta) + (A\xi + D_{S^\perp}\eta)\,. \end{aligned} \tag{6.7}$$

Since $D\zeta$ is a $(1,0)$-form with values in E, $(D_S\xi+B\eta)$ and $(A\xi+D_{S^\perp}\eta)$ are also $(1,0)$-forms with values in S and $S^\perp$, respectively. Since $A\xi$ is a $(1,0)$-form by (6.4), it follows that $D_{S^\perp}\eta$ is a $(1,0)$-form. This shows that the corresponding connection D_Q is the Hermitian connection of (Q, h_Q).

Finally, if $\xi\in A^0(S)$ and $\eta\in A^0(S^\perp)$, then

$$\begin{aligned} 0&=dh(\xi,\eta)=h(D\xi,\eta)+h(\xi,D\eta)\\ &=h(D_S\xi+A\xi,\eta)+h(\xi,B\eta+D_{S^\perp}\eta)=h(A\xi,\eta)+h(\xi,B\eta)\,. \end{aligned}$$

This also shows that B is a $(0,1)$-form since A is a $(1,0)$-form. Q.E.D.

Let $e_1,\cdots,e_p,e_{p+1},\cdots,e_r$ be a local C^∞ unitary frame field for E such that $e_1,\cdots,e_p$ is a local frame field for S. Then $e_{p+1},\cdots,e_r$ is a local frame field for $S^\perp=Q$. We shall use the following convention on the range of indices:

$$1\leqq a,b,c\leqq p\,,\qquad p+1\leqq\lambda,\mu,\nu\leqq r\,,\qquad 1\leqq i,j,k\leqq r\,.$$

If (ω^i_j) denotes the connection form of the Hermitian connection D of (E,h) with respect to $e_1,\cdots,e_r$, then

$$\text{(6.8)}\qquad De_a=\sum\omega^b_a e_b+\sum\omega^\lambda_a e_\lambda\,,\qquad De_\lambda=\sum\omega^a_\lambda e_a+\sum\omega^\mu_\lambda e_\mu\,.$$

It follows then that

$$\text{(6.9)}\qquad \begin{aligned} D_Se_a&=\sum\omega^b_a e_b\,, & Ae_a&=\sum\omega^\lambda_a e_\lambda\,,\\ Be_\lambda&=\sum\omega^a_\lambda e_a\,, & D_Qe_\lambda&=\sum\omega^\mu_\lambda e_\mu\,. \end{aligned}$$

Since $e_1,\cdots,e_r$ is not holomorphic, ω^b_a and ω^μ_λ may not be $(1,0)$-forms. However, by (6.4) and (6.6), ω^λ_a are $(1,0)$-forms while $\omega^a_\lambda=-\bar\omega^\lambda_a$ are $(0,1)$-forms. We see from (6.9) that (ω^b_a) is the connection form for (S,h_S) with respect to the frame field $e_1,\cdots,e_p$ and that (ω^μ_λ) is the connection form for (Q,h_Q) with respect to the frame field $e_{p+1},\cdots,e_r$.

From (6.8) we obtain

$$\text{(6.10)}\qquad \begin{aligned} DDe_a&=\sum(d\omega^b_a+\sum\omega^b_c\wedge\omega^c_a+\sum\omega^b_\lambda\wedge\omega^\lambda_a)e_b\\ &\quad+\sum(d\omega^\lambda_a+\sum\omega^\lambda_c\wedge\omega^c_a+\sum\omega^\lambda_\mu\wedge\omega^\mu_a)e_\lambda\,,\\ DDe_\lambda&=\sum(d\omega^a_\lambda+\sum\omega^a_c\wedge\omega^c_\lambda+\sum\omega^a_\mu\wedge\omega^\mu_\lambda)e_a\\ &\quad+\sum(d\omega^\mu_\lambda+\sum\omega^\mu_c\wedge\omega^c_\lambda+\sum\omega^\mu_\nu\wedge\omega^\nu_\lambda)e_\mu\,. \end{aligned}$$

Let $R=D\circ D$, $R_S=D_S\circ D_S$ and $R_Q=D_Q\circ D_Q$ be the curvatures of the Hermitian connections D, D_S and D_Q in E, S and Q, respectively. They are of degree $(1,1)$.

Making use of (6.9) we can rewrite (6.10) as follows:

$$(6.10') \qquad \begin{aligned} Re_a &= (R_S + B \wedge A + DA)e_a\,, \\ Re_\lambda &= (R_Q + A \wedge B + DB)e_\lambda\,. \end{aligned}$$

Considering only the terms of degree (1, 1), we obtain

$$(6.11) \qquad \begin{aligned} Re_a &= (R_S + B \wedge A + D''A)e_a = (R_S - B \wedge B^* - D''B^*)e_a\,, \\ Re_\lambda &= (R_Q + A \wedge B + D'B)e_\lambda = (R_Q - B^* \wedge B + D'B)e_\lambda\,. \end{aligned}$$

In matrix notation we can write (6.11) as follows:

$$(6.12) \qquad R = \begin{pmatrix} R_S - B \wedge B^* & D'B \\ -D''B^* & R_Q - B^* \wedge B \end{pmatrix} = \begin{pmatrix} R_S - A^* \wedge A & D'A^* \\ -D''A & R_Q - A \wedge A^* \end{pmatrix}$$

These are vector bundle analogues of the *Gauss-Codazzi equations*.

Let g be an Hermitian metric on M. In terms of local unitary coframes $\theta^1, \cdots, \theta^n$ for (M, g), we can express part of (6.11) as follows: Setting the (1,0)-forms ω_a^λ

$$\omega_a^\lambda = \sum A_{a\alpha}^\lambda \theta^\alpha\,,$$

we have

$$(6.13) \qquad \begin{aligned} R_{S\,a\alpha\bar{\beta}}^{\;b} &= R_{a\alpha\bar{\beta}}^{b} - \sum A_{a\alpha}^{\lambda} \bar{A}_{b\beta}^{\lambda}\,, \\ R_{Q\,\lambda\alpha\bar{\beta}}^{\;\mu} &= R_{\lambda\alpha\bar{\beta}}^{\mu} + \sum A_{c\alpha}^{\mu} \bar{A}_{c\beta}^{\lambda}\,. \end{aligned}$$

From (4.18) we obtain

(6.14) **Proposition** *If the second fundamental form A vanishes identically, then $S^\perp$ is a holomorphic subbundle and the orthogonal decomposition*

$$E = S \oplus S^\perp$$

is holomorphic.

§7 Hermitian manifolds

Let M be a complex manifold of dimension n. Its tangent bundle TM will be considered as a holomorphic vector bundle of rank n in the following manner. When M is regarded as a real manifold, TM is a real vector bundle of rank $2n$. The complex structure of M defines an almost complex structure $J \in A^0(\mathrm{End}(TM))$, $J \circ J = -I$, where I denotes the identity transformation of TM. Let T_CM denote the complexification of TM, i.e., $T_CM = TM \otimes C$. It is a complex vector bundle of rank

$2n$. Extend J to a complex endomorphism of $T_C M$. Since $J^2=-I$, J has eigenvalues $\sqrt{-1}$ and $-\sqrt{-1}$. Let

$$T_C M = T'M + T''M \tag{7.1}$$

be the decomposition of $T_C M$ into the eigen-spaces of J, where $T'M$ (resp. $T''M$) is the eigen-space for $\sqrt{-1}$ (resp. $-\sqrt{-1}$). Then $T'M$ is a holomorphic vector bundle of rank n and $T''M$ is the conjugate bundle of $T'M$, i.e., $T''M=\bar{T}'M$. We identify TM with $T'M$ by sending $X\in TM$ to $\frac{1}{2}(X-\sqrt{-1}JX)\in T'M$.

Perhaps, it would be instructive to repeat the construction above using local coordinates. Let $z^1, \cdots, z^n$ be a local coordinate system for M as a complex manifold. Let $z^j=x^j+\sqrt{-1}y^j$ so that $x^1, y^1, \cdots, x^n, y^n$ form a local coordinate system for M as a real manifold. Then $(\partial/\partial x^1, \partial/\partial y^1, \cdots, \partial/\partial x^n, \partial/\partial y^n)$ is a basis for TM over $\boldsymbol{R}$ and also a basis for $T_C M$ over $\boldsymbol{C}$. The almost complex structure J is determined by

$$J(\partial/\partial x^j)=\partial/\partial y^j\,, \qquad J(\partial/\partial y^j)=-(\partial/\partial x^j)\,. \tag{7.2}$$

Let

$$\frac{\partial}{\partial z^j}=\frac{1}{2}\left(\frac{\partial}{\partial x^j}-\sqrt{-1}\frac{\partial}{\partial y^j}\right), \qquad \frac{\partial}{\partial \bar{z}^j}=\frac{1}{2}\left(\frac{\partial}{\partial x^j}+\sqrt{-1}\frac{\partial}{\partial y^j}\right). \tag{7.3}$$

Then $(\partial/\partial z^1, \cdots, \partial/\partial z^n)$ is a basis for $T'M$ while $(\partial/\partial \bar{z}^1, \cdots, \partial/\partial \bar{z}^n)$ is a basis for $T''M$. The identification of TM with $T'M$ is given by

$$\partial/\partial x^j \longrightarrow \partial/\partial z^j\,.$$

Let.

$$dz^j=dx^j+\sqrt{-1}dy^j\,.$$

Then $(dz^1, \cdots, dz^n)$ is the holomorphic local frame field for the holomorphic cotangent bundle T^*M dual to the frame field $(\partial/\partial z^1, \cdots, \partial/\partial z^n)$.

To a connection D in the holomorphic tangent bundle TM, we can associate, in addition to the curvature R, another invariant called the torsion T of D defined by

$$T(\xi, \eta)=D_\xi\eta-D_\eta\xi-[\xi, \eta] \qquad \text{for} \quad \xi, \eta\in A^0(TM)\,. \tag{7.4}$$

Then T is A^0-bilinear and can be considered as an element of $A^2(TM)$ since it is skew-symmetric in ξ and η.

Let $s=(s_1, \cdots, s_n)$ be a holomorphic local frame field for TM and $\theta=(\theta^1, \cdots, \theta^n)$ the dual frame field for the holomorphic cotangent bundle T^*M. Let $\omega=(\omega^i_j)$ be the connection form of D with respect to s. Then

$$(7.5) \qquad T = s(d\theta + \omega\wedge\theta) = \sum s_i(d\theta^i + \sum \omega^i_j \wedge \theta^j)\,.$$

To prove (7.5), observe first that

$$\xi = \sum \xi^i s_i\,, \qquad \text{where} \quad \xi^i = \theta^i(\xi) \quad \text{so that} \quad \xi = s\cdot\theta(\xi)\,,$$
$$\eta = \sum \eta^i s_i\,, \qquad \text{where} \quad \eta^i = \theta^i(\eta) \quad \text{so that} \quad \eta = s\cdot\theta(\eta)\,.$$

Hence,

$$D\eta = s(d(\theta(\eta)) + \omega\cdot\theta(\eta))\,, \qquad D_\xi\eta = s(\xi(\theta(\eta)) + \omega(\xi)\cdot\theta(\eta))\,,$$
$$D\xi = s(d(\theta(\xi)) + \omega\cdot\theta(\xi))\,, \qquad D_\eta\xi = s(\eta(\theta(\xi)) + \omega(\eta)\cdot\theta(\xi))\,.$$

Substituting these into (7.4), we obtain

$$\begin{aligned} T(\xi,\eta) &= s(\xi(\theta(\eta)) + \omega(\xi)\theta(\eta) - \eta(\theta(\xi)) - \omega(\eta)\theta(\xi) - \theta([\xi,\eta])) \\ &= s(d\theta(\xi,\eta) + (\omega\wedge\theta)(\xi,\eta))\,. \end{aligned}$$

Q.E.D.

From (7.5) and (c) of (3.9) we obtain

(7.6) **Proposition** *A connection D in TM satisfies $D'' = d''$ if and only if its torsion T is of degree* $(2,0)$ *i.e.,* $T\in A^{2,0}(TM)$.

Let g be an Hermitian structure in TM. It is called an *Hermitian metric* on M. A complex manifold M with an Hermitian metric g is called an *Hermitian manifold.* Then we can write (see (4.2))

$$(7.7) \qquad g = \sum g_{i\bar{j}}\, dz^i d\bar{z}^j\,, \qquad \text{where} \quad g_{i\bar{j}} = g(\partial/\partial z^i, \partial/\partial\bar{z}^j)\,.$$

(Following the tradition, we write $dz^i d\bar{z}^j$ instead of $dz^i \otimes d\bar{z}^j$). Let D be the Hermitian connection of g. It is characterized by the following two properties (1) and (2). The latter is equivalent to (2)′, (see (4.9) and (7.6)).

(1) $Dg = 0$,

(2) $D'' = d''$,

(2)′ $T\in A^{2,0}(TM)$.

Let R be the curvature of D; it is in $A^{1,1}(\mathrm{End}(TM))$, (see (4.15)). In terms of the frame field $(\partial/\partial z^1, \cdots, \partial/\partial z^n)$ and its dual $(dz^1, \cdots, dz^n)$, the curvature can be expressed as

$$(7.8) \qquad R = \sum \Omega^i_j\, dz^j \otimes \frac{\partial}{\partial z^i}\,, \qquad \text{where} \quad \Omega^i_j = \sum R^{\,i}_{j k\bar{h}}\, dz^k \wedge d\bar{z}^h\,.$$

Expressing (4.13) in terms of local coordinates, we have

(7.9)
$$R^i_{ja\bar b} = -\sum g^{i\bar k}\frac{\partial^2 g_{j\bar k}}{\partial z^a \partial \bar z^b} + \sum g^{i\bar k} g^{p\bar q}\frac{\partial g_{p\bar k}}{\partial z^a}\frac{\partial g_{j\bar q}}{\partial \bar z^b}.$$

If we set

(7.10)
$$R_{j\bar k a\bar b} = \sum g_{i\bar k} R^i_{ja\bar b}$$

then (7.9) reads as follows:

(7.11)
$$R_{j\bar k a\bar b} = -\frac{\partial^2 g_{j\bar k}}{\partial z^a \partial \bar z^b} + \sum g^{p\bar q}\frac{\partial g_{p\bar k}}{\partial z^a}\frac{\partial g_{j\bar q}}{\partial \bar z^b}.$$

We define two Hermitian tensor fields contracting the curvature tensor R in two different ways. We set

(7.12)
$$R_{k\bar h} = \sum R^i_{ik\bar h} = \sum g^{i\bar j} R_{i\bar j k\bar h}, \qquad Ric = \sum R_{k\bar h} dz^k \otimes d\bar z^h,$$

(7.13)
$$K_{i\bar j} = \sum g^{k\bar h} R_{i\bar j k\bar h}, \qquad \hat K = \sum K_{i\bar j} dz^i \otimes d\bar z^j.$$

To each of these, we can associate a real $(1,1)$-form:

(7.14)
$$\rho = \sqrt{-1}\sum R_{k\bar h} dz^k \wedge d\bar z^h, \qquad \kappa = \sqrt{-1}\sum K_{i\bar j} dz^i \wedge d\bar z^j.$$

The Ricci tensor Ric has a simple geometric interpretation. The Hermitian metric of M induces an Hermitian structure in the determinant bundle $\det(TM)$, whose curvature is the trace of R, (see Section 5). The Ricci tensor is therefore the curvature of $\det(TM)$. Clearly,

(7.14)′
$$\rho = \sqrt{-1}\sum \Omega^i_i.$$

Since $\sum \Omega^i_i$ is the curvature form of $\det(TM)$, it can be calculated by setting $H = \det(g_{i\bar j})$ in (4.14). Thus,

(7.15)
$$\rho = \sqrt{-1}\, d''d' \log(\det(g_{i\bar j})).$$

The real $(1,1)$-form ρ is closed and, as we shall see later, represents the cohomology class $2\pi c_1$, where c_1 denotes the first Chern class of M.

The tensor field $\hat K$ is a special case of the *mean curvature* of an Hermitian vector bundle, which will play an essential role in vanishing theorems for holomorphic sections. We define the *scalar curvature* of the Hermitian manifold M by

(7.16)
$$\sigma = \sum g^{k\bar h} R_{k\bar h} = \sum g^{i\bar j} K_{i\bar j}.$$

Associated to each Hermitian metric g is the *fundamental 2-form* or the *Kähler 2-form* Φ defined by

(7.17) $$\Phi=\sqrt{-1}\sum g_{i\bar{j}}dz^i \wedge d\bar{z}^j .$$

Then

(7.18) $$\Phi^n=(\sqrt{-1})^n n!\det(g_{i\bar{j}})dz^1 \wedge d\bar{z}^1 \wedge \cdots \wedge dz^n \wedge d\bar{z}^n .$$

The real (1, 1)-form Φ may or may not be closed. If it is closed, then g is called a *Kähler metric* and (M, g) is called a *Kähler manifold.*

(7.19) **Proposition** *For an Hermitian metric g, the following conditions are mutually equivalent*:

(1) *g is Kähler, i.e., $d\Phi=0$;*

(2) $\dfrac{\partial g_{i\bar{j}}}{\partial z^k}=\dfrac{\partial g_{k\bar{j}}}{\partial z^i}$;

(3) $\dfrac{\partial g_{i\bar{j}}}{\partial \bar{z}^k}=\dfrac{\partial g_{i\bar{k}}}{\partial \bar{z}^j}$;

(4) *There exists a locally defined real function f such that $\Phi=\sqrt{-1}d'd''f$, i.e.,*

$$g_{i\bar{j}}=\frac{\partial^2 f}{\partial z^i \partial \bar{z}^j} .$$

(5) *The Hermitian connection has vanishing torsion.*

Proof Since (2) is equivalent to $d'\Phi=0$ while (3) is equivalent to $d''\Phi=0$ and since $d''\Phi=\overline{d'\Phi}$, we see that (1), (2) and (3) are mutually equivalent. Clearly, (4) implies (1). To prove the converse, assume $d\Phi=0$. By Poincaré lemma, there exists a locally defined real 1-form ψ such that $\Phi=d\psi$. We can write $\psi=\varphi+\bar{\varphi}$, where φ is a (1, 0)-form and $\bar{\varphi}$ is its conjugate. Since Φ is of degree (1, 1), we have

$$d'\varphi=0 , \quad d''\bar{\varphi}=0 , \qquad \Phi=d''\varphi+d'\bar{\varphi} .$$

Then there exists a locally defined function p such that $\varphi=d'p$. We have $\bar{\varphi}=d''\bar{p}$. Hence,

$$\Phi=d''\varphi+d'\bar{\varphi}=d''d'p+d'd''\bar{p}=d'd''(\bar{p}-p)=\sqrt{-1}d'd''f ,$$

where

$$f=\sqrt{-1}(p-\bar{p}) .$$

To see that (2) and (5) are equivalent, we write, as in (4.10), in terms of coordinates. Then

$$\omega^i_j = \sum g^{i\bar{k}} d' g_{j\bar{k}} . \tag{7.20}$$

We substitute (7.20) into (7.5). Since $d\theta^i = d(dz^i) = 0$, we obtain

$$T = \sum \left(\sum g^{i\bar{k}} \frac{\partial g_{j\bar{k}}}{\partial z^h} dz^h \wedge dz^j \right) \frac{\partial}{\partial z^i} .$$

It is now clear that (2) and (5) are equivalent. Q.E.D.

By (7.11) and (7.19), the curvature components of a Kähler metric can be written locally as follows:

$$R_{i\bar{j}k\bar{h}} = -\frac{\partial^4 f}{\partial z^i \partial \bar{z}^j \partial z^k \partial \bar{z}^h} + \sum g^{p\bar{q}} \frac{\partial^3 f}{\partial z^p \partial \bar{z}^j \partial z^k} \frac{\partial^3 f}{\partial \bar{z}^q \partial z^i \partial \bar{z}^h} . \tag{7.21}$$

Immediate from (7.21) is the following symmetry for the curvature of a Kähler metric:

$$R_{i\bar{j}k\bar{h}} = R_{k\bar{h}i\bar{j}} . \tag{7.22}$$

This implies that, in the Kähler case, the two tensor fields Ric and $\hat{K}$ introduced in (7.12) and (7.13) coincide:

$$R_{i\bar{j}} = K_{i\bar{j}} . \tag{7.23}$$

In the next chapter we use a special case ($p=1$) of the following

(7.24) **Proposition** *A closed real (p, p)-form ω on a compact Kähler manifold M is cohomologous to zero if and only if $\omega = id'd''\varphi$ for some real $(p-1, p-1)$-form φ.*

Proof Assume that ω is cohomologous to zero so that $\omega = d\alpha$, where α is a real $(2p-1)$-form. Write $\alpha = \beta + \bar{\beta}$, where β is a $(p-1, p)$-form and $\bar{\beta}$ is its complex conjugate. Then

$$\omega = d\alpha = d'\bar{\beta} + (d''\bar{\beta} + d'\beta) + d''\beta .$$

Comparing the degrees of the forms involved, we obtain

$$\omega = d\alpha = d''\bar{\beta} + d'\beta , \qquad d'\bar{\beta} = 0 , \qquad d''\beta = 0 .$$

We may write

$$\beta = H\beta + d''\gamma ,$$

where $H\beta$ denotes the harmonic part of β and γ is a $(p-1, p-1)$-form. Then

$$\bar{\beta} = \overline{H\beta} + d'\bar{\gamma} .$$

Hence,

$$\omega = d''\bar{\beta} + d'\beta = d''d'\bar{\gamma} + d'd''\gamma = d'd''(\gamma - \bar{\gamma}) = id'd''\varphi ,$$

where $\varphi = -i(\gamma - \bar{\gamma})$.

The converse is a trivial, local statement. Q.E.D.

Chapter II

Chern classes

In order to minimize topological prerequisites, we take the axiomatic approach to Chern classes. This enables us to separate differential geometric aspects of Chern classes from their topological aspects; for the latter the reader is referred to Milnor-Stasheff [1], Hirzebruch [1] and Husemoller [1]. Section 2 is taken from Kobayashi-Nomizu [1; Chapter XII]. For the purpose of reading this book, the reader may take as definition of Chern classes their expressions in terms of curvature. The original approach using Grassmannian manifolds can be found in Chern's book [1].

The Riemann-Roch formula of Hirzebruch, recalled in Section 4, is used only in Section 10 of Chapter V and Section 8 of Chapter VII. Section 5 on symplectic vector bundles will not be used except in Sections 5 and 7 of Chapter VII.

§1 Chern classes

We recall the axiomatic definition of Chern classes (cf. Hirzebruch [1]). We consider the category of complex vector bundles over real manifolds.

Axiom 1 For each complex vector bundle E over M and for each integer $i \geqq 0$, the i-th *Chern class* $c_i(E) \in H^{2i}(M; \boldsymbol{R})$ is given, and $c_0(E)=1$.

We set $c(E)=\sum_{i=0}^{\infty} c_i(E)$ and call $c(E)$ the *total Chern class* of E.

Axiom 2 (Naturality). Let E be a complex vector bundle over M and $f\colon N\to M$ a C^∞ map. Then

$$c(f^*E)=f^*(c(E)) \in H^*(M; \boldsymbol{R}),$$

where f^*E is the pull-back bundle over N.

Axiom 3 (Whitney sum formula). Let $E_1, \cdots, E_q$ be complex line bundles over M. Let $E_1 \oplus \cdots \oplus E_q$ be their Whitney sum. Then

$$c(E_1 \oplus \cdots \oplus E_q)=c(E_1) \cdots c(E_q).$$

To state Axiom 4, we need to define the so-called tautological line bundle $\mathcal{O}(-1)$ over the complex projective space P_nC. It will be defined as a line subbundle of the product vector bundle $P_nC \times C^{n+1}$. A point x of P_nC is a 1-dimensional complex subspace, denoted by L_x, of C^{n+1}. To each $x \in P_nC$ we assign the corresponding line L_x as the fibre over x, we obtain a line subbundle L of $P_nC \times C^{n+1}$. It is customary in algebraic geometry to denote the line bundle L by $\mathcal{O}(-1)$.

Axiom 4 (Normalization). If L is the tautological line bundle over P_1C, then $-c_1(L)$ is the generator of $H^2(P_1C; Z)$; in other words, $c_1(L)$ evaluated (or integrated) on the fundamental 2-cycle P_1C is equal to -1.

In topology, $c_i(E)$ is defined for a topological complex vector bundle E as an element of $H^{2i}(M; Z)$. However, the usual topological proof of the existence and uniqueness of the Chern classes holds in our differentiable case without any modification, (see, for example, Husemoller [1]).

Before we explain another definition of the Chern classes, we need to generalize the construction of the tautological line bundle L.

Let E be a complex vector bundle of rank r over M. At each point x of M, let $P(E_x)$ be the $(r-1)$-dimensional projective space of lines through the origin in the fibre E_x. Let $P(E)$ be the fibre bundle over M whose fibre at x is $P(E_x)$. In other words,

$$P(E) = (E - \text{zero section})/C^* . \tag{1.1}$$

Using the projection $p: P(E) \to M$, we pull back the bundle E to obtain the vector bundle p^*E of rank r over $P(E)$. We define the *tautological line* bundle $L(E)$ over $P(E)$ as a subbundle of p^*E as follows. The fibre $L(E)_\xi$ at $\xi \in P(E)$ is the complex line in $E_{p(\xi)}$ represented by ξ.

We need also the following theorem of Leray-Hirsch in topology.

(1.2) **Theorem** *Let P be a fibre bundle over M with fibre F and projection p. For each $x \in M$, let $j_x: P_x = p^{-1}(x) \to P$ be the injection. Suppose that there exist homogeneous elements $a_1, \cdots, a_r \in H^*(P; R)$ such that for every $x \in M$, $j_x^* a_1, \cdots, j_x^* a_r$ form a basis of $H^*(P_x; R)$. Then $H^*(P; R)$ is a free $H^*(M; R)$-module with basis $a_1, \cdots, a_r$ under the action defined by $p^*: H^*(M; R) \to H^*(P; R)$, i.e.,*

$$H^*(P; R) = p^*(H^*(M; R)) \cdot a_1 \oplus \cdots \oplus p^*(H^*(M; R)) \cdot a_r .$$

Proof For each open set U in M, let $P_U = p^{-1}(U)$ and $j_U : P_U \to P$ be the injection. By the Künneth formula, the theorem is true over an open set U such that P_U is a product $U \times F$. We shall show that if the theorem holds over open sets U, V and $U \cap V$, then it holds over $U \cup V$. Let $d_i = \deg(a_i)$ and denote by x_i an indeterminate of degree d_i. We define two functors K^m and L^m on the open subsets of M by

$$K^m(U) = \sum_{i=1}^{r} H^{m-d_i}(U; \boldsymbol{R}) \cdot x_i ,$$

$$L^m(U) = H^m(P_U; \boldsymbol{R}) .$$

Define a morphism $\alpha_U : K^m(U) \to L^m(U)$ by

$$\alpha_U(\sum s_i x_i) = \sum p^*(s_i) a_i \qquad \text{for} \quad s_i \in H^{m-d_i}(U; \boldsymbol{R}) .$$

Then we have the following commutative diagram:

$$\begin{array}{ccccccccc}
\leftarrow K^m(U \cap V) & \leftarrow & K^m(U) \oplus K^m(V) & \leftarrow & K^m(U \cup V) & \leftarrow & K^{m-1}(U \cap V) & \leftarrow & K^{m-1}(U) \oplus K^{m-1}(V) \\
\downarrow \alpha_{U \cap V} & & \downarrow \alpha_U \oplus \alpha_V & & \downarrow \alpha_{U \cup V} & & \downarrow \alpha_{U \cap V} & & \downarrow \alpha_U \oplus \alpha_V \\
\leftarrow L^m(U \cap V) & \leftarrow & L^m(U) \oplus L^m(V) & \leftarrow & L^m(U \cup V) & \leftarrow & L^{m-1}(U \cap V) & \leftarrow & L^{m-1}(U) \oplus L^{m-1}(V)
\end{array}$$

where the rows are exact Mayer-Vietoris sequence. By our assumption, both $\alpha_{U \cap V}$ and $\alpha_U \oplus \alpha_V$ are isomorphisms. By the "5-lemma", $\alpha_{U \cup V}$ is an isomorphism, which proves that the theorem holds over $U \cup V$. Q.E.D.

Let E be a complex vector bundle over M of rank r. Let E^* be its dual bundle. We apply (1.2) to the projective bundle $P(E^*)$ over M with fibre $P_{r-1}\boldsymbol{C}$. Let $L(E^*)$ be the tautological line bundle over $P(E^*)$. Let g denote the first Chern class of the line bundle $L(E^*)^{-1}$, i.e.,

$$g = -c_1(L(E^*)) \in H^2(P(E^*); \boldsymbol{R}) .$$

From the definition of g it follows that g, restricted to each fibre $P(E^*_x) = P_{r-1}\boldsymbol{C}$, generates the cohomology ring $H^*(P(E^*_x); \boldsymbol{R})$, that is, $1, g, g^2, \cdots, g^{r-1}$, restricted to $P(E^*_x)$, is a basis of $H^*(P(E^*_x); \boldsymbol{R})$. Hence, by (1.2), $H^*(P(E^*); \boldsymbol{R})$ is a free $H^*(M; \boldsymbol{R})$-module with basis $1, g, g^2, \cdots, g^{r-1}$. This implies that there exist uniquely determined cohomology classes $c_i \in H^{2i}(M; \boldsymbol{R})$, $i = 1, \cdots, r$, such that

$$(1.3) \qquad g^r - c_1 g^{r-1} + c_2 g^{r-2} - \cdots + (-1)^r c_r = 0 .$$

Then c_i is the i-th Chern class of E, i.e.,

$$c_i = c_i(E)\,. \tag{1.4}$$

For the proof, see, for example, Husemoller [1]. As Grothendieck did, this can be taken as a definition of the Chern classes. In order to use (1.3) and (1.4) as a definition of $c_i(E)$, we have first to define the first Chern class $c_1(L(E^*))$ of the line bundle $L(E^*)$. The first Chern class of a line bundle can be defined easily in the following manner. Let $\mathscr{A}$ (resp. $\mathscr{A}^*$) be the sheaf of germs of C^∞ complex functions (resp. nowhere vanishing complex functions) over M. Consider the exact sequence of sheaves:

$$0 \longrightarrow \boldsymbol{Z} \xrightarrow{\ j\ } \mathscr{A} \xrightarrow{\ e\ } \mathscr{A}^* \longrightarrow 0\,, \tag{1.5}$$

where j is the natural injection and e is defined by

$$e(f) = e^{2\pi i f} \qquad \text{for} \quad f \in \mathscr{A}\,.$$

This induces the exact sequence of cohomology groups:

$$H^1(M;\mathscr{A}) \xrightarrow{\ e\ } H^1(M;\mathscr{A}^*) \xrightarrow{\ \delta\ } H^2(M;\boldsymbol{Z}) \xrightarrow{\ j\ } H^2(M;\mathscr{A})\,. \tag{1.6}$$

Identifying $H^1(M;\mathscr{A}^*)$ with the group of (equivalence classes of) line bundles over M, we define the first Chern class of a line bundle F by

$$c_1(F) = \delta(F)\,, \qquad F \in H^1(M;\mathscr{A}^*)\,. \tag{1.7}$$

We note that $\delta\colon H^1(M;\mathscr{A}^*) \to H^2(M;\boldsymbol{Z})$ is an isomorphism since $H^k(M;\mathscr{A}) = 0$, $(k \geq 1)$, for any fine sheaf $\mathscr{A}$.

Now the *Chern character* $\mathrm{ch}(E)$ of a complex vector bundle E is defined as follows by means of the formal factorization of the total Chern class:

$$\text{If} \quad \sum c_i(E)x^i = \prod (1+\xi_i x)\,, \quad \text{then} \quad \mathrm{ch}(E) = \sum \exp \xi_i\,. \tag{1.8}$$

While the Chern class $c(E)$ is in $H^*(M;\boldsymbol{Z})$, the Chern character $\mathrm{ch}(E)$ is in $H^*(M;\boldsymbol{Q})$. The Chern character satisfies (see Hirzebruch [1])

$$\begin{aligned} \mathrm{ch}(E \oplus E') &= \mathrm{ch}(E) + \mathrm{ch}(E') \\ \mathrm{ch}(E \otimes E') &= \mathrm{ch}(E)\,\mathrm{ch}(E')\,. \end{aligned} \tag{1.9}$$

The first few terms of the Chern character $\mathrm{ch}(E)$ can be expressed in terms of Chern classes as follows:

$$\mathrm{ch}(E) = r + c_1(E) + \frac{1}{2}(c_1(E)^2 - 2c_2(E)) + \cdots\,, \tag{1.10}$$

where r is the rank of E.

The Chern classes of the conjugate vector bundle $\bar{E}$ are related to those of E by the following formulas (see Milnor-Stasheff [1]):

$$(1.11) \qquad c_i(\bar{E}) = (-1)^i c_i(E), \qquad \mathrm{ch}_i(\bar{E}) = (-1)^i \mathrm{ch}_i(E).$$

An Hermitian structure h in E defines an isomorphism of $\bar{E}$ onto the dual bundle E^* of E; since $h(\xi, \eta)$ is linear in ξ and conjugate linear in η, the map $\eta \longmapsto h(\cdot, \eta)$ defines an isomorphism of $\bar{E}$ onto E^*. This is only a C^∞ isomorphism even when E is a holomorphic vector bundle. From this isomorphism and (1.11) we obtain

$$(1.12) \qquad c_i(E^*) = (-1)^i c_i(E), \qquad \mathrm{ch}_i(E^*) = (-1)^i \mathrm{ch}_i(E).$$

Hence,

$$\begin{aligned}(1.13) \qquad \mathrm{ch}(\mathrm{End}(E)) &= \mathrm{ch}(E \otimes E^*) = \mathrm{ch}(E)\,\mathrm{ch}(E^*) \\ &= (r + \mathrm{ch}_1(E) + \mathrm{ch}_2(E) + \cdots)(r - \mathrm{ch}_1(E) + \mathrm{ch}_2(E) - \cdots) \\ &= (r^2 + 2r \cdot \mathrm{ch}_2(E) - (\mathrm{ch}_1(E))^2 + \cdots).\end{aligned}$$

In particular, from (1.10) and (1.13) we obtain

$$(1.14) \qquad c_1(\mathrm{End}(E)) = 0, \qquad c_2(\mathrm{End}(E)) = 2rc_2(E) - (r-1)c_1(E)^2.$$

We conclude this section by a few comments on Chern classes of coherent sheaves. The results stated below will not be used except in the discussion on Gieseker stability in Section 10 of Chapter V. We assume that the reader is familiar with basic properties of coherent sheaves.

For a fixed complex manifold M, let $S(M)$ (resp. $B(M)$) be the category of coherent sheaves (resp. holomorphic vector bundles, i.e., locally free coherent sheaves) over M. We indentify a holomorphic vector bundle E with the sheaf $\mathcal{O}(E)$ of germs of holomorphic sections of E. Let $\boldsymbol{Z}[S(M)]$ (resp. $\boldsymbol{Z}[B(M)]$) denote the free abelian group generated by the isomorphism classes $[\mathcal{S}]$ of sheaves $\mathcal{S}$ in $S(M)$ (resp. in $B(M)$). For each exact sequence

$$0 \longrightarrow \mathcal{S}' \longrightarrow \mathcal{S} \longrightarrow \mathcal{S}'' \longrightarrow 0$$

of sheaves of $S(M)$ (resp. $B(M)$), we form the element

$$-[\mathcal{S}'] + [\mathcal{S}] - [\mathcal{S}'']$$

of $\boldsymbol{Z}[S(M)]$ (resp. $\boldsymbol{Z}[B(M)]$). Let $A(S(M))$ (resp. $A(B(M))$) be the subgroup of $\boldsymbol{Z}[S(M)]$ (resp. $\boldsymbol{Z}[B(M)]$) generated by such elements. We define the Grothendieck groups

$$K_S(M)=\boldsymbol{Z}[S(M)]/A(S(M)),$$
(1.15)
$$K_B(M)=\boldsymbol{Z}[B(M)]/A(B(M)).$$

We define a multiplication in $\boldsymbol{Z}[S(M)]$ using the tensor product $\mathscr{S}_1 \otimes \mathscr{S}_2$ of sheaves. Then $A(B(M))$ is an ideal of the ring $\boldsymbol{Z}[B(M)]$, and we obtain a commutative ring structure in $K_B(M)$. On the other hand, $A(S(M))$ is not an ideal of $\boldsymbol{Z}[S(M)]$. We have only

$$\boldsymbol{Z}[B(M)]\cdot A(S(M)) \subset A(S(M)), \qquad \boldsymbol{Z}[S(M)]\cdot A(B(M)) \subset A(S(M)).$$

Hence, $K_S(M)$ is a $K_B(M)$-module. The natural injection $\boldsymbol{Z}[B(M)] \rightarrow \boldsymbol{Z}[S(M)]$ induces a $K_B(M)$-module homomorphism

(1.16)
$$i\colon K_B(M) \longrightarrow K_S(M).$$

It follows from (1.9) that the Chern character ch is a ring homomorphism of $K_B(M)$ into $H^*(M; \boldsymbol{Q})$. The following theorem shows that the Chern character is defined also for coherent sheaves when M is projective algebraic.

(1.17) **Theorem** *If M is projective algebraic, the homomorphism i of* (1.16) *is an isomorphism.*

We shall not prove this theorem in detail. We show only how to construct the inverse of i. Given a coherent sheaf $\mathscr{S}$ over M, there is a resolution of $\mathscr{S}$ by vector bundles, i.e., an exact sequence

$$0 \longrightarrow \mathscr{E}_n \longrightarrow \cdots \longrightarrow \mathscr{E}_1 \longrightarrow \mathscr{E}_0 \longrightarrow \mathscr{S} \longrightarrow 0,$$

where $\mathscr{E}_i$ are all locally free coherent sheaves. For a complex manifold M in general, such a resolution exists only locally. If M is projective algebraic, a global resolution exists, (see Serre [1]). Then the inverse of i is given by

$$j\colon [\mathscr{S}] \longmapsto \sum (-1)^i [\mathscr{E}_i].$$

The first Chern class $c_1(\mathscr{S})$ of a coherent sheaf $\mathscr{S}$ can be defined more directly and relatively easily, see Section 6 of Chapter V, and we shall need only $c_1(\mathscr{S})$ in this book except in Section 10 of Chapter V. Recently, a definition of Chern classes for coherent sheaves on general compact complex manifolds has been found, see O'Brian-Toledo-Tong [1].

§2 Chern classes in terms of curvatures

We defined the i-th Chern class $c_i(E)$ of a complex vector bundle E as an element of $H^{2i}(M; \boldsymbol{R})$. Via the de Rham theory, we should be able to represent $c_i(E)$ by a closed $2i$-form γ_i. In this section, we shall construct such a γ_i using the curvature form of a connection in E. For convenience, we imbed $H^{2i}(M; \boldsymbol{R})$ in $H^{2i}(M; \boldsymbol{C})$ and represent $c_i(E)$ by a closed complex $2i$-form γ_i.

We begin with simple algebraic preliminaries. Let V be a complex vector space and let

$$f: V \times \cdots \times V \longrightarrow \boldsymbol{C}$$

be a symmetric multilinear form of degree k, i.e.,

$$f(X_1, \cdots, X_k) \in \boldsymbol{C} \qquad \text{for} \quad X_1, \cdots, X_k \in V,$$

where f is linear in each variable X_i and symmetric in $X_1, \cdots, X_k$. We write

$$f(X) = f(X, \cdots, X) \qquad \text{for} \quad X \in V.$$

Then the mapping $X \rightarrow f(X)$ may be considered as a homogeneous polynomial of degree k. Thus, every symmetric multilinear form of degree k on V gives rise to a homogeneous polynomial function of degree k on V. It is easy to see (for instance, using a basis) that this gives an isomorphism between the algebra of symmetric multilinear forms on V and the algebra of polynomials on V.

If a group G acts linearly on V, then we say that a symmetric multilinear form f is G-invariant if

$$f(a(X_1), \cdots, a(X_k)) = f(X_1, \cdots, X_k) \qquad \text{for} \quad a \in G.$$

Similarly, a homogeneous polynomial f is G-invariant if

$$f(a(X)) = f(X) \qquad \text{for} \quad a \in G.$$

Then every G-invariant form gives rise to a G-invariant polynomial, and vice-versa.

We shall now consider the following special case. Let V be the Lie algebra $\mathfrak{gl}(r; \boldsymbol{C})$ of the general linear group $GL(r; \boldsymbol{C})$, i.e., the Lie algebra of all $r \times r$ complex matrices. Let G be $GL(r; \boldsymbol{C})$ acting on $\mathfrak{gl}(r; \boldsymbol{C})$ by the adjoint action, i.e.,

$$X \in \mathfrak{gl}(r; \boldsymbol{C}) \longrightarrow aXa^{-1} \in \mathfrak{gl}(r; \boldsymbol{C}), \qquad a \in GL(r; \boldsymbol{C}).$$

Now we define homogeneous polynomials f_k on $\mathfrak{gl}(r; \boldsymbol{C})$ of degree $k = 1, 2, \cdots, r$ by

$$\det\left(I_r - \frac{1}{2\pi i}X\right) = 1 + f_1(X) + f_2(X) + \cdots + f_r(X), \qquad X \in \mathfrak{gl}(r; \boldsymbol{C}). \tag{2.1}$$

Since

$$\det\left(I_r - \frac{1}{2\pi i}aXa^{-1}\right) = \det\left(a\left(I_r - \frac{1}{2\pi i}X\right)a^{-1}\right) = \det\left(I_r - \frac{1}{2\pi i}X\right),$$

the polynomials $f_1, \cdots, f_r$ are $GL(r; \boldsymbol{C})$-invariant. It is known that these polynomials generate the algebra of $GL(r; \boldsymbol{C})$-invariant polynomials on $\mathfrak{gl}(r; \boldsymbol{C})$.

Since $GL(r; \boldsymbol{C})$ is a connected Lie group, the $GL(r; \boldsymbol{C})$-invariance can be expressed infinitesimally. Namely, a symmetric multilinear form f on $\mathfrak{gl}(r; \boldsymbol{C})$ is $GL(r; \boldsymbol{C})$-invariant if and only if

$$\begin{aligned} f([Y, X_1], X_2, \cdots, X_k) + f(X_1, [Y, X_2], \cdots, X_k) + \cdots \\ + f(X_1, X_2, \cdots, [Y, X_k]) = 0 \qquad \text{for} \quad X_1, \cdots, X_k, \ Y \in \mathfrak{gl}(r; \boldsymbol{C}). \end{aligned} \tag{2.2}$$

This can be verified by differentiating

$$f(e^{tY}X_1e^{-tY}, \cdots, e^{tY}X_ke^{-tY}) = f(X_1, \cdots, X_k)$$

with respect to t at $t=0$.

Let E be a complex vector bundle of rank r over M. Let D be a connection in E and R its curvature. Choosing a local frame field $s = (s_1, \cdots, s_r)$, we denote the connection form and the curvature form of D by ω and Ω, respectively. Given a $GL(r; \boldsymbol{C})$-invariant symmetric multilinear form f of degree k on $\mathfrak{gl}(r; \boldsymbol{C})$, we set

$$\gamma = f(\Omega) = f(\Omega, \cdots, \Omega). \tag{2.3}$$

If $s' = sa^{-1}$ is another frame field, then the corresponding curvature form Ω' is given by $a\Omega a^{-1}$, (see (I.1.14)). Since f is $GL(r; \boldsymbol{C})$-invariant, it follows that γ is independent of the choice of s and hence is a globally defined differential form of degree $2k$. We claim that γ is closed, i.e.,

$$d\gamma = 0. \tag{2.4}$$

This can be proved by using the Bianchi identity (see (I.5.17)) as follows.

$$d(f(\Omega, \cdots, \Omega)) = D(f(\Omega, \cdots, \Omega)) = f(D\Omega, \cdots, \Omega) + \cdots + f(\Omega, \cdots, D\Omega) = 0$$

since $D\Omega = 0$. Or (see (I.1.13))

$$\begin{aligned} d(f(\Omega, \cdots, \Omega)) &= f(d\Omega, \cdots, \Omega) + \cdots + f(\Omega, \cdots, d\Omega) \\ &= f([\Omega, \omega], \cdots, \Omega) + \cdots + f(\Omega, \cdots, [\Omega, \omega]) = 0 \end{aligned}$$

by (2.2).

Since γ is closed, it represents a cohomology class in $H^{2k}(M; \boldsymbol{C})$. We shall show that the cohomology class does not depend on the choice of connection D. We consider two connections D_0 and D_1 in E and connect them by a line of connections:

$$D_t = (1-t)D_0 + tD_1\,, \qquad 0 \leqq t \leqq 1\,. \tag{2.5}$$

Let ω_t and Ω_t be the connection form and the curvature form of D_t with respect to s. Then

$$\omega_t = \omega_0 + t\alpha\,, \qquad \text{where} \quad \alpha = \omega_1 - \omega_0\,, \tag{2.6}$$

$$\Omega_t = d\omega_t + \omega_t \wedge \omega_t \tag{2.7}$$

so that (see (I.5.16))

$$\frac{d\Omega_t}{dt} = d\alpha + \alpha \wedge \omega_t + \omega_t \wedge \alpha = D_t \alpha\,. \tag{2.8}$$

We set

$$\varphi = k \int_0^1 f(\alpha, \Omega_t, \cdots, \Omega_t) dt\,. \tag{2.9}$$

From (I.1.7) we see that the difference α of two connection forms transforms in the same way as the curvature form under a transformation of the frame field s. It follows that $f(\alpha, \Omega_t, \cdots, \Omega_t)$ is independent of s and hence a globally defined $(2k-1)$-form on M. Therefore, φ is a $(2k-1)$-form on M.

From $D_t \Omega_t = 0$, we obtain

$$\begin{aligned} kdf(\alpha, \Omega_t, \cdots, \Omega_t) &= kD_t f(\alpha, \Omega_t, \cdots, \Omega_t) = kf(D_t\alpha, \Omega_t, \cdots, \Omega_t) \\ &= kf\left(\frac{d\Omega_t}{dt}, \Omega_t, \cdots, \Omega_t\right) = \frac{d}{dt} f(\Omega_t, \Omega_t, \cdots, \Omega_t)\,. \end{aligned}$$

Hence,

$$d\varphi = k \int_0^1 \frac{d}{dt} f(\Omega_t, \cdots, \Omega_t) dt = f(\Omega_1, \cdots, \Omega_1) - f(\Omega_0, \cdots, \Omega_0)\,, \tag{2.10}$$

which proves that the cohomology class of γ does not depend on the connection D.

Using $GL(r; \boldsymbol{C})$-invariant polynomials f_k defined by (2.1), we define

$$\gamma_k = f_k(\Omega)\,, \qquad k = 1, \cdots, r\,. \tag{2.11}$$

In other words,

$$\det\left(I_r - \frac{1}{2\pi i}\Omega\right) = 1 + \gamma_1 + \gamma_2 + \cdots + \gamma_r . \tag{2.12}$$

A simple linear algebra calculation shows

$$\gamma_k = \frac{(-1)^k}{(2\pi i)^k k!} \sum \delta^{j_1 \cdots j_k}_{i_1 \cdots i_k} \Omega^{i_1}_{j_1} \wedge \cdots \wedge \Omega^{i_k}_{j_k} . \tag{2.13}$$

In particular,

$$\gamma_1 = -\frac{1}{2\pi i} \sum \Omega^j_j , \tag{2.14}$$

$$\gamma_2 = -\frac{1}{8\pi^2} \sum (\Omega^j_j \wedge \Omega^k_k - \Omega^j_k \wedge \Omega^k_j) . \tag{2.15}$$

(2.16) **Theorem** *The k-th Chern class $c_k(E)$ of a complex vector bundle E is represented by the closed 2k-form γ_k defined by* (2.11) *or* (2.12).

Proof We shall show that the cohomology classes represented by the γ_k's satisfy the four axioms given in Section 1. Axiom 1 is trivially satisfied. (We simply set $\gamma_0 = 1$.) For Axiom 2, in the bundle f^*E induced from E by $f: N \to M$, we use the connection f^*D induced from a connection D in E. Then its curvature form is given by $f^*\Omega$, (see Section 5, Chapter I). Since $f_k(f^*\Omega) = f^*(f_k(\Omega)) = f^*\gamma_k$, Axiom 2 is satisfied. To verify Axiom 3, let $D_1, \cdots, D_q$ be connections in line bundles $E_1, \cdots, E_q$, respectively. Let $\Omega_1, \cdots, \Omega_q$ be their curvature forms. We use the connection $D = D_1 \oplus \cdots \oplus D_q$ in $E = E_1 \oplus \cdots \oplus E_q$. Then its curvature form Ω is diagonal with diagonal entries $\Omega_1, \cdots, \Omega_q$, (see (I.5.14)). Hence,

$$\det\left(I_r - \frac{1}{2\pi i}\Omega\right) = \left(1 - \frac{1}{2\pi i}\Omega_1\right) \wedge \cdots \wedge \left(1 - \frac{1}{2\pi i}\Omega_q\right),$$

which establishes Axiom 3. Finally, to verify Axiom 4, we take a natural Hermitian structure in the tautological line bundle L over P_1C, i.e., the one arising from the natural inner product in C^2. Since a fibre of L is a complex line through the origin in C^2, each element $\zeta \in L$ is represented by a vector (ζ^0, ζ^1) in C^2. Then the Hermitian structure h is defined by

$$h(\zeta, \zeta) = |\zeta^0| + |\zeta^1|^2 .$$

Considering (ζ^0, ζ^1) as a homogeneous coordinate in P_1C, let $z = \zeta^1/\zeta^0$ be the inhomogeneous coordinate in $U = P_1C - \{(0, 1)\}$. Let s be the frame field of L over U defined by

$$s(z) = (1, z) \in L_z \in C^2 .$$

With respect to s, h is given by the function

$$H(z) = h(s(z), s(z)) = 1 + |z|^2 .$$

The connection form ω and the curvature form Ω are given by (see (I.4.11 & 13))

$$\omega = \frac{\bar{z}dz}{1 + |z|^2}$$

$$\Omega = \frac{-dz \wedge d\bar{z}}{(1 + |z|^2)^2} .$$

Hence,

$$\gamma_1 = \frac{dz \wedge d\bar{z}}{2\pi i(1 + |z|^2)^2} .$$

Using the polar coordinate (r, t) defined by $z = re^{2\pi it}$, we write

$$\gamma_1 = \frac{-2rdr \wedge dt}{(1 + r^2)^2} \qquad \text{on} \quad U .$$

Then integrating γ_1 over P_1C, we obtain

$$\int_{P_1C} \gamma_1 = \int_U \gamma_1 = -\int_0^1 \left(\int_0^\infty \frac{2rdr}{(1 + r^2)^2} \right) dt = -1 .$$

This verifies Axiom 4. Q.E.D.

We shall now express the Chern character $\mathrm{ch}(E)$ in terms of the curvature. Going back to (2.1), we see that if $X \in \mathfrak{gl}(r; C)$ is a diagonal matrix with diagonal entries $i\xi_1, \cdots, i\xi_r$, then

$$\det\left(I_r - \frac{1}{2\pi i} X \right) = \left(1 - \frac{\xi_1}{2\pi} \right) \cdots \left(1 - \frac{\xi_r}{2\pi} \right). \tag{2.17}$$

Consider now the $GL(r; C)$-invariant function defined by the power series

$$\mathrm{Tr}\left(\exp\left(\frac{-1}{2\pi i} X \right) \right) = \mathrm{Tr}\left(\sum_{k=0}^{\infty} \frac{(-1)^k}{k!\,(2\pi i)^k} X^k \right) \qquad \text{for} \quad X \in \mathfrak{gl}(r; C) . \tag{2.18}$$

Again, when X is a diagonal matrix with diagonal entries $i\xi_1, \cdots, i\xi_r$, (2.18) reduces to

$$\sum_{j=1}^{r} \exp\frac{\xi_j}{2\pi} = \sum_{k=0}^{\infty}\left(\frac{1}{k!(2\pi)^k}\sum_{j=1}^{r}(\xi_j)^k\right). \tag{2.19}$$

Comparing (1.8), (2.12), (2.17) and (2.19), we see that the Chern character $\mathrm{ch}(E)$ can be obtained by substituting the curvature form Ω into X in (2.18). In other words, $\mathrm{ch}(E) \in H^*(M; \boldsymbol{R})$ is the cohomology class represented by the closed form

$$\mathrm{Tr}\left(\exp\frac{-1}{2\pi i}\Omega\right). \tag{2.20}$$

In other words, $\mathrm{ch}_k(E)$ is represented by

$$\frac{1}{k!}\left(\frac{-1}{2\pi i}\right)^k \sum \Omega^{j_1}_{j_2} \wedge \Omega^{j_2}_{j_3} \wedge \cdots \wedge \Omega^{j_k}_{j_1}. \tag{2.21}$$

We shall denote the k-th Chern form γ_k of (2.13) by $c_k(E, D)$ when the curvature form Ω comes from a connection D. Similarly, we denote the form (2.21) by $\mathrm{ch}_k(E, D)$. If E is a holomorphic vector bundle with an Hermitian structure h and if D is the Hermitian connection, we write also $c_k(E, h)$ and $\mathrm{ch}_k(E, h)$ for $c_k(E, D)$ and $\mathrm{ch}_k(E, D)$.

Let E be a holomorphic vector bundle over M with an Hermitian structure h. Then the k-th Chern form $c_k(E, h)$ is a (k, k)-form since the curvature is of degree $(1, 1)$. In particular, the first Chern class $c_1(E)$ is represented by (see (I.5.25))

$$c_1(E, h) = -\frac{1}{2\pi i}\sum \Omega^k_k = -\frac{1}{2\pi i}\sum R_{\alpha\bar{\beta}} dz^\alpha \wedge d\bar{z}^\beta. \tag{2.22}$$

Conversely,

(2.23) **Proposition** *Given any closed real $(1, 1)$-form φ representing $c_1(E)$, there is an Hermitian structure h in E such that $\varphi = c_1(E, h)$ provided that M is compact Kähler. (In fact, given any h', a suitable conformal change of h' yields a desired Hermitian structure.)*

Proof Given

$$\varphi = -\frac{1}{2\pi i}\sum S_{\alpha\bar{\beta}} dz^\alpha \wedge d\bar{z}^\beta,$$

we wish to find an Hermitian structure h whose Ricci form is $\sum S_{\alpha\bar{\beta}} dz^\alpha \wedge d\bar{z}^\beta$. Let

$\sum R_{\alpha\bar\beta} dz^\alpha \wedge d\bar z^\beta$ be the Ricci form of any given Hermitian structure h'. Since M is compact Kähler, there is a real function f on M such that (see (I.7.24))

$$\sum R_{\alpha\bar\beta} dz^\alpha \wedge d\bar z^\beta - \sum S_{\alpha\bar\beta} dz^\alpha \wedge d\bar z^\beta = d'd''f,$$

or

$$R_{\alpha\bar\beta} - S_{\alpha\bar\beta} = \partial_\alpha \partial_{\bar\beta} f .$$

On the other hand, we know (see (I.5.25))

$$R_{\alpha\bar\beta} = -\partial_\alpha \partial_{\bar\beta}(\log \det(h'_{i\bar j})) .$$

By setting $h = e^{f/r} h'$ so that $\det(h_{i\bar j}) = e^f \det(h'_{i\bar j})$, we obtain a desired Hermitian structure h. Q.E.D.

§3 Chern classes for flat bundles

In Section 2 of Chapter I, we discussed flat and projectively flat complex vector bundles. Such bundles must satisfy certain simple topological conditions.

(3.1) **Proposition** *Let E be a complex vector bundle of rank r over a manifold M.*

(a) *If E is flat* (*i.e., satisfies one of the three equivalent conditions in* (I.2.5)), *then all its Chern classes $c_i(E) \in H^{2i}(M; \mathbf{R})$ are zero for $i \geqq 1$;*

(b) *If E is projectively flat* (*i.e., satisfies one of the three equivalent conditions in* (I.2.7)), *then its total Chern class $c(E)$ can be expressed in terms of the first Chern class $c_1(E)$ as follows:*

$$c(E) = \left(1 + \frac{c_1(E)}{r}\right)^r .$$

Proof (a) By Condition (2) of (I.2.5), E admits a connection with vanishing curvature. Our assertion follows from (2.15).

(b) By (I.2.8), E admits a connection whose curvature form $\Omega = (\Omega^i_j)$ satisfies the following condition:

$$\Omega = a I_r , \qquad \text{or} \quad \Omega^i_j = a \delta^i_j ,$$

where a is a 2-form on M. By (2.16), the total Chern class $c(E)$ can be represented by a closed form

$$(1 - \frac{a}{2\pi i})^r. \tag{3.2}$$

In particular, $c_1(E)$ is represented by the closed 2-form $(-r/2\pi i)a$. Substituting this back into (3.2), we obtain our assertion (b). Q.E.D.

From (b) of (3.1) it follows that the Chern character $\mathrm{ch}(E)$ of a projectively flat bundle E and the Chern character $\mathrm{ch}(E^*)$ of the dual bundle E^* are given by

$$\mathrm{ch}(E) = r \cdot \exp\left(\frac{1}{r} c_1(E)\right), \qquad \mathrm{ch}(E^*) = r \cdot \exp\left(-\frac{1}{r} c_1(E)\right). \tag{3.3}$$

Hence, if E is projectively flat,

$$\mathrm{ch}(\mathrm{End}(E)) = \mathrm{ch}(E) \cdot \mathrm{ch}(E^*) = r^2. \tag{3.4}$$

This is consistent with the fact that $\mathrm{End}(E)$ is flat if E is projectively flat, (see (I.2.9)). In particular, if E is projectively flat, then

$$2r \cdot c_2(E) - (r-1)c_1(E)^2 = 0. \tag{3.5}$$

Let $\Omega = (\Omega^i_j)$ be the curvature form of D with respect to a local frame field s of E. Let D denote also the naturally induced connection in the dual bundle E^*. Then its curvature form with respect to the dual frame field t is given by $-\Omega$, (see (I.5.5) and (I.5.6)). It follows that for a projectively flat connection, we have the formulas of (3.1), (3.3), (3.4) and (3.5) at the level of differential forms, i.e.,

$$c(E, D) = \left(1 + \frac{1}{r} c_1(E, D)\right)^r, \tag{3.6}$$

$$\mathrm{ch}(E, D) = r \cdot \exp\left(\frac{1}{r} c_1(E, D)\right), \qquad \mathrm{ch}(E^*) = r \cdot \exp\left(-\frac{1}{r} c_1(E, D)\right), \tag{3.7}$$

$$\mathrm{ch}(\mathrm{End}(E), D) = \mathrm{ch}(E, D) \cdot \mathrm{ch}(E^*, D) = r^2, \tag{3.8}$$

$$2r \cdot c_2(E, D) - (r-1)c_1(E, D)^2 = 0. \tag{3.9}$$

§4 Formula of Riemann-Roch

Let E be a holomorphic vector bundle over a compact complex manifold M. Let $H^i(M, E)$ denote the i-th cohomology of M with coefficients in the sheaf $\mathcal{O}(E)$ of germs of holomorphic sections of E. We often write $H^i(M, E)$ instead of $H^i(M, \mathcal{O}(E))$. The Euler characteristic $\chi(M, E)$ is defined by

$$\chi(M, E) = \sum(-1)^i \dim H^i(M, E)\,. \tag{4.1}$$

Let $c(M)$ be the total Chern class of (the tangent bundle of) M, $c(M)=c(TM)$. We defined the Chern character $\mathrm{ch}(E)$ using the formal factorization of the total Chern class $c(E)$. Similarly, we define the total Todd class $\mathrm{td}(M)$ of (the tangent bundle of) M as follows:

$$\mathrm{td}(M) = \prod \frac{\xi_i}{1-e^{-\xi_i}} \qquad \text{if } \sum c_k(M)x^k = \prod(1+\xi_i x)\,. \tag{4.2}$$

The first few terms of $\mathrm{td}(M)$ are given by

$$\mathrm{td}(M) = 1 + \frac{1}{2}c_1(M) + \frac{1}{12}(c_1(M)^2 + c_2(M)) + \cdots\,. \tag{4.3}$$

Now, the Riemann-Roch formula states

$$\chi(M, E) = \int_M \mathrm{td}(M)\mathrm{ch}(E)\,. \tag{4.4}$$

The formula was proved by Hirzebruch when M is an algebraic manifold. It was generalized by Atiyah and Singer to the case where M is a compact complex manifold. The Chern classes and character can be defined for coherent analytic sheaves. The Riemann-Roch formula holds for any coherent analytic sheaf $\mathscr{F}$:

$$\chi(M, \mathscr{F}) = \int_M \mathrm{td}(M)\mathrm{ch}(\mathscr{F})\,. \tag{4.5}$$

For all these, see Hirzebruch [1]. The formula for coherent sheaves has been proved by O'Brian-Toledo-Tong [2].

§ 5 Symplectic vector bundles

We recall first some basic facts about the symplectic group $Sp(m)$. On the vector space $V = \boldsymbol{C}^{2m}$, consider a skew-symmetric bilinear form given by the matrix

$$S = \begin{pmatrix} 0 & I_m \\ -I_m & 0 \end{pmatrix}. \tag{5.1}$$

Then the complex symplectic group $Sp(m; \boldsymbol{C})$ consists of linear transformations of V leaving S invariant. Thus,

$$Sp(m; \boldsymbol{C}) = \{X \in GL(2m; \boldsymbol{C});\ {}^tXSX = S\}\,. \tag{5.2}$$

Its Lie algebra $\mathfrak{sp}(m; \boldsymbol{C})$ is given by

$$\text{(5.3)} \qquad \mathfrak{sp}(m; \boldsymbol{C}) = \{X \in \mathfrak{gl}(2m; \boldsymbol{C});\ {}^t XS + SX = 0\}\,.$$

Now, $Sp(m)$ is a maximal compact subgroup of $Sp(m; \boldsymbol{C})$ defined by

$$\text{(5.4)} \qquad Sp(m) = \{X \in U(2m);\ {}^t XSX = S\} = Sp(m; \boldsymbol{C}) \cap U(2m)\,.$$

Its Lie algebra is given by

$$\text{(5.5)} \qquad \mathfrak{sp}(m) = \{X \in \mathfrak{u}(2m);\ {}^t XS + SX = 0\} = \mathfrak{sp}(m; \boldsymbol{C}) \cap \mathfrak{u}(2m)\,.$$

By simple calculation we obtain

$$\text{(5.6)} \qquad Sp(m) = \left\{ \begin{pmatrix} A & B \\ -\bar{B} & \bar{A} \end{pmatrix};\ {}^t\bar{A}A + {}^t B\bar{B} = I_m,\ {}^t\bar{A}B = {}^t B\bar{A} \right\},$$

$$\text{(5.7)} \qquad \mathfrak{sp}(m) = \left\{ \begin{pmatrix} A & B \\ -\bar{B} & \bar{A} \end{pmatrix};\ A = -{}^t\bar{A},\ B = {}^t B \right\}.$$

We can regard V as a $4m$-dimensional real vector space. We consider the real endomorphisms J_1, J_2, J_3 of V defined by

$$\text{(5.8)} \qquad J_1\begin{pmatrix} z \\ w \end{pmatrix} = \begin{pmatrix} iz \\ iw \end{pmatrix}, \qquad J_2\begin{pmatrix} z \\ w \end{pmatrix} = \begin{pmatrix} -\bar{w} \\ \bar{z} \end{pmatrix}, \qquad J_3\begin{pmatrix} z \\ w \end{pmatrix} = \begin{pmatrix} -i\bar{w} \\ i\bar{z} \end{pmatrix},$$

where $z, w \in \boldsymbol{C}^m$. Each of J_1, J_2, J_3 commutes with every element of $Sp(m)$ and $\mathfrak{sp}(m)$.

Let

$$\boldsymbol{H} = \{a_0 + a_1 i + a_2 j + a_3 k;\ a_0, a_1, a_2, a_3 \in \boldsymbol{R}\}$$

denote the field of quaternions. Then the map $\boldsymbol{H}^* = \boldsymbol{H} - \{0\} \to GL(4m; \boldsymbol{R})$ given by

$$\text{(5.9)} \qquad a_0 + a_1 i + a_2 j + a_3 k \longrightarrow a_0 I + a_1 J_1 + a_2 J_2 + a_3 J_3$$

is a representation. Every element of the form

$$\text{(5.10)} \qquad J = a_1 J_1 + a_2 J_2 + a_3 J_3\,, \qquad a_1^2 + a_2^2 + a_3^2 = 1$$

(which corresponds to a purely imaginary unit quaternion) satisfies $J^2 = -I$ and defines a complex structure in $V = \boldsymbol{R}^{4m}$. In particular, J_1 defines the given complex structure of $V = \boldsymbol{C}^{2m}$ and $-J_1$ defines the conjugate complex structure.

Let E be a complex vector bundle of rank $2m$ over a (real) manifold M. If its structure group $GL(2m; \boldsymbol{C})$ can be reduced to $Sp(m; \boldsymbol{C})$, we call E a *symplectic vector bundle*. On a symplectic vector bundle E we have a non-degenerate skew-symmetric form S:

$$S_x\colon E_x \times E_x \longrightarrow \boldsymbol{C}\,, \qquad x \in M\,,$$

and vice-versa. If the structure group is reduced to $Sp(m; \boldsymbol{C})$, then an Hermitian structure h in E reduces it further to $Sp(m)$.

There are simple topological obstructions to reducing the structure group to $Sp(m)$.

(5.11) **Theorem** *If E is a symplectic vector bundle, then its Chern classes $c_{2j+1}(E) \in H^{4j+2}(M, \boldsymbol{R})$ vanish for all j.*

Proof Let $\bar{E}$ denote the conjugate complex vector bundle. Let E^* be the complex vector bundle dual to E. Since an Hermitian structure h in E defines a conjugate linear isomorphism between E and E^*, it defines a complex linear isomorphism between $\bar{E}$ and E^*. We have, in general, $c_j(E^*) = (-1)^j c_j(E)$, (see (1.12)). Hence, $c_j(\bar{E}) = (-1)^j c_j(E)$. Let J_1 denote the given complex structure of E. Then the complex structure of $\bar{E}$ is given by $-J_1$. But these two complex structures are contained in a connected family of complex structures which is homeomorphic to a 2-sphere, (see (5.10)). Since the Chern classes are integral cohomology classes, they remain fixed under a continuous deformation of complex structures. Hence, $c_j(E) = c_j(\bar{E})$. This combined with $c_j(\bar{E}) = (-1)^j c_j(E)$ yields the stated result.

Q.E.D.

We note that we have shown $2c_{2j+1}(E) = 0$ in $H^{4j+2}(M, \boldsymbol{Z})$. The following differential geometric argument may be of some interest although it says nothing about $c_{2j+1}(E)$ as integral classes. We consider a connection in the associated principal bundle P with the structure group $Sp(m)$. It suffices to show (see (2.16)) that the invariant polynomials corresponding to c_{2j+1} are all zero. Let

$$\text{(5.12)} \qquad F(t, X) = \det\left(tI - \frac{1}{2\pi i} X\right) \quad \text{for} \quad X \in \mathfrak{sp}(m)\,,$$

$$= t^{2m} + f_1(X) t^{2m-1} + \cdots + f_{2m}(X)\,.$$

Clearly, we have

$$\text{(5.13)} \qquad F(-t, X) = t^{2m} - f_1(X) t^{2m-1} + \cdots + f_{2m}(X)\,.$$

On the other hand, using ${}^tX = -SXS^{-1}$ (see (5.5)), we obtain

$$\text{(5.14)} \qquad F(-t, X) = F(t, -X) = F(t, -{}^tX) = F(t, SXS^{-1}) = F(t, X)\,.$$

Hence,

$$f_{2j+1}(X)=0 \qquad \text{for} \quad X\in \mathfrak{sp}(m)\,. \tag{5.15}$$

This argument proves vanishing of the Chern forms

$$c_{2j+1}(E,h)=0 \tag{5.16}$$

provided we use a connection in P (so that both the Hermitian structure h and the symplectic structure are parallel).

What we have shown applies, in particular, to tangent bundles. Let M be a complex manifold of dimension $2m$. A *holomorphic symplectic structure* or *form* is a closed holomorphic 2-form ω on M of maximal rank, i.e., $\omega^m \neq 0$. If we take an Hermitian metric g on M and a connection which makes ω also parallel, then all $(2j+1)$st Chern forms vanish.

The remainder of this section makes use of vanishing theorems and vector bundle cohomology which will be explained in the next Chapter.

Assume that M is a compact Kähler manifold of dimension $2m$. Then every holomorphic form is automatically closed. In particular, every holomorphic 2-form ω of maximal rank defines a holomorphic symplectic structure on M. Since the first Chern class $c_1(M)$ vanishes for such a manifold M, by the theorems of Yau there is a Kähler metric g with vanishing Ricci tensor. By (III.1.34), every holomorphic form, in particular ω, is parallel with respect to this Ricci-flat Kähler metric g. The argument above shows that, for such a metric g, not only the first Chern form $c_1(M,g)$ but also all Chern forms $c_{2j+1}(M,g)$ must vanish.

There are other obstructions to the existence of a holomorphic symplectic form ω. The form ω defines a holomorphic isomorphism between the (holomorphic) tangent bundle and cotangent bundle of M. This has some consequences on the cohomology of M. For example, we have an isomorphism

$$H^p(M,\Lambda^q T)\approx H^p(M,\Lambda^q T^*)\,, \tag{5.17}$$

where T and T^* denote the holomorphic tangent and cotangent bundles of M. On the other hand, since the canonical line bundle of M is trivial, the Serre duality (see (III.2.60)) implies

$$H^p(M,\Lambda^q T)\underset{\text{dual}}{\approx} H^{2m-p}(M,\Lambda^q T^*)\,. \tag{5.18}$$

Combining (5.17) and (5.18), we obtain

$$h^{q,p}=h^{q,2m-p}\,. \tag{5.19}$$

There are other restrictions on Hodge numbers. In the classical argument on primitive forms involving the operators L and Λ (see Section 2 of Chapter III, also Weil [1]), let the holomorphic symplectic form ω play the role of the Kähler form Φ (which is a real symplectic form). Then we obtain inequalities

$$h^{p,q} \leqq h^{p+2,q} \quad \text{for} \quad p+2 \leqq m\,. \tag{5.20}$$

If M is irreducible, its holonomy group is $Sp(m)$ according to Berger's classification of the holonomy groups, (Berger [1]). Since every holomorphic form is parallel, from the representation theory of $Sp(m)$ we see that, for an irreducible M (i.e., with holonomy group $Sp(m)$),

$$\begin{aligned} &h^{1,0} = h^{3,0} = h^{5,0} = \cdots = 0\,, \\ &h^{0,0} = h^{2,0} = h^{4,0} = \cdots = 1\,. \end{aligned} \tag{5.21}$$

On a 2-dimensional compact Kähler manifold M, a holomorphic symplectic form exists if and only if the canonical line bundle is trivial. There are two classes of such Kähler surfaces: (i) complex tori and (ii) $K3$ surfaces.

Clearly, every complex torus of even dimension admits a holomorphic symplectic form. The first higher dimensional example of a simply connected compact Kähler manifold with holomorphic symplectic structure was given by Fujiki. Given a $K3$ surface S, consider $V = (S \times S)/\boldsymbol{Z}_2$, where $\boldsymbol{Z}_2$ acts by interchanging the factors. Since the diagonal of $S \times S$ is fixed, V is a singular variety. By blowing up the diagonal we obtain a 4-dimensional nonsingular Kähler manifold which carries a holomorphic symplectic form. For a systematic study of holomorphic symplectic structures, see Beauville [1], [2], Fujiki [1] and Wakakuwa [1].

Chapter III
Vanishing theorems

In Section 1 we prove Bochner's vanishing theorems and their variants. For Bochner's original theorems, see Yano-Bochner [1]. For a modern exposition, see Wu [1]. These theorems are on vanishing of holomorphic sections or 0-th cohomology groups of holomorphic bundles under some "negativity" conditions on bundles. Vanishing theorems are proved also for Einstein-Hermitian vector bundles which will play a central role in subsequent chapters.

In Section 2 we collect definitions of and formulas relating various operators on Kähler manifolds. The reader who needs more details should consult, for example, Weil [1]. These operators and formulas are used in Section 3 to prove vanishing theorems of Kodaira, Nakano, Vesentini, Girbau and Gigante, i.e., vanishing of higher dimensional cohomology groups for "semi-negative" line bundles.

In Section 4 we prove Bott's vanishing theorem for line bundles over projective spaces. This is needed in Section 5 to relate vector bundle cohomology to line bundle cohomology, (Le Potier's isomorphism theorem). This isomorphism together with vanishing theorems for line bundle cohomology yields the corresponding vanishing theorems for vector bundle cohomology.

Finally, in Section 6 we examine the concept of negativity for vector bundles from view points of algebraic geometry and function theory as well as differential geometry.

Results from Section 1 will be used extensively in subsequent chapters. Although Kodaira's original vanishing theorem will be used a little in Chapter VII, other more general vanishing theorems are not needed in the remainder of the book. The reader whose main interest lies in Einstein-Hermitian vector bundles and stability may skip Sections $2 \sim 6$. The reader who is more interested in cohomology vanishing theorems should consult also the recent monograph by Shiffman-Sommese [1]. It contains also a more extensive bibliography on the subject.

§1 Vanishing theorem for holomorphic sections

Throughout this section, let E be a holomorphic vector bundle of rank r over an

n-dimensional complex manifold M. We recall (see (I.4.9)) that given an Hermitian structure h in E, there is a unique connection D, i.e., the Hermitian connection, such that

$$Dh = 0.$$

Let $s = (s_1, \cdots, s_r)$ be a holomorphic local frame field for E and $t = (t^1, \cdots, t^r)$ the dual frame field for E^*. Given a C^∞ section

$$\xi = \sum \xi^i s_i \in A^0(E),$$

we can write

$$D\xi = D'\xi + d''\xi, \qquad D'\xi = \sum (d'\xi^i + \sum \omega^i_j \xi^j) s_i, \qquad d''\xi = \sum d''\xi^i s_i.$$

Since $D'\xi$ is a $(1,0)$-form and $d''\xi$ is a $(0,1)$-form, we may write

$$(1.1) \qquad d'\xi^i + \sum \omega^i_j \xi^j = \sum \nabla_\alpha \xi^i dz^\alpha,$$

$$(1.2) \qquad d''\xi^i = \sum \nabla_{\bar\beta} \xi^i d\bar z^\beta$$

in terms of a local coordinate system $(z^1, \cdots, z^n)$ of M. The equation above may be regarded as a definition of $\nabla_\alpha \xi^i$ and $\nabla_{\bar\beta} \xi^i$.

Let R be the curvature of D, i.e., $R = D \circ D$. The curvature form $\Omega = (\Omega^i_j)$ with respect to the frame field s is given by (cf. (I.1.11))

$$(1.3) \qquad R(s_j) = \sum \Omega^i_j s_i,$$

Since Ω is a $(1,1)$-form (see (I.4.15))

$$(1.4) \qquad \Omega^i_j = \sum R^i_{j\alpha\bar\beta} dz^\alpha \wedge d\bar z^\beta.$$

We prove the following formula.

(1.5) **Proposition** *If ξ is a holomorphic section of an Hermitian vector bundle (E, h), then*

$$d'd''h(\xi, \xi) = h(D'\xi, D'\xi) - h(R(\xi), \xi),$$

or, in terms of local coordinates,

$$\frac{\partial^2 h(\xi, \xi)}{\partial z^\alpha \partial \bar z^\beta} = \sum h_{i\bar j} \nabla_\alpha \xi^i \overline{\nabla_\beta \xi^j} - \sum h_{i\bar k} R^i_{j\alpha\bar\beta} \xi^j \bar\xi^k.$$

Proof If f is a function, it is a section of the trivial line bundle and hence

$$d''d'f = D''D'f.$$

We apply this to $f=h(\xi, \xi)$. Since ξ is holomorphic,

$$D''\xi=0\,, \qquad \text{or equivalently} \quad D'\bar{\xi}=0\,.$$

Hence,

$$\begin{aligned} d'd''h(\xi, \xi) &= -d''d'h(\xi, \xi) = -D''D'h(\xi, \xi) = -D''h(D'\xi, \xi) \\ &= -h(D''D'\xi, \xi) + h(D'\xi, D'\xi)\,, \end{aligned}$$

where, to derive the third and fourth equalities, we used the fact that h is Hermitian and hence

$$\left.\begin{aligned} D'h(\varphi, \psi) &= h(D'\varphi, \psi) \pm h(\varphi, D''\psi)\,, \\ D''h(\varphi, \psi) &= h(D''\varphi, \psi) \pm h(\varphi, D'\psi)\,. \end{aligned}\right. \qquad \varphi \in A^p(E)\,, \quad \psi \in A^q(E)$$

(The signs $\pm$ depend on the parity of the degree p). Since

$$D''D'\xi = D''D'\xi + D'D''\xi = DD\xi = R(\xi)\,,$$

we obtain

$$d'd''h(\xi, \xi) = -h(R(\xi), \xi) + h(D'\xi, D'\xi)\,. \qquad \text{Q.E.D.}$$

Let g be an Hermitian metric on M and write

$$g = \sum g_{\alpha\bar{\beta}}dz^\alpha d\bar{z}^\beta\,.$$

As usual, we denote the inverse matrix of $(g_{\alpha\bar{\beta}})$ by $(g^{\alpha\bar{\beta}})$. Let

$$\text{(1.6)} \qquad K^i_j = \sum g^{\alpha\bar{\beta}} R^i_{j\alpha\bar{\beta}}\,, \qquad K_{j\bar{k}} = \sum h_{i\bar{k}} K^i_j\,.$$

Then $K=(K^i_j)$ defines an endomorphism of E and $\hat{K}=(K_{j\bar{k}})$ defines the corresponding Hermitian form in E by

$$\text{(1.7)} \qquad K(\xi) = \sum K^i_j \xi^j s_i\,, \qquad \hat{K}(\xi, \eta) = \sum K_{j\bar{k}} \xi^j \bar{\eta}^k$$

for $\xi = \sum \xi^i s_i$ and $\eta = \sum \eta^i s_i$. We call K the *mean curvature transformation* and $\hat{K}$ the *mean curvature form* of E (or more precisely, (E, h, M, g)). We call also both K and $\hat{K}$ simply the *mean curvature of E*.

Taking the trace of the formula in (1.5) with respect to g, we obtain the so-called Weitzenböck's formula:

(1.8) **Proposition** *Let (E, h) be an Hermitian vector bundle over an Hermitian manifold (M, g). If ξ is a holomorphic section of E, then*

$$\sum g^{\alpha\bar{\beta}} \frac{\partial^2 h(\xi, \xi)}{\partial z^\alpha \partial \bar{z}^\beta} = \|D'\xi\|^2 - \hat{K}(\xi, \xi),$$

where

$$\|D'\xi\|^2 = \sum h_{i\bar{j}} g^{\alpha\bar{\beta}} \nabla_\alpha \xi^i \nabla_{\bar{\beta}} \bar{\xi}^j .$$

We are now in a position to prove the following vanishing theorem of Bochner type.

(1.9) **Theorem** *Let (E, h) be an Hermitian vector bundle over a compact Hermitian manifold (M, g). Let D be the Hermitian connection of E and R its curvature. Let $\hat{K}$ be the mean curvature of E.*

(i) *If $\hat{K}$ is negative semi-definite everywhere on M, then every holomorphic section ξ of E is parallel, i.e.,*

$$D\xi = 0$$

and satisfies

$$\hat{K}(\xi, \xi) = 0 .$$

(ii) *If $\hat{K}$ is negative semi-definite everywhere on M and negative definite at some point of M, then E admits no nonzero holomorphic sections.*

The proof is based on the following maximum principle of E. Hopf.

(1.10) **Theorem** *Let U be a domain in $\boldsymbol{R}^m$. Let f, g^{ij}, h^i $(1 \leqq i, j \leqq m)$, be C^∞ real functions on U such that the matrix (g^{ij}) is symmetric and positive definite everywhere on U. If*

$$L(f) := \sum g^{ij} \frac{\partial^2 f}{\partial x^i \partial x^j} + \sum h^i \frac{\partial f}{\partial x^i} \geqq 0 \qquad \text{on} \quad U$$

and f has a relative maximum in the interior of U, then f is a constant function.

Proof of (1.9)

(i) Let $f = h(\xi, \xi)$ and apply (1.10). Let A be the maximum value of f on M. The set $f^{-1}(A) = \{x \in M; f(x) = A\}$ is evidently closed. We shall show that it is also open. Let x_0 be any point such that $f(x_0) = A$, and let U be a coordinate neighborhood around x_0. Since $\hat{K} \leqq 0$, we have

$$L(f)=\sum g^{\alpha\bar{\beta}}\partial_\alpha\partial_{\bar{\beta}}h(\xi,\xi)=\|D'\xi\|^2-\hat{K}(\xi,\xi)\geqq 0\,.$$

By (1.10), $f=A$ in U. Hence, $f^{-1}(A)$ is open. Thus we have shown that $f=A$ on M. This implies $L(f)=0$, which in turn implies $D'\xi=0$ and $\hat{K}(\xi,\xi)=0$. Since ξ is holomorphic, $D''\xi=d''\xi=0$. Hence, $D\xi=0$.

(ii) If ξ is a nonzero holomorphic section of E, then it never vanishes on M since it is parallel by (i). Since $\hat{K}(\xi,\xi)=0$, $\hat{K}$ cannot be definite anywhere on M. Q.E.D.

We shall now derive several consequences of (1.9).

(1.11) **Corollary** *Let (E, h_E) and (F, h_F) be two Hermitian vector bundles over a compact Hermitian manifold (M, g). Let K_E and K_F be the mean curvatures of E and F, respectively.*

(i) *If both $\hat{K}_E$ and $\hat{K}_F$ are negative semi-definite everywhere on M, then every holomorphic section ξ of $E\otimes F$ is parallel, i.e.,*

$$D_{E\otimes F}\xi=0$$

and satisfies

$$\hat{K}_{E\otimes F}(\xi,\xi)=0\,.$$

(ii) *If both $\hat{K}_E$ and $\hat{K}_F$ are negative semi-definite everywhere and either one is negative definite somewhere in M, then $E\otimes F$ admits no nonzero holomorphic sections.*

Proof We know (see (I.5.13)) that the curvature $R_{E\otimes F}$ of $E\otimes F$ is given by

$$R_{E\otimes F}=R_E\otimes I_F+I_E\otimes R_F\,.$$

Taking the trace with respect to g, we obtain

$$K_{E\otimes F}=K_E\otimes I_F+I_E\otimes K_F\,. \tag{1.12}$$

Choosing suitable orthonormal bases in fibres E_x and F_x, we can represent K_E and K_F by diagonal matrices. If $a_1,\cdots,a_r$ and $b_1,\cdots,b_p$ are the diagonal elements of K_E and K_F, respectively, then $K_{E\otimes F}$ is also a diagonal matrix with diagonal elements a_i+b_j. Now our assertion follows from (1.9). Q.E.D.

(1.13) *Remark* From the proof above it is clear that (i) holds if $a_i+b_j\leqq 0$ for all i,j everywhere on M and that (ii) holds if, in addition, $a_i+b_j<0$ for all i,j at some point of M.

The following results are immediate from (1.11).

(1.14) **Corollary** *Let (E, h) be an Hermitian vector bundle over a compact Hermitian manifold (M, g).*

(i) *If $\hat{K}$ is negative semi-definite everywhere on M, then every holomorphic section ξ of a tensor power $E^{\otimes m}$ is parallel and satisfies $\hat{K}_{E^{\otimes m}}(\xi, \xi)=0$.*

(ii) *If, moreover, $\hat{K}$ is negative definite at some point of M, then $E^{\otimes m}$ admits no nonzero holomorphic sections.*

We identify the sheaf $\Omega^p(E)$ of germs of holomorphic p-forms with values in E with the corresponding vector bundle $E\otimes(\bigwedge^p T^*M)$. The latter is an Hermitian vector bundle with the Hermitian structure induced by (E, h) and (M, g). We denote its mean curvature by $K_{E\otimes\wedge^p T^*}$.

(1.15) **Corollary** *Let (E, h) and (M, g) be as in* (1.14).

(i) *If $\hat{K}_E$ is negative semi-definite and $\hat{K}_{TM}$ is positive semi-definite everywhere on M, every holomorphic section ξ of $\Omega^p(E)$ is parallel and satisfies $\hat{K}_{E\otimes\ {}^{p}T^{*}}(\xi, \xi)=0$.*

(ii) *If, moreover, either $\hat{K}_E$ is negative definite or $\hat{K}_{TM}$ is positive definite at some point of M, then $\Omega^p(E)$ has no nonzero sections, i.e.,*

$$H^{p,0}(M, E)=0 \qquad \text{for} \quad p>0 .$$

If F is an Hermitian line bundle, its curvature form Ω can be written locally

$$\Omega=\sum R_{\alpha\bar{\beta}}dz^{\alpha}\wedge d\bar{z}^{\beta} .$$

If, at each point of M, at least one of the eigen-values of $(R_{\alpha\bar{\beta}})$ is negative, then we can choose an Hermitian metric g on M such that the mean curvature $K=\sum g^{\alpha\bar{\beta}}R_{\alpha\bar{\beta}}$ is negative everywhere on M. From (1.9) we obtain

(1.16) **Corollary** *Let (F, h) be an Hermitian line bundle over a compact complex manifold M. If, at each point x of M, the Chern form $c_1(F, h)$ is negative in one direction, i.e., $c_1(F, h)(X, X)<0$ for some vector $X\in T_xM$, then F admits no nonzero holomorphic section.*

If M is Kähler, we can express the condition of (1.16) in terms of the Chern class $c_1(F)$. Let g be a Kähler metric on M and Φ the associated Kähler form. We define the *degree* $\deg(E)$ of a holomorphic vector bundle E by

(1.17) $$\deg(E)=\int_M c_1(E)\wedge\Phi^{n-1}.$$

We note that $\deg(E)$ depends only on the cohomology classes $c_1(E)$ and $[\Phi]$, not on the forms representing them.

Let $\theta^1, \cdots, \theta^n$ be a local orthonormal coframe field on M so that

$$\Phi=i\sum\theta^\alpha\wedge\bar\theta^\alpha.$$

In general, for a $(1,1)$-form $\alpha=i\sum a_{\alpha\bar\beta}\theta^\alpha\wedge\bar\theta^\beta$, we have

(1.18) $$\alpha\wedge\Phi^{n-1}=\frac{1}{n}\left(\sum a_{\alpha\bar\alpha}\right)\Phi^n.$$

Let ξ be a holomorphic section of E. If we multiply the equation of (1.5) by Φ^{n-1}, we obtain

(1.19) $$id'd''h(\xi,\xi)\wedge\Phi^{n-1}=\frac{1}{n}\left(\|D'\xi\|^2-\hat K(\xi,\xi)\right)\Phi^n,$$

which is nothing but the formula of Weitzenböck in (1.8).

Going back to the case of an Hermitian line bundle F, let

$$c_1(F,h)=\frac{i}{2\pi}\sum R_{\alpha\bar\beta}\theta^\alpha\wedge\bar\theta^\beta$$

be its Chern form. Then from (1.18) we obtain

$$c_1(F,h)\wedge\Phi^{n-1}=\frac{1}{2n\pi}\left(\sum R_{\alpha\bar\alpha}\right)\Phi^n.$$

In this case, we can rewrite (1.19) as follows:

(1.20) $$id'd''h(\xi,\xi)\wedge\Phi^{n-1}=\frac{1}{n}\left(\|D'\xi\|^2\Phi^n-2n\pi\|\xi\|^2c_1(F,h)\wedge\Phi^{n-1}\right).$$

If we choose a suitable Hermitian structure h, then $c_1(F,h)$ is harmonic (see (II.2.23)) so that $\sum R_{\alpha\bar\alpha}$ is a constant function on M. Set

$$c=\sum R_{\alpha\bar\alpha}.$$

Then

(1.21) $$2n\pi c_1(F,h)\wedge\Phi^{n-1}=\left(\sum R_{\alpha\bar\alpha}\right)\Phi^n=c\Phi^n.$$

Integrating (1.21) we obtain

(1.22) $$2n\pi\cdot\deg(F)=\int_M c\Phi^n\,.$$

Substituting (1.22) into (1.20) and integrating the resulting equation, we obtain

(1.23) $$0=\int_M (\|D'\xi\|^2-c\|\xi\|^2)\Phi^n\,.$$

From (1.22) and (1.23) we obtain

(1.24) **Theorem** *Let F be a holomorphic line bundle over a compact Kähler manifold M.*

(a) *If* $\deg(F)<0$, *then F admits no nonzero holomorphic sections;*

(b) *If* $\deg(F)=0$, *then every holomorphic section of F has no zeros unless it vanishes identically.*

In fact, a holomorphic section ξ in the case $c=0$ is parallel, i.e., $D'\xi=0$.

(1.25) **Corollary** *Let M be a compact Kähler manifold such that the degree of its tangent bundle is positive. Then its pluri-canonical genera* $P_m=\dim H^0(M, K_M^m)$ *are all zero. (Here* $K_M=\bigwedge^n T^*M$ *is the canonical line bundle of M, and* $K_M^m=K_M^{\otimes m}$.*) If the degree is zero, then* $P_m\leqq 1$.

Proof This follows from (1.24) and the fact that $c_1(M)=-c_1(K_M)$. Q.E.D.

In order to generalize (1.25) to Hermitian manifolds, we shall show negativity of $\hat{K}$ in (1.9) may be replaced by negativity of the average of the maximum eigenvalue of K. For this purpose, we consider conformal changes of h. Let a be a real positive function on M. Given an Hermitian structure h in a holomorphic vector bundle E, we consider a new Hermitian structure $h'=ah$. With respect to a fixed local holomorphic frame field $s=(s_1,\cdots,s_r)$ of E, we calculate the connection forms $\omega=(\omega^i_j)$ and $\omega'=(\omega'^i_j)$ of h and h', respectively. By (I.4.10) we have

$$\omega^i_j=\sum h^{i\bar{k}}d'h_{j\bar{k}}\,,$$

$$\omega'^i_j=\sum\frac{1}{a}h^{i\bar{k}}d'(ah_{j\bar{k}})=\sum h^{i\bar{k}}d'h_{j\bar{k}}+\frac{d'a}{a}\delta^i_j\,.$$

Hence,

(1.26) $$\omega'^i_j=\omega^i_j+d'(\log a)\delta^i_j\,.$$

The relation between the curvature forms $\Omega=(\Omega^i_j)$ and $\Omega'=(\Omega'^i_j)$ of h and h' can be obtained by applying d'' to (1.26). Thus,

$$\Omega'^i_j=\Omega^i_j+d''d'(\log a)\delta^i_j\,. \tag{1.27}$$

In terms of local coordinates $z^1,\cdots,z^n$ of M, (1.27) can be expressed as follows:

$$R'^i_{j\alpha\bar\beta}=R^i_{j\alpha\bar\beta}-\delta^i_j\frac{\partial^2\log a}{\partial z^\alpha\partial\bar z^\beta}\,. \tag{1.28}$$

Taking the trace of (1.28) with respect to the Hermitian metric g, we obtain

$$K'^i_j=K^i_j+\delta^i_j\Box(\log a)\,,\qquad \text{where}\quad \Box=-\sum g^{\alpha\bar\beta}\frac{\partial^2}{\partial z^\alpha\partial\bar z^\beta}\,. \tag{1.29}$$

(1.30) **Theorem** *Let (E,h) be an Hermitian vector bundle over a compact Hermitian manifold (M,g). Let K be the mean curvature of E and $\lambda_1<\cdots<\lambda_r$ be the eigen-values of K. If*

$$\int_M\lambda_r\Phi^n<0\,,\qquad (\Phi^n=\textit{the volume form of } M)\,,$$

then E admits no nonzero holomorphic section.

This follows from (1.9) and the following lemma.

(1.31) **Lemma** *Under the same assumption as in* (1.30), *there is a real positive function a on M such that the mean curvature K' of the new Hermitian structure $h'=ah$ is negative definite.*

Proof Let f be a C^∞ function on M such that $\lambda_r<f$ and $\int_M f\Phi^n=0$. Let u be a solution of

$$\Box u=-f\,. \tag{1.32}$$

From Hodge theory we know that (1.32) has a solution if and only if f is orthogonal to all $\Box$-harmonic functions. But every $\Box$-harmonic function on a compact manifold is constant. Hence, (1.32) has a solution if and only if $\int f\Phi^n=0$. We set $a=e^u$ so that $\Box(\log a)=-f$. Since $\lambda_r<f$, K' is negative definite by (1.29). Q.E.D.

(1.33) **Corollary** *Let M be a compact Hermitian manifold and σ its scalar curvature. If*

$$\int_M \sigma \cdot \Phi^n < 0,$$

then its pluri-canonical genera $P_m = \dim H^0(M, K_M^m)$ *are all zero.*

Proof We apply (1.30) to the Hermitian line bundle $K_M^m = (\bigwedge^n T^*M)^m$. Then its mean curvature is given by $-m\sigma$. Q.E.D.

We consider a compact Kähler manifold (M, g), its tangent bundle (TM, g) and related vector bundles. Since the Ricci tensor of M coincides with the mean curvature $\hat{K}$ of the tangent bundle (TM, g), (see (I.7.23)), (1.9) applied to $E = (T^*M)^{\otimes p}$ yields the following result of Bochner, (see Yano-Bochner [1]):

(1.34) **Theorem** *If M is a compact Kähler manifold with positive semi-definite Ricci tensor, every holomorphic section of $(T^*M)^{\otimes p}$ (in particular, every holomorphic p-form on M) is parallel.*

If, moreover, the Ricci tensor is positive definite at some point, there is no nonzero holomorphic section of $(T^*M)^{\otimes p}$ (in particular, no nonzero holomorphic p-form on M).

If the Ricci tensor is negative semi-definite, applying (1.9) to $E = (TM)^{\otimes p}$ we obtain the dual statement. The case $p = 1$ is of geometric interest.

(1.35) *If M is a compact Kähler manifold with negative semi-definite Ricci tensor, every holomorphic vector field on M is parallel.*

If, moreover, the Ricci tensor is negative definite at some point, then M admits no nonzero holomorphic vector field.

Applying (1.9) to $E = \bigwedge^p T^*M$, we can obtain vanishing of holomorphic p-forms under an assumption weaker than that in (1.34).

(1.36) **Theorem** *Let M be a compact Kähler manifold and fix an integer p, $1 \leqq p \leqq n$. If the eigen-values $r_1, \cdots, r_n$ of the Ricci tensor satisfies the inequality*

$$r_{i_1} + \cdots + r_{i_p} > 0 \qquad \text{for all} \quad i_1 < \cdots < i_p$$

at each point of M, then M admits no nonzero holomorphic p-form.

Instead of negativity of the mean curvature K we shall now consider the following Einstein condition. We say that an Hermitian vector bundle (E, h) over an Hermitian manifold (M, g) satisfies the *weak Einstein condition with factor* φ if

$$(1.37) \qquad K = \varphi I_E, \qquad i.e., \quad K^i_j = \varphi \delta^i_j,$$

where φ is a function on M. If φ is a constant, we say that (E, h) satisfies the *Einstein condition.* In Chapter IV we shall study Einstein-Hermitian vector bundles systematically. We shall see then that if (E, h) satisfies the weak Einstein condition and if (M, g) is a compact Kähler manifold, then with a suitable conformal change $h \to h' = ah$ of the Hermitian structure h the Hermitian vector bundle (E, h') satisfies the Einstein condition. Here, we shall prove only the following

(1.38) **Theorem** *Let (E, h) be an Hermitian vector bundle over a compact Hermitian manifold (M, g). If it satisfies the weak Einstein condition, then every holomorphic section of $E^{\otimes m} \otimes E^{*\otimes m} = (\mathrm{End}(E))^{\otimes m}$, $m \geqq 0$, is parallel.*

Proof The curvature form of E^* is related to that of E by (I.5.5) and (I.5.6). Since the mean curvature of E is given by φI_E, that of E^* is given by $-\varphi I_{E^*}$. From (1.12) it follows that the mean curvature of $E^{\otimes m} \otimes E^{*\otimes m}$ vanishes identically. Now, the theorem follows from (1.9). Q.E.D.

We shall make a few concluding remarks. Myers' theorem states that the fundamental group of a compact Riemannian manifold with positive definite Ricci tensor is finite. Bochner's first vanishing theorem showed that, under the same assumption, the first Betti number of the manifold is zero. Although Bochner's result is weaker than Myers' theorem in this special case, Bochner's technique is more widely applicable and has led to Kodaira's vanishing theorem which will be discussed in Section 3. For a survey on Bochner's technique, see Wu [1]. Combining vanishing of holomorphic p-forms by Bochner (see (1.34)) with the Riemann-Roch-Hirzebruch formula and making use of Myers' theorem, we see that a compact Kähler manifold with positive Ricci tensor is simply connected, (Kobayashi [1]). Yau's solution of the Calabi conjecture allows us to replace the assumption on positivity of the Ricci tensor by the assumption that $c_1(M)$ is positive.

Bochner's vanishing theorem on holomorphic p-forms will be generalized to the cohomology vanishing theorem in Section 3 where the assumption is stated in terms of the Chern class rather than the Ricci tensor. For vanishing theorems of general holomorphic tensors such as (1.34), see Lichnerowicz [1]. The reader interested in vanishing theorems for (non-Kähler) Hermitian manifolds should consult Gauduchon [1].

We may restate (1.35), (also proved first by Bochner), as a theorem on the group of holomorphic transformations of M. *If M is a compact Kähler manifold with negative Ricci tensor, then the group of holomorphic transformations is discrete.* However, we know now that this group is actually finite even under the assumption that M is a projective algebraic manifold of general type, (see Kobayashi [2; pp. 82–88]).

The concept of Einstein-Hermitian manifold was introduced to understand differential geometrically Bogomolov's semistability through (1.38), (see Kobayashi [3]). If (E, h) is the tangent bundle (TM, g) over a Kähler manifold (M, g), the Einstein condition means that (M, g) is an Einstein-Kähler manifold. Hence, the term "Einstein-Hermitian vector bundle".

§2 Operators on Kähler manifolds

Let M be an n-dimensional compact complex manifold with a Kähler metric g. Let Φ be its Kähler form. As before, $A^{p,q}$ denotes the space of C^∞ (p, q)-forms on M. We choose (1,0)-forms $\theta^1, \cdots, \theta^n$ locally to form a unitary frame field for the (holomorphic) cotangent bundle T^*M so that

$$g=\sum \theta^\alpha \bar{\theta}^\alpha \tag{2.1}$$

and

$$\Phi=\sqrt{-1}\sum \theta^\alpha \wedge \bar{\theta}^\alpha . \tag{2.2}$$

For each ordered set of indices $A=\{\alpha_1, \cdots, \alpha_p\}$, we write

$$\theta^A=\theta^{\alpha_1}\wedge \cdots \wedge \theta^{\alpha_p}, \qquad \bar{\theta}^A=\bar{\theta}^{\alpha_1}\wedge \cdots \wedge \bar{\theta}^{\alpha_p}, \tag{2.3}$$

and denote by $A'=\{\alpha_{p+1}, \cdots, \alpha_n\}$ a complementary ordered set of indices. We might as well assume that $\alpha_{p+1}<\cdots<\alpha_n$ although this is not essential. We define the star operator $*$

$$*: A^{p,q} \longrightarrow A^{n-q,n-p} \tag{2.4}$$

as a linear map (over $A^0=A^{0,0}$) satisfying

$$*(\theta^A \wedge \bar{\theta}^B)=(\sqrt{-1})^n \varepsilon(A, B)\theta^{B'} \wedge \bar{\theta}^{A'} , \tag{2.5}$$

where $\varepsilon(A, B)=\pm 1$ is determined by

$$\varepsilon(A, B)=(-1)^{np+n(n+1)/2}\sigma(AA')\cdot\sigma(BB') . \tag{2.6}$$

Here, $\sigma(AA')$ denotes the sign of the permutation (AA'), i.e., it is 1 or -1 according as (AA') is an even or odd permutation of $(1, 2, \cdots, n)$. The reason for the complicated definition (2.6) will become clear soon. It is simple to verify

$$(2.7) \qquad **\varphi=(-1)^{p+q}\varphi \qquad \text{for} \quad \varphi\in A^{p,q}\,.$$

Next, we define the dual operator

$$(2.8) \qquad \bar{*}: A^{p,q} \longrightarrow A^{n-p,n-q} \qquad \text{by} \quad \bar{*}\varphi=*\bar{\varphi}=\overline{*\varphi}\,.$$

The sign in (2.6) is so chosen that we have

$$(2.9) \qquad (\theta^A\wedge\bar{\theta}^B)\wedge\bar{*}(\theta^A\wedge\bar{\theta}^B)=(\sqrt{-1})^n\theta^1\wedge\bar{\theta}^1\wedge\cdots\wedge\theta^n\wedge\bar{\theta}^n = \frac{1}{n!}\Phi^n$$

Given (p, q)-forms

$$\varphi=\frac{1}{p!\,q!}\sum\varphi_{A\bar{B}}\theta^A\wedge\bar{\theta}^B \quad \text{and} \quad \psi=\frac{1}{p!\,q!}\sum\psi_{A\bar{B}}\theta^A\wedge\bar{\theta}^B\,,$$

we have

$$(2.10) \qquad \varphi\wedge\bar{*}\psi=\left(\frac{1}{p!\,q!}\sum\varphi_{A\bar{B}}\bar{\psi}_{A\bar{B}}\right)\frac{1}{n!}\Phi^n\,.$$

We define an inner product in the space of (p, q)-forms on M by setting

$$(2.11) \qquad (\varphi, \psi)=\int_M \varphi\wedge\bar{*}\psi\,.$$

Then

$$(\varphi, \psi)=\overline{(\psi, \varphi)}\,.$$

In general, $e(\psi)$ denotes the exterior multiplication by ψ, i.e.,

$$(2.12) \qquad e(\psi)\cdot\varphi=\psi\wedge\varphi\,.$$

In particular, we write $L=e(\Phi)$, i.e.,

$$(2.13) \qquad L\varphi=\Phi\wedge\varphi$$

so that

$$L: A^{p,q} \longrightarrow A^{p+1,q+1}\,.$$

We set

(2.14) $$\Lambda = *^{-1} \circ L \circ * : A^{p,q} \longrightarrow A^{p-1,q-1}.$$

If φ is a (p, q)-form and ψ is a $(p+1, q+1)$-form, then

(2.15) $$\varphi \wedge \bar{*} \Lambda \psi = L\varphi \wedge \bar{*} \psi.$$

Integrating (2.15) we obtain

(2.16) $$(\varphi, \Lambda\psi) = (L\varphi, \psi),$$

showing that Λ is the adjoint of L. We note that both L and Λ as well as the star operator $*$ are algebraic operators in the sense that they are defined at each point of M and are linear over A^0.

Given a (p, q)-form

$$\varphi = \frac{1}{p!\,q!} \sum \varphi_{A\bar{B}} \theta^A \wedge \bar{\theta}^B,$$

we have

(2.17) $$\Lambda\varphi = \frac{1}{(p-1)!\,(q-1)!} \sum \varphi'_{A^*\bar{B}^*} \theta^{A^*} \wedge \bar{\theta}^{B^*},$$

where

$$\varphi'_{A^*\bar{B}^*} = (-1)^p \sqrt{-1} \sum_{\alpha} \varphi_{\alpha A^* \bar{\alpha} \bar{B}^*}.$$

(The notation αA^* stands for $(\alpha, \alpha_1, \cdots, \alpha_{p-1})$ if $A^* = (\alpha_1, \cdots, \alpha_{p-1})$. This formula may be derived, for instance, from (2.15).

The commutation relation between L and Λ is given by

(2.18) $$(\Lambda L - L\Lambda)\varphi = (n-p-q)\varphi \qquad \text{for} \quad \varphi \in A^{p,q}.$$

We define

(2.19) $$\begin{aligned} \delta' &= -\bar{*} d' \bar{*} = -* d'' * : A^{p,q} \longrightarrow A^{p-1,q}, \\ \delta'' &= -\bar{*} d'' \bar{*} = -* d' * : A^{p,q} \longrightarrow A^{p,q-1}, \\ \delta &= \delta' + \delta''. \end{aligned}$$

Then for differential forms φ and ψ defined on M of appropriate degrees, we have

(2.20) $$(d\varphi, \psi) = (\varphi, \delta\psi), \qquad (d'\varphi, \psi) = (\varphi, \delta'\psi), \qquad (d''\varphi, \psi) = (\varphi, \delta''\psi).$$

Let $(e_1, \cdots, e_n)$ be the local unitary frame field of the (holomorphic) tangent bundle TM dual to $(\theta^1, \cdots, \theta^n)$. Corresponding to the formula

$$\begin{aligned} d' &= \sum e(\theta^\alpha)\nabla_{e_\alpha} \qquad (\text{i.e., } d'\varphi = \sum \theta^\alpha \wedge \nabla_{e_\alpha}\varphi), \\ d'' &= \sum e(\bar\theta^\alpha)\nabla_{\bar e_\alpha} \qquad (\text{i.e., } d''\varphi = \sum \bar\theta^\alpha \wedge \nabla_{\bar e_\alpha}\varphi), \end{aligned} \tag{2.21}$$

we have

$$\begin{aligned} \delta' &= -\sum \iota(e_\alpha)\nabla_{\bar e_\alpha}, \\ \delta'' &= -\sum \iota(\bar e_\alpha)\nabla_{e_\alpha}. \end{aligned} \tag{2.22}$$

We have the following commutation relations among the various operators introduced above.

$$\begin{aligned} & d'L - Ld' = 0, \qquad d''L - Ld'' = 0, \\ & \delta'\Lambda - \Lambda\delta' = 0, \qquad \delta''\Lambda - \Lambda\delta'' = 0, \\ & L\delta' - \delta' L = \sqrt{-1}d'', \qquad L\delta'' - \delta'' L = -\sqrt{-1}d', \\ & \Lambda d' - d'\Lambda = \sqrt{-1}\delta'', \qquad \Lambda d'' - d''\Lambda = -\sqrt{-1}\delta'. \end{aligned} \tag{2.23}$$

Corresponding to the well known formulas

$$d'd' = 0, \qquad d''d'' = 0, \qquad d'd'' + d''d' = 0, \tag{2.24}$$

which follow from $dd = 0$, we have

$$\delta'\delta' = 0, \qquad \delta''\delta'' = 0, \qquad \delta'\delta'' + \delta''\delta' = 0. \tag{2.25}$$

Further, we have

$$\begin{aligned} d'\delta'' &= -\delta'' d' = -\sqrt{-1}\delta'' L\delta'' = -\sqrt{-1}d'\Lambda d'', \\ d''\delta' &= -\delta' d'' = \sqrt{-1}\delta' L\delta' = \sqrt{-1}d''\Lambda d'. \end{aligned} \tag{2.26}$$

We recall that the Laplacian Δ is defined by

$$\Delta = d\delta + \delta d. \tag{2.27}$$

If we set

$$\square = d''\delta'' + \delta'' d'', \tag{2.28}$$

then

$$\square = d'\delta' + \delta' d' = \frac{1}{2}\Delta \tag{2.29}$$

and

$$\Delta L = L\Delta, \qquad \Box L = L\Box. \tag{2.30}$$

Let E be a holomorphic vector bundle of rank r on a compact Kähler manifold M. Let $A^{p,q}(E)$ denote the space of C^∞ (p, q)-forms with values in E. The operators $*$, L, and d'' defined above extend to $A^{p,q}(E)$. Consequently, $\delta' = -*d''*$ also extends to $A^{p,q}(E)$. In order to generalize d' and δ'' as operators on $A^{p,q}(E)$, we choose an Hermitian structure h in E and use the Hermitian connection D in E. We recall (I.4.9) that

$$D = D' + d'', \tag{2.31}$$

where

$$D' : A^{p,q}(E) \longrightarrow A^{p+1,q}(E).$$

Using D' in place of d', we define

$$\delta''_h = -*D'* : A^{p,q}(E) \longrightarrow A^{p,q-1}(E). \tag{2.32}$$

Using a local frame field $s = (s_1, \cdots, s_r)$ of E, let $\varphi = \sum \varphi^i s_i$. Then

$$\delta''_h(\sum \varphi^i s_i) = \sum (\delta'' \varphi^i - *\sum \omega^i_k \wedge *\varphi^k) s_i. \tag{2.33}$$

In fact, if φ is an r-form, then

$$\begin{aligned}\delta''_h(\sum \varphi^i s_i) &= -*D'*(\sum \varphi^i s_i) = -\sum (*d'*\varphi^i)s_i - \sum (-1)^r *(*\varphi^i \wedge D's_i)\\ &= \sum (-*d'*\varphi^i - (-1)^r * \sum *\varphi^k \wedge \omega^i_k) s_i\\ &= \sum (\delta''\varphi^i - *\sum \omega^i_k \wedge *\varphi^k) s_i.\end{aligned}$$

We define the local inner product $h(\varphi, \psi)$ of $\varphi = \sum \varphi^i s_i$ and $\psi = \sum \psi^j s_j$ by

$$h(\varphi, \psi) = \sum h_{i\bar{j}} \varphi^i \wedge *\bar{\psi}^j \tag{2.34}$$

and the global inner product (φ, ψ) by

$$(\varphi, \psi) = \int_M h(\varphi, \psi). \tag{2.35}$$

Then

$$\begin{aligned} &(\varphi, \psi) = \overline{(\psi, \varphi)}, && (L\varphi, \psi) = (\varphi, \Lambda\psi),\\ &(D'\varphi, \psi) = (\varphi, \delta'\psi), && (d''\varphi, \psi) = (\varphi, \delta''_h\psi). \end{aligned} \tag{2.36}$$

For example, to prove the last equality, let φ be a $(p, q-1)$-form and ψ a (p, q)-form. Then

$$\begin{aligned} d(h(\varphi, \psi)) &= d''(h(\varphi, \psi)) \\ &= \sum d''h_{i\bar{j}} \wedge \varphi^i \wedge *\bar{\psi}^j + \sum h_{i\bar{j}} d''\varphi^i \wedge *\bar{\psi}^j + \sum h_{i\bar{j}} \varphi^i \wedge **d''*\bar{\psi}^j \\ &= h(d''\varphi, \psi) + \sum h_{i\bar{j}} \varphi^i \wedge **h^{k\bar{j}} d'' h_{k\bar{m}} \wedge *\bar{\psi}^m + \sum h_{i\bar{j}} \varphi^i \wedge **d''*\bar{\psi}^j \\ &= h(d''\varphi, \psi) - \sum h_{i\bar{j}} \varphi^i \wedge *(-\overline{*d'*\psi^j} - \sum *\bar{\omega}^j_m \wedge *\bar{\psi}^m) \\ &= h(d''\varphi, \psi) - h(\varphi, \delta''_h \psi), \qquad \text{(using (2.33))}. \end{aligned}$$

We now list various commuting relations between these operators acting on $A^{p,q}(E)$. They all generalize the formulas for the operators acting on $A^{p,q}$.

$$(2.37) \qquad (\Lambda L - L\Lambda)\varphi = (n-p-q)\varphi \qquad \text{for} \quad \varphi \in A^{p,q}(E).$$

$$(2.38) \qquad \begin{aligned} D'L - LD' &= 0, \qquad & d''L - Ld'' &= 0, \\ \delta'\Lambda - \Lambda\delta' &= 0, \qquad & \delta''_h\Lambda - \Lambda\delta''_h &= 0. \end{aligned}$$

$$(2.39) \qquad \begin{aligned} L\delta' - \delta'L &= \sqrt{-1}d'', \qquad & L\delta''_h - \delta''_h L &= -\sqrt{-1}D', \\ \Lambda D' - D'\Lambda &= \sqrt{-1}\delta''_h, \qquad & \Lambda d'' - d''\Lambda &= -\sqrt{-1}\delta'. \end{aligned}$$

Clearly, (2.38) and (2.39) generalize (2.23). The formulas involving only, L, Λ, d'' and δ' follow immediately from the corresponding formulas in (2.23). The remaining formulas can be obtained from the corresponding formulas in (2.23) with the aid of normal frame fields (I.4.19). We can see their validity also by observing that their adjoints involve only L, Λ, d'' and δ'. For example,

$$((D'L - LD')\varphi, \psi) = (\varphi, (\Lambda\delta' - \delta'\Lambda)\psi),$$

thus reducing the formula $D'L - LD' = 0$ to the formula $\Lambda\delta' - \delta'\Lambda = 0$.

Since

$$R = D \circ D = (D' + d'')(D' + d'') = D'D' + (D'd'' + d''D') + d''d''$$

and since the curvature R is of degree $(1, 1)$, we obtain

$$(2.40) \qquad D'D' = 0, \qquad d''d'' = 0, \qquad D'd'' + d''D' = R,$$

which generalizes (2.24). The last formula should be understood as follows. If $\Omega = (\Omega^i_j)$ denotes the curvature form with respect to a local frame field $s = (s_1, \cdots, s_r)$, then

$$(D'd''+d''D')\varphi=\sum \Omega^i_j\wedge\varphi^j s_i \qquad \text{for} \quad \varphi=\sum\varphi^i s_i\,.$$

Thus, the curvature R in (2.40) represents a combination of the exterior multiplication and the linear transformation by R. Denoting this operator by $e(R)$, we write the last formula of (2.40) as follows:

$$D'd''+d''D'=e(R)\,. \tag{2.41}$$

From (2.40) (and (2.41)) we obtain the following

$$\delta''_h\delta''_h=0\,, \qquad \delta'\delta'=0\,, \qquad \delta''_h\delta'+\delta'\delta''_h=-*^{-1}e(R)*\,. \tag{2.42}$$

For example, the last formula in (2.42) follows from (2.41) with the aid of (2.7). In fact, if φ is an r-form with values in E, then

$$\begin{aligned}(\delta''_h\delta'+\delta'\delta''_h)\varphi&=(*D'**d''*+*d''**D'*)\varphi=(-1)^{r+1}*(D'd''+d''D')*\varphi\\ &=(-1)^{r+1}*e(R)*\varphi=-*^{-1}e(R)*\varphi\,.\end{aligned}$$

This last formula is known as *Nakano's formula*.

Generalizing (2.28) to vector bundle valued forms, we define

$$\Box_h=d''\delta''_h+\delta''_h d''\,. \tag{2.43}$$

Let $\boldsymbol{H}^{p,q}(E)$ denote the space of $\Box_h$-harmonic (p,q)-forms with values in E, i.e.,

$$\boldsymbol{H}^{p,q}(E)=\mathrm{Ker}(\Box_h\colon A^{p,q}(E)\longrightarrow A^{p,q}(E))\,. \tag{2.44}$$

We recall that the *d''-cohomology* or the *Dolbeault cohomology* of M with coefficients in E is defined to be

$$H^{p,q}(M,E)=\frac{\mathrm{Ker}(d''\colon A^{p,q}(E)\longrightarrow A^{p,q+1}(E))}{\mathrm{Image}(d''\colon A^{p,q-1}(E)\longrightarrow A^{p,q}(E))}\,. \tag{2.45}$$

The *Dolbeault isomorphism theorem* states

$$H^{p,q}(M,E)\approx H^q(M,\Omega^p(E))\,. \tag{2.46}$$

This is the d''-analogue of the de Rham isomorphism theorem.

The *Hodge theorem* states:

(2.47) **Theorem** (i) $\dim \boldsymbol{H}^{p,q}(E)<\infty$;
This allows us to define the orthogonal projection

$$H\colon A^{p,q}(E)\longrightarrow \boldsymbol{H}^{p,q}(E)\,;$$

(ii) *There is a unique operator (called the Green's operator)*

$$G: A^{p,q}(E) \longrightarrow A^{p,q}(E)$$

such that $G(H^{p,q}(E))=0$, $d''\circ G=G\circ d''$, $\delta''_h\circ G=\delta''_h\circ G$ *and*

$$H+\Box_h\circ G=I,$$

or more explicitly

$$\varphi=H\varphi+d''(\delta''_h G\varphi)+\delta''_h(d''G\varphi) \qquad \text{for} \quad \varphi\in A^{p,q}(E).$$

This implies immediately the following isomorphism:

(2.48) $$H^{p,q}(M,E)\approx \boldsymbol{H}^{p,q}(E).$$

For the proofs of the Dolbeault theorem and the Hodge theorem, see for example Griffiths-Harris [1], where the theorems are proved when E is a trivial line bundle.

We can define a product

$$A^{p,q}(E)\times A^{r,s}(E^*) \longrightarrow A^{p+r,q+s}$$
$$(\varphi,\psi)\longmapsto \varphi\wedge\psi$$

in a natural way. If $r=n-p$ and $s=n-q$, then we can define a pairing

$$\iota: A^{p,q}(E)\times A^{n-p,n-q}(E^*)\longrightarrow \boldsymbol{C}$$
$$(\varphi,\psi)\longmapsto \int_M \varphi\wedge\psi.$$

This pairing induces a pairing

(2.49) $$\iota: H^{p,q}(M,E)\times H^{n-p,n-q}(M,E^*)\longrightarrow \boldsymbol{C}$$

(2.50) Duality Theorem of Serre. *The pairing* ι *in* (2.49) *is a dual pairing.*

For the proof, see Serre [2] which contains also the proof of the Dolbeault isomorphism theorem (2.46) for vector bundle cohomology. The reader can obtain concise outlines of the proofs of (2.46), (2.48) and (2.49) in Hirzebruch [1].

Summarizing the three preceding theorems, we have

sheaf cohomology *d''-cohomology* $\square_h$-*harmonic forms*

$$H^q(M, \Omega^p(E)) \overset{\text{Dolbeault}}{\approx} H^{p,q}(M, E) \overset{\text{Hodge-Kodaira}}{\approx} \boldsymbol{H}^{p,q}(E)$$

Serre duality (between $H^{p,q}(M, E)$ and $H^{n-p,n-q}(M, E)$)

$$H^{n-p,n-q}(M, E)$$

§3 Vanishing theorems for line bundle cohomology

Let F be a holomorphic line bundle over a compact complex manifold M of dimension n. Let $c_1(F) \in H^2(M; \boldsymbol{R})$ denote the (real) first Chern class of F.

We say that $c_1(F)$ is *negative* (resp. *semi-negative*, *positive*, *semi-positive*, *of rank* $\geqq k$) and write $c_1(F) < 0$ (resp. $c_1(F) \leqq 0$, $c_1(F) > 0$, $c_1(F) \geqq 0$, rank $c_1(F) \geqq k$) if the cohomology class $c_1(F)$ can be represented by a closed real (1, 1)-form

$$\varphi = -\frac{1}{2\pi i} \sum \varphi_{\alpha\bar{\beta}} dz^\alpha \wedge d\bar{z}^\beta$$

such that at each point x of M the Hermitian matrix $(\varphi_{\alpha\bar{\beta}}(x))$ is negative definite (resp. negative semi-definite, positive definite, positive semi-definite, of rank $\geqq k$). We say that F is *negative* (resp. *positive*) if $c_1(F)$ is negative (resp. positive).

Vanishing theorems for higher dimensional cohomology began with Kodaira [1].

(3.1) **Vanishing theorem of Kodaira** *If* $c_1(F) < 0$, *then*

$$H^q(M; \Omega^0(F)) = 0 \qquad \textit{for} \quad q \leqq n-1\,.$$

His theorem has been generalized as follows.

(3.2) **Vanishing theorem of Nakano** (Akizuki-Nakano [1]) *If* $c_1(F) < 0$, *then*

$$H^q(M; \Omega^p(F)) = 0 \qquad \textit{for} \quad p+q \leqq n-1\,.$$

(3.3) **Vanishing theorem of Vesentini** [1] *If M is Kähler and* $c_1(F) \leqq 0$ *with rank* $c_1(F) \geqq k$, *then*

$$H^q(M, \Omega^0(F)) = 0 \qquad \textit{for} \quad q \leqq k-1\,,$$

$$H^0(M, \Omega^p(F)) = 0 \qquad \textit{for} \quad p \leqq k-1\,.$$

(3.4) **Vanishing theorem of Gigante** [1] **and Girbau** [1] *If M is Kähler and $c_1(F) \leqq 0$ with rank $c_1(F) \geqq k$, then*

$$H^q(M, \Omega^p(F)) = 0 \qquad \textit{for} \quad p+q \leqq k-1 .$$

Since (3.4) implies the preceding three theorems, we shall prove (3.4). By (II.2.23), there is an Hermitian structure h in F such that $\varphi_{\alpha\bar{\beta}} = R_{\alpha\bar{\beta}}$ so that $(R_{\alpha\bar{\beta}})$ is negative semi-definite and of rank $\geqq k$ everywhere on M. Throughout the proof, we fix such an Hermitian structure h in F. We fix also an arbitrarily chosen Kähler metric g on M and use various operators and formulas explained in Section 2. In particular, $\boldsymbol{H}^{p,q}(F)$ denotes the space of $\square_h$-harmonic (p, q)-forms with values in F.

(3.5) **Nakano's inequality** *If $\varphi \in \boldsymbol{H}^{p,q}(F)$, then*

$$\sqrt{-1}((\Lambda e(R) - e(R)\Lambda)\varphi, \varphi) = \|D'\varphi\|^2 + \|\delta'\varphi\|^2 \geqq 0 .$$

Proof From $\square_h \varphi = 0$, we obtain

$$d''\varphi = 0 , \qquad \delta''_h \varphi = 0 .$$

From (2.41), we obtain

$$e(R)\varphi = d''D'\varphi .$$

Making use of (2.39), we obtain

$$\begin{aligned} \sqrt{-1}(\Lambda e(R)\varphi, \varphi) &= \sqrt{-1}(\Lambda d''D'\varphi, \varphi) \\ &= \sqrt{-1}(d''\Lambda D'\varphi, \varphi) + (\delta'D'\varphi, \varphi) \\ &= (D'\varphi, D'\varphi) . \end{aligned}$$

Similarly, we have

$$\begin{aligned} -\sqrt{-1}(e(R)\Lambda\varphi, \varphi) &= -\sqrt{-1}(d''D'\Lambda\varphi, \varphi) - \sqrt{-1}(D'd''\Lambda\varphi, \varphi) \\ &= (D'\delta'\varphi, \varphi) - \sqrt{-1}(D'\Lambda d''\varphi, \varphi) \\ &= (\delta'\varphi, \delta'\varphi) . \end{aligned}$$

This completes the proof of (3.5).

Before we start the proof of (3.4) we should remark that (3.2) follows immediately from (3.5). In fact, if $(R_{\alpha\bar{\beta}})$ is negative definite, we can use $-\sum R_{\alpha\bar{\beta}} dz^\alpha d\bar{z}^\beta$ as our Kähler metric g. Then $L = \sqrt{-1}e(R)$ and the left hand side of

(3.5) is equal to $-((\Lambda L-L\Lambda)\varphi, \varphi)$. Hence, (3.5) in this case reads as follows:

$$((\Lambda L-L\Lambda)\varphi, \varphi)\leqq 0\,.$$

On the other hand, by (2.18) we have

$$((\Lambda L-L\Lambda)\varphi, \varphi)=(n-p-q)(\varphi, \varphi)\,.$$

Hence, $\varphi=0$ if $p+q<n$.

To prove (3.4), at each point x of M, we choose (1, 0)-forms $\theta^1, \cdots, \theta^n$ to form a unitary frame for the cotangent space T^*_xM as in Section 2. We may choose them in such a way that the curvature R of h is diagonal with respect to this frame, i.e.,

$$R_{\alpha\bar\beta}(x)=r_\alpha\delta_{\alpha\beta} \qquad \alpha, \beta=1, \cdots, n\,,$$

where $r_1, \cdots, r_n$ are real numbers since $(R_{\alpha\bar\beta})$ is Hermitian. For an ordered set of indices $A=\{\alpha_1, \cdots, \alpha_p\}$, we write $\theta^A=\theta^{\alpha_1}\wedge\cdots\wedge\theta^{\alpha_p}$ and $\bar\theta^A=\bar\theta^{\alpha_1}\wedge\cdots\wedge\bar\theta^{\alpha_p}$ as in (2.3). Let $\varphi\in A^{p,q}(F)$ and write at x

$$\varphi=\sum\varphi_{A\bar B}\theta^A\wedge\bar\theta^B\,, \qquad \varphi_{A\bar B}\in F_x\,.$$

Then we have the following formula due to Gigante.

$$(3.6) \qquad \sqrt{-1}(\Lambda e(R)\varphi-e(R)\Lambda\varphi)_{A\bar B}=\Bigl(-\sum_{\alpha\in A\cap B}r_\alpha+\sum_{\beta\notin A\cup B}r_\beta\Bigr)\varphi_{A\bar B} \qquad \text{at} \quad x\,.$$

Proof of (3.6) Given two (p, q)-forms φ and ψ at x, we denote their inner product at x by (φ, ψ). Thus,

$$(\varphi, \psi)\frac{\Phi^n}{n!}=\varphi\wedge\bar{*}\psi\,.$$

Omitting exterior product symbols $\wedge$ in the following calculation, we have

$$\begin{aligned}\sqrt{-1}(\Lambda e(R)\theta^A\bar\theta^B, \theta^C\bar\theta^D)&=\sqrt{-1}(e(R)\theta^A\bar\theta^B, L(\theta^C\bar\theta^D))\\ &=\sum_{\alpha,\beta}r_\alpha(\theta^\alpha\bar\theta^\alpha\theta^A\bar\theta^B, \theta^\beta\bar\theta^\beta\theta^C\bar\theta^D)\,.\end{aligned}$$

In this summation, a term does not vanish only when α and β satisfy one of the following two conditions:

(i) $\alpha\notin A\cup B, \beta\notin C\cup D, \alpha\neq\beta$ and $\theta^\alpha\bar\theta^\alpha\theta^A\bar\theta^B=\varepsilon\theta^\beta\bar\theta^\beta\theta^C\bar\theta^D$ with $\varepsilon=\pm 1$,

or

(ii) $\alpha=\beta\notin A\cup B$, and $\theta^A\bar\theta^B=\varepsilon\theta^C\bar\theta^D$ with $\varepsilon=\pm 1$.

On the other hand, using

$$-\sqrt{-1}(e(R)\Lambda\theta^A\bar{\theta}^B, \theta^C\bar{\theta}^D)\frac{\Phi^n}{n!} = -\sqrt{-1}e(R)(\bar{*}^{-1}L\bar{*}\theta^A\theta^B)\bar{*}(\theta^C\bar{\theta}^D)$$

$$= -\sqrt{-1}(e(R)\bar{*}\theta^C\bar{\theta}^D, L\bar{*}\theta^A\bar{\theta}^B)\frac{\Phi^n}{n!},$$

we obtain

$$-\sqrt{-1}(e(R)\Lambda\theta^A\bar{\theta}^B, \theta^C\bar{\theta}^D) = -\sum_{\alpha,\beta} r_\alpha(\theta^\alpha\bar{\theta}^\alpha\bar{*}\theta^C\bar{\theta}^D, \theta^\beta\bar{\theta}^\beta\bar{*}\theta^A\bar{\theta}^B).$$

In this summation, a term does not vanish only when α and β satisfy one of the following two conditions:

(iii) $\alpha\in C\cap D$, $\beta\in A\cap B$, $\alpha\neq\beta$ and $\theta^\alpha\bar{\theta}^\alpha\bar{*}\theta^C\bar{\theta}^D = \varepsilon\theta^\beta\bar{\theta}^\beta\bar{*}\theta^A\bar{\theta}^B$, $\varepsilon=\pm 1$,

or

(iv) $\alpha=\beta\in A\cap B$ and $\theta^A\bar{\theta}^B = \varepsilon\theta^C\bar{\theta}^D$, $\varepsilon=\pm 1$.

Now we claim that the terms coming from (i) and the terms coming from (iii) cancel each other out. In fact, we observe first that in both (i) and (iii) we have (up to permutations)

$$(\alpha, A) = (\beta, C), \qquad (\alpha, B) = (\beta, D).$$

Calculation using (2.5) shows that the sign ε in (i) agrees with ε in (iii). Thus,

$$\theta^\alpha\bar{\theta}^\alpha\theta^A\bar{\theta}^B = \varepsilon\theta^\beta\bar{\theta}^\beta\theta^C\bar{\theta}^D, \qquad \theta^\alpha\bar{\theta}^\alpha\bar{*}\theta^C\bar{\theta}^D = \varepsilon\theta^\beta\bar{\theta}^\beta\bar{*}\theta^A\bar{\theta}^B.$$

Hence,

$$(\theta^\alpha\bar{\theta}^\alpha\theta^A\bar{\theta}^B, \theta^\beta\bar{\theta}^\beta\theta^C\bar{\theta}^D) = \varepsilon, \qquad (\theta^\alpha\bar{\theta}^\alpha\bar{*}\theta^C\bar{\theta}^D, \theta^\beta\bar{\theta}^\beta\bar{*}\theta^A\bar{\theta}^B) = \varepsilon.$$

This proves our claim. Considering only the terms coming from (ii) and (iv), we obtain

$$-1((\Lambda e(R) - e(R)\Lambda)\theta^A\bar{\theta}^B, \theta^C\bar{\theta}^D) = \left(\sum_{\beta\notin A\cup B} r_\beta - \sum_{\alpha\in A\cap B} r_\alpha\right)(\theta^A\bar{\theta}^B, \theta^C\bar{\theta}^D).$$

This proves (3.6).

To complete the proof of Theorem (3.4), we need the following

(3.7) **Lemma** *Given real numbers $r_1, \cdots, r_n$ such that*

$$r_1 \leqq r_2 \leqq \cdots \leqq r_n \leqq 0, \qquad r_1, \cdots, r_k < 0$$

and a positive number μ, let

$$r'_\alpha = r_\alpha/(1-\mu r_\alpha), \qquad \alpha = 1, \cdots, n.$$

If μ is sufficiently large, for any indices $\alpha_1 < \cdots < \alpha_s$ and $\beta_1 < \cdots < \beta_{n-t}$ with $s+t<k$ we have

$$-(r'_{\alpha_1} + \cdots + r'_{\alpha_s}) + (r'_{\beta_1} + \cdots + r'_{\beta_{n-t}}) < 0 .$$

Proof It follows easily that

$$r'_1 \leqq r'_2 \leqq \cdots \leqq r'_n \leqq 0 , \qquad r'_1, \cdots, r'_k < 0 .$$

Hence it suffices to prove the following extreme case:

$$-(r'_1 + \cdots + r'_s) + (r'_{t+1} + \cdots + r'_n) < 0 .$$

But

$$\begin{aligned}(r'_1 + \cdots + r'_s) - (r'_{t+1} + \cdots + r'_{t+s}) \geqq s(r'_1 - r'_{t+s}) \geqq s(r'_1 - r'_k) \\ = s(r_1 - r_k)/(1 - \mu r_1)(1 - \mu r_k) ,\end{aligned}$$

and

$$-(r'_{t+s+1} + \cdots + r'_n) \geqq -(k-s-t) r'_k = -(k-s-t) r_k/(1 - \mu r_k) .$$

Adding these two inequalities, we obtain

$$(r'_1 + \cdots + r'_s) - (r'_{t+1} + \cdots + r'_n) \geqq \frac{-(k-s-t) r_k}{(1 - \mu r_k)} + \frac{s(r_1 - r_k)}{(1 - \mu r_1)(1 - \mu r_k)} .$$

The first term on the right is positive and the second term on the right is negative. By taking μ sufficiently large, we can make the absolute value of the second term smaller than the first term. Q.E.D.

We shall now complete the proof of Theorem (3.4). We fix a point x of M. We choose $\theta^1, \cdots, \theta^n$ as above so that $(R_{\alpha\bar{\beta}})$ is diagonal at x with diagonal entries $r_1, \cdots, r_n$. Since $(R_{\alpha\bar{\beta}})$ is assumed to be negative semi-definite of rank $\geqq k$, we may assume that

$$r_1 \leqq \cdots \leqq r_n \leqq 0 , \qquad r_1, \cdots, r_k < 0 .$$

Using the curvature form Ω of the Hermitian structure h and the Kähler form Φ of g, we define

$$\Phi' = \Phi - \mu\sqrt{-1}\Omega ,$$

where μ is a positive constant. Since the curvature Ω is negative semi-definite, Φ' is a positive (1, 1)-form. It is obviously closed. Let g' be the corresponding Kähler metric on M. Making use of a unitary frame $\theta'^1, \cdots, \theta'^n$ of the cotangent space $T^*_x M$ with respect to the new metric g', we write

$$\Omega = \sum R'_{\alpha\bar{\beta}} \theta'^\alpha \wedge \bar{\theta}'^\beta \qquad \text{at} \quad x\,.$$

Then the eigen-values of $(R'_{\alpha\bar{\beta}})$ are given by

$$r'_\alpha = r_\alpha/(1 - \mu r_\alpha)\,, \qquad \alpha = 1, \cdots, n\,.$$

Let Λ' denote the adjoint of $L' = e(\Phi')$. Then it follows from (3.6) and (3.7) that if $p+q<k$, then the eigen-values of the linear transformation

$$\sqrt{-1}(\Lambda' e(R) - e(R)\Lambda') \colon A^{p,q}(F)_x \longrightarrow A^{p,q}(F)_x$$

are all negative. This holds, by continuity, in a neighborhood of x. Since M is compact, by covering M with a finite number of such neighborhoods and taking μ sufficiently large, we obtain in the range $p+q<k$ the following inequality:

$$\sqrt{-1}((\Lambda' e(R) - e(R)\Lambda')\varphi, \varphi) < 0 \qquad \text{for} \quad 0 \neq \varphi \in A^{p,q}(F)\,.$$

This contradicts Nakano's inequality (3.5). Hence, $\varphi = 0$, showing that $\boldsymbol{H}^{p,q}(F) = 0$ for $p+q<k$. This completes the proof of (3.4).

By the Serre duality theorem, (3.4) can be dualized as follows.

(3.8) **Corollary** *If M is Kähler and $c_1(F) \geqq 0$ with rank $c_1(F) \geqq k$, then*

$$H^q(M, \Omega^p(F)) = 0 \qquad \text{for} \quad p+q \geqq 2n-k+1\,.$$

(3.9) **Corollary** *Let M be a compact Kähler manifold and F a line bundle over M such that $F^{\otimes m}$ is generated by global sections for large m. Let*

$$\Phi_m \colon M \longrightarrow P_N\boldsymbol{C} \qquad (N+1 = \dim H^0(M, F^{\otimes m}))$$

be the holomorphic mapping defined by the sections of $F^{\otimes m}$. If $\dim \Phi_m(M) = k$, then

$$H^q(M, \Omega^p(F^{-1})) = 0 \qquad \text{for} \quad p+q \leqq k-1\,.$$

This follows from the fact that the Chern form of $F^{\otimes m}$ can be obtained by pulling back (a suitable positive multiple of) the Fubini-Study Kähler form of $P_N\boldsymbol{C}$ by Φ_m. (3.9) is due to Ramanujam [1] and Umemura [1].

There are various generalizations of the vanishing theorem proved in this section. The following vanishing theorem, due to Kawamata [1] and Viehweg [1], generalizes the vanishing theorem of Kodaira.

(3.10) **Theorem** *Let M be a projective algebraic manifold of dimension n and F a line bundle over M. If*

(i) $\int_C c_1(F) \geqq 0$ *for every curve* C *in* M

(ii) $\int_M c_1(F)^n > 0$,

then

$$H^q(M, \Omega^n(F)) = 0 \quad \textit{for} \quad q \geqq 1 .$$

Let D be a divisor defining the line bundle F. Then (i) says that D is numerically effective while (ii) states that the highest self-intersection D^n is positive. We note that if $c_1(F)$ is semipositive, then (i) is satisfied. The converse is probably true.

Nakano's vanishing theorem has been generalized to certain non-compact manifolds. A complex manifold M is said to be *weakly* 1-*complete* if there is a smooth real function f on M such that

(i) f is plurisubharmonic, i.e., its complex Hessian $(\partial^2 f/\partial z^\alpha \partial \bar{z}^\beta)$ is positive semidefinite;

(ii) $\{x \in M; f(x) < c\}$ is a relatively compact subset of M for every $c \in \boldsymbol{R}$.

Every compact complex manifold is weakly 1-complete since a constant function satisfies the conditions above. On the other hand, it follows from Remmert's proper embedding theorem that every holomorphically complete manifold is weakly 1-complete. Sometimes, the term "*pseudoconvex*" is used for "weakly 1-complete".

A holomorphic line bundle over a (possibly noncompact) complex manifold M is said to be *positive* if there is an Hermitian structure h with positive definite curvature. A *semipositive* line bundle of rank $\geqq k$ can be defined in a similar manner.

The following generalization of (the dual of) (3.2) is also due to Nakano [1], [2].

(3.11) **Theorem** *If F is a positive line bundle over a weakly* 1-*complete complex manifold M of dimension n, then*

$$H^q(M, \Omega^p(F)) = 0 \quad \textit{for} \quad p+q \geqq n+1 .$$

The strongest result in this direction, due to Takegoshi-Ohsawa [1], generalizes (the dual of) (3.4):

(3.12) **Theorem** *Let M be a weakly* 1-*complete Kähler manifold of dimension n and F a semipositive line bundle whose curvature has at least $n-k+1$ positive eigenvalues outside a proper compact subset K of M. Then*

$$H^q(M, \Omega^p(F))=0 \qquad \textit{for} \quad p+q \geqq n+k\,.$$

Comparing (3.8) with (3.10), it is reasonable to conjecture the following:

Let M be a projective algebraic manifold of dimension n and F a line bundle over M. If

(i) $\int_C c_1(F) \geqq 0$ for every curve C in M,

(ii) $\int_M c_1(F)^k \wedge \Phi^{n-k} > 0$,

then

$$H^q(M, \Omega^p(F))=0 \qquad \text{for} \quad p+q \geqq 2n-k+1\,.$$

§4 Vanishing theorem of Bott

We recall first the construction of the tautological line bundle L over the projective space $P_n = P_n C$ given in Section 1, Chapter II and calculate the curvature of L with respect to the natural Hermitian structure to see that $c_1(L)$ is negative.

Let $V = C^{n+1}$. We recall that a point x of P_n is a line through 0 of V and that the fibre L_x over x is the line represented by x. Thus, L may be considered as a line subbundle of the product vector bundle $P_n \times V$ over P_n in a natural manner. We use the Hermitian structure induced from the natural inner product in V. Let $(\zeta^0, \zeta^1, \cdots, \zeta^n)$ be the natural coordinate system in V, which is used as the homogeneous coordinate system for P_n. In the open set U_0 of P_n defined by $\zeta^0 \neq 0$, we use the inhomogeneous coordinate system

$$z^1 = \zeta^1/\zeta^0, \ \cdots, \ z^n = \zeta^n/\zeta^0$$

and a local holomorphic frame field $s\colon U_0 \to L$ given by

$$s(z^1, \cdots, z^n) = (1, z^1, \cdots, z^n) \in V\,. \tag{4.1}$$

With respect to this frame field, the Hermitian structure h of L is given by

$$h(s, s) = 1 + |z^1|^2 + \cdots + |z^n|^2\,. \tag{4.2}$$

Its curvature form ρ is given by

$$\begin{aligned} \rho &= d''d'(\log(1+\textstyle\sum |z^k|^2)) \\ &= -\sum \frac{\partial^2 \log(1+\sum |z^k|^2)}{\partial z^\alpha \partial \bar{z}^\beta} dz^\alpha \wedge d\bar{z}^\beta\,. \end{aligned} \tag{4.3}$$

It follows from (4.3) that $c_1(L)$ is negative according to the definition given in Section 3. We note that the Fubini-Study metric of P_n is given by

$$\sum \frac{\partial^2 \log(1+\sum|z^k|^2)}{\partial z^\alpha \partial \bar{z}^\beta} dz^\alpha d\bar{z}^\beta \tag{4.4}$$

(4.5) **Lemma**

$$H^q(P_n, \Omega^p) = \begin{cases} \boldsymbol{C}\cdot c_1(L)^p & \textit{for} \quad p=q\,, \\ 0 & \textit{for} \quad p\neq q\,. \end{cases}$$

Proof By the Hodge decomposition theorem for Kähler manifolds, we have

$$H^r(P_n, \boldsymbol{C}) = \bigoplus_{p+q=r} H^q(P_n, \Omega^p)\,.$$

Since $c_1(L)$ is negative, $c_1(L)^p$ defines a nonzero element of $H^p(P_n, \Omega^p) = H^{p,p}(P_n, \boldsymbol{C})$. On the other hand, we know that

$$\dim H^r(P_n, \boldsymbol{C}) = \begin{cases} 1 & \text{if } r \text{ is even}\,, \\ 0 & \text{if } r \text{ is odd}\,. \end{cases}$$

Now (4.5) follows immediately. Q.E.D.

We constructed the tautological line bundle L as a line subbundle of the product bundle $P_n \times V$. Now we shall show that the quotient bundle $(P_n \times V)/L$ is isomorphic to $TP_n \otimes L$, where TP_n is the tangent bundle of P_n. In other words, we shall construct an exact sequence

$$0 \longrightarrow L \longrightarrow P_n \times V \longrightarrow TP_n \otimes L \longrightarrow 0\,. \tag{4.6}$$

Setting $V = P_n \times V$ and $T = TP_n$ for simplicity's sake, we write (4.6) as

$$0 \longrightarrow L \longrightarrow V \longrightarrow T \otimes L \longrightarrow 0\,.$$

Tensoring this with L^*, we have

$$0 \longrightarrow 1 \longrightarrow L^{*\oplus n+1} \longrightarrow T \longrightarrow 0\,. \tag{4.7}$$

It suffices therefore to prove (4.7). Let $f^i \in L^*_x$, $i=0, 1, \cdots, n$. Then each f^i is a linear functional on the line L_x in V. If $(\zeta^0, \zeta^1, \cdots, \zeta^n)$ denotes the coordinate system in $V = \boldsymbol{C}^{n+1}$, then each $\partial/\partial\zeta^i$ is a holomorphic vector field on V. Hence, each $f^i(\partial/\partial\zeta^i)$ is a vector field defined along the line L_x. This vector field is invariant under the scalar multiplication by $\boldsymbol{C}^*$ and hence projects down to a tangent vector of P_n at x by the projection $V - \{0\} \to P_n = (V - \{0\})/\boldsymbol{C}^*$. We denote the projected tangent vector of P_n by Z_i. Define

$$\alpha\colon L^{*\oplus n+1} \longrightarrow T, \qquad \alpha(f^0, f^1, \cdots, f^n) = \sum Z_i .$$

Since $V - \{0\} \to P_n$ is surjective, it is not hard to see that α is also surjective. Hence, the kernel of α must be a line bundle. Since the vector field $\sum \zeta^i(\partial/\partial\zeta^i)$ is radial, it projects down to the zero vector field on P_n. Hence,

$$\alpha(\zeta^0, \zeta^1, \cdots, \zeta^n) = 0 .$$

Since, for each $x \in P_n$, at least one of ζ^i is nonzero on L_x, it follows that $(\zeta^0, \zeta^1, \cdots, \zeta^n)$ defines a nowhere vanishing section of $L^{*\oplus n+1}$ and hence generates a trivial line subbundle which is exactly the kernel of α. This proves (4.7).

Considering the determinant of the sequence (4.7), we obtain

$$L^{*\otimes n+1} = \det T .$$

Dualizing this, we can write

(4.8) $$L^{\otimes n+1} = K_{P_n} ,$$

where K_{P_n} denotes the canonical line bundle $(\det T)^*$ of P_n.

Let W be a hyperplane of $V = \boldsymbol{C}^{n+1}$ and $P(W)$ the $(n-1)$-dimensional projective space defined by W. With the natural imbedding $i\colon P(W) \to P_n$, $P(W)$ is a hyperplane of P_n. Let $L(W)$ denote the tautological line bundle over $P(W)$.

(4.9) **Lemma** (i) *The line bundle $L(W)$ is the restriction of L to $P(W)$, and the curvature form ρ of the natural Hermitian structure h of L, restricted to $P(W)$, coincides with the curvature form ρ_W of $h|_{L(W)}$.*

(ii) *The dual bundle L^* of L is isomorphic to the line bundle determined by the divisor $P(W)$ of P_n. (In other words, L^* has a holomorphic section whose zeros define the divisor $P(W)$.)*

Proof (i) is obvious. To prove (ii), let $f\colon V \to \boldsymbol{C}$ be a linear functional with kernel W. Then f can be considered as a holomorphic section of L^* since it is linear on each line L_x, $x \in P_n$. As a section, it vanishes exactly on $P(W)$ to the first order. Q.E.D.

For each integer k, we set

$$L^k = \begin{cases} L^{\otimes k} & \text{if } k>0\,, \\ \text{trivial line bundle} & \text{if } k=0\,, \\ (L^*)^{\otimes(-k)} & \text{if } k<0\,. \end{cases}$$

We are now in a position to prove the vanishing theorem of Bott [1]. The following proof using the induction on the dimension of P_n is due to Enoki.

(4.10) **Theorem** *If L is the tautological line bundle over P_n, then*

$$H^q(P_n, \Omega^p(L^{-k}))=0 \qquad \textit{for} \quad p, q \geqq 0 \quad \textit{and} \quad k \in \boldsymbol{Z}\,,$$

with the following exceptions:

(1) $p=q$ *and* $k=0$,

(2) $q=0$ *and* $k>p$,

(3) $q=n$ *and* $k<p-n$.

Proof (a). $n=1$.

Since

$$H^1(P_1, \Omega^p(L^{-k})) \overset{\text{dual}}{\sim} H^0(P_1, \Omega^{1-p}(L^k))$$

by the Serre duality, it suffices to consider the case $q=0$.

For $p=0$, $H^0(P_1, \Omega^0(L^{-k}))=0$ for $k<0$ by (1.14) or (3.1) since L is negative. (Or more directly, by (ii) of (4.9) the sheaf $\Omega^0(L)$ can be identified with the sheaf $\Omega^0(-p)$ of germs of holomorphic functions vanishing at one point p $(=P(W))$ of P_1. Hence, if $k<0$, $\Omega^0(L^{-k})$ has no nonzero sections).

For $p=1$, noting $\Omega^1=\Omega^0(K_{P_1})$ and using (4.8), we obtain

$$H^0(P_1, \Omega^1(L^{-k}))=H^0(P_1, \Omega^0(L^{2-k}))=0 \qquad \text{if} \quad k \leqq 1\,.$$

(b). $n-1 \Rightarrow n$.

Again, by the duality theorem, we have

$$H^q(P_n, \Omega^p(L^k)) \sim H^{n-q}(P_n, \Omega^{n-p}(L^{-k}))\,.$$

It suffices therefore to consider the case $k \geq 0$. The case $k=0$ is proved in (4.5). In order to complete the proof by induction on k, we fix a hyperplane $D=P(W)$ in P_n. By (ii) of (4.9), $\Omega^p(L^{-m-1})$ can be identified with the sheaf of L^{-m}-valued meromorphic p-forms with pole of order 1 at D. Then we have the following exact sequence of sheaves:

$$(4.11)\qquad 0 \longrightarrow \Omega^p(L^{-m}) \longrightarrow \Omega^p(L^{-m-1}) \xrightarrow{R} \Omega_D^{p-1}(L^{-m}) \longrightarrow 0\,,$$

where R is the residue map which may be defined as follows. If $f=0$ defines D locally, write $\varphi \in \Omega^p(L^{-m-1})_x$ as

$$\varphi = \frac{1}{f}(df \wedge \psi + \cdots)\,, \qquad \psi \in \Omega^{p-1}(L^{-m})_x\,,$$

where the dots indicate the terms not involving df. Then

$$R(\varphi) = i^*\psi\,,$$

where $i\colon D \to P_n$ is the imbedding. From (4.11) we obtain the following long exact sequence:

$$(4.12)\qquad H^{q-1}(D, \Omega^{p-1}(L^{-k})) \xrightarrow{\delta} H^q(P_n, \Omega^p(L^{-k})) \longrightarrow H^q(P_n, \Omega^p(L^{-k-1})) \xrightarrow{R} H^q(D, \Omega^{p-1}(L^{-k}))$$

To prove (4.10) for $k=1$, we set $k=0$ in (4.12) and show that

$$H^{q-1}(D, \Omega^{p-1}) \xrightarrow{\delta} H^q(P_n, \Omega^p) \quad \text{and} \quad H^q(D, \Omega^{p-1}) \xrightarrow{\delta} H^{q+1}(P_n, \Omega^p)$$

are both isomorphisms. We consider the first one; the second one is obtained from the first by shifting q by 1. By (4.5), the problem reduces to showing that

$$(4.13)\qquad \mathbf{C}\cdot c_1(L)^{p-1} = H^{p-1}(D, \Omega^{p-1}) \xrightarrow{\delta} H^p(P_n, \Omega^p) = \mathbf{C}\cdot c_1(L)^p$$

is an isomorphism. (Here, L denotes the tautological line bundle of P_n and its restriction to $D=P(W)$ at the same time). We recall the construction of the connecting homomorphism δ. Given a cohomology class in $H^{p-1}(D, \Omega^{p-1})$, represent it by a d''-closed $(p-1, p-1)$-form ω on D. Then find an L^{-1}-valued $(p, p-1)$-form $\tilde{\omega}$ on P_n such that $R(\tilde{\omega}) = \omega$. Then $\delta[\omega] = [d''\tilde{\omega}]$. We apply this construction to $\omega = \rho_W^{p-1}$, where ρ_W denotes the curvature of the tautological line bundle (L_D, h) over $D=P(W)$. By (1) of (4.9), $\rho_W = \rho|_D$. By (ii) of (4.9), there is a holomorphic section f of L^{-1} which vanishes exactly on D to the first order. If we set

$$\tilde{\omega} = h(f,f)^{-1} d'(h(f,f)) \wedge \rho^{p-1}\,,$$

then $R(\tilde{\omega}) = \rho_W^{p-1}$. Hence,

$$\delta[\rho_W^{p-1}] = [d''\tilde{\omega}] = [\rho^p]\,.$$

Since ρ represents $c_1(L)$ up to a constant factor, this shows that (4.13) is an isomorphism.

Now assume (4.10) for some $k \geqq 1$. Then (4.12) implies immediately (4.10) for $k+1$. This completes the induction. Q.E.D.

What we proved here is a very special case of Bott's general results on homogeneous vector bundles, see Bott [1]. This special case will play an important role in the next section.

§ 5 Vanishing theorems for vector bundle cohomology

Let E be a holomorphic vector bundle of rank r over a complex manifold M of dimension n. We defined in (II.1.1) the bundle

$$P(E) = \bigcup_{x \in M} P(E_x) = (E - \text{zero section})/\boldsymbol{C}^*$$

and the tautological line bundle

$$L(E) = \bigcup_{x \in M} L(E_x)$$

over $P(E)$.

In order to obtain vanishing theorems for vector bundle cohomology, we prove the following theorem of Le Potier which relates vector bundle cohomology to line bundle cohomology. Here we follow Schneider's [1] proof which simplifies the original proof of Le Potier [1], (cf. Verdier [1]).

(5.1) **Theorem** *Let E^* denote the dual of a holomorphic vector bundle E over M. Then there is a natural isomorphism*

$$H^q(M, \Omega_M^p(E^*)) \approx H^q(P(E), \Omega_{P(E)}^p(L(E)^*)) .$$

Proof Let $\pi: P(E) \to M$ be the projection. We consider a subsheaf

$$\mathscr{F}^p = \Omega_{P(E)}^0(\pi^*(\Lambda^p T^* M) \otimes L(E)^*)$$

of $\Omega_{P(E)}^p(L(E)^*)$.

(5.2) **Lemma** (i) *For each open set U of M, the inclusion $\mathscr{F}^p \subset \Omega_{P(E)}^p(L(E)^*)$ induces an isomorphism*

$$H^0(\pi^{-1}(U), \mathscr{F}^p) \approx H^0(\pi^{-1}(U), \Omega_{P(E)}^p(L(E)^*)) .$$

(ii) *There is a natural isomorphism*

$$H^0(\pi^{-1}(U), \mathscr{F}^p) \approx H^0(U, \Omega^p_M(E^*)) .$$

Proof (i) For each x in U, $L(E)|_{\pi^{-1}(x)} = L(E_x)$ and the bundle $(\pi^* \bigwedge^s T^*M)|_{\pi^{-1}(x)}$ is a product bundle. From the decomposition

$$(\bigwedge^p T^*P(E))|_{\pi^{-1}(x)} = \bigoplus_{s+t=p} (\pi^* \bigwedge^s T^*_x M \otimes \bigwedge^t T^*P(E_x)),$$

we obtain

$$\begin{aligned} H^0(\pi^{-1}(x), \Omega^p(L(E)^*)) &= \bigoplus_{s+t=p} (\bigwedge^s T^*_x M) \otimes H^0(P(E_x), \Omega^t(L(E_x)^*)) \\ &= (\bigwedge^p T^*_x M) \otimes H^0(P(E_x), \Omega^0(L(E_x)^*)), \end{aligned}$$

where the last equality is a consequence of the vanishing theorem of Bott (see (4.10)):

$$H^0(P(E_x), \Omega^t(L(E_x)^*)) = 0, \qquad t > 0 .$$

(ii) Using the identifications

$$\pi^*(\bigwedge^p T^*M) \otimes L(E)^* = \mathrm{Hom}(L(E), \pi^* \bigwedge^p T^*M)$$

and

$$\bigwedge^p T^*M \otimes E^* = \mathrm{Hom}(E, \bigwedge^p T^*M),$$

we define an isomorphism f: $H^0(\pi^{-1}(U), \mathscr{F}^p) \to H^0(U, \Omega^p_M(E^*))$ by setting

$$\pi^*(f(\eta)(x)(e)) = \begin{cases} \eta([e])(e) & \text{for } e \neq 0 \\ 0 & \text{for } e = 0, \end{cases} \tag{5.3}$$

where $\eta \in H^0(\pi^{-1}(U), \mathscr{F}^p)$, $x \in U$, $e \in E_x$. (On the right hand side, e is regarded as an element of $L(E)_{[e]}$.) Q.E.D.

(5.4) **Lemma** *If $U \subset M$ is a Stein manifold and $E|_U$ is holomorphically a product bundle, then*

(i) $H^q(\pi^{-1}(U), \mathscr{F}^p) = 0$ *for* $p \geqq 0$, $q > 0$,

(ii) $H^q(\pi^{-1}(U), \Omega^p_{P(E)}(L(E)^*)) = 0$ *for* $p \geqq 0$, $q > 0$.

Proof Fixing a trivialization $\pi^{-1}(U) \approx U \times P_{r-1}\boldsymbol{C}$, we let

$$\rho : \pi^{-1}(U) \longrightarrow P_{r-1}\boldsymbol{C}$$

be the projection onto the second factor. Then $L(E)|_{\pi^{-1}(U)} \approx \rho^* L$, where L denotes the tautological line bundle over $P_{r-1}C$.

(i) Since

$$\mathscr{F}^p = \mathcal{O}((\pi^* \textstyle\bigwedge^p T^*M) \otimes \rho^* L) \quad \text{on} \quad \pi^{-1}(U),$$

we have (by Künneth formula for sheaf cohomology (see Kaup [1], [2]))

$$H^q(\pi^{-1}(U), \mathscr{F}^p) = \bigoplus_{i+j=q} H^i(U, \Omega^p_M) \hat{\otimes} H^j(P_{r-1}C, \mathcal{O}(L^*)),$$

where $\hat{\otimes}$ denotes the topological tensor product. Since U is Stein, $H^i(U, \Omega^p_M) = 0$ for $i > 0$. On the other hand, by the vanishing theorem of Bott (see (4.10)), $H^j(P_{r-1}, \mathcal{O}(L^*)) = 0$ for $j > 0$. Hence, $H^q(\pi^{-1}(U), \mathscr{F}^p) = 0$ for $q > 0$.

(ii) Since

$$\pi^*(\textstyle\bigwedge^p T^*M) \otimes L(E)^* \approx \displaystyle\bigoplus_{s+t=p} \pi^*(\textstyle\bigwedge^s T^*M) \otimes \rho^*((\textstyle\bigwedge^t T^* P_{r-1}) \otimes L^*)$$

$$\text{on} \quad \pi^{-1}(U),$$

as in (i) we have

$$H^q(\pi^{-1}(U), \Omega^p(L(E)^*)) \approx \bigoplus_{i+j=q} \bigoplus_{s+t=p} H^i(U, \Omega^s_M) \otimes H^j(P_{r-1}, \Omega^t(L^*)) = 0.$$

Q.E.D.

We need also the following theorem of Leray (for the proof, see for example, Godement [1]).

(5.5) **Theorem** *Let X and Y be topological spaces and $\pi: X \to Y$ a proper map. Given a sheaf $\mathscr{F}$ of abelian groups over X, $R^q\pi_*\mathscr{F}$ denotes the sheaf over Y defined by the presheaf*

$$U \subset Y \longmapsto H^q(\pi^{-1}(U), \mathscr{F}).$$

If, for some fixed p,

$$R^q\pi_*\mathscr{F} = 0 \qquad \textit{for all} \quad q \neq p,$$

then for every i there is a natural isomorphism

$$H^i(X, \mathscr{F}) \approx H^{i-p}(Y, R^p\pi_*\mathscr{F}).$$

Using the preceding two lemmas and the theorem of Leray, we shall now complete the proof of (5.1). By (ii) of (5.2) and (i) of (5.4), we have

$$R^q\pi_*\mathscr{F}^p = \begin{cases} \Omega^p_M(E^*) & \text{for } q=0\,, \\ 0 & \text{for } q\neq 0\,. \end{cases}$$

By (5.5), we have a natural isomorphism

$$(5.6) \qquad H^q(M, \Omega^p(E^*)) \approx H^q(P(E), \mathscr{F}^p)\,.$$

Take a Stein open cover $\{U_i\}$ of M such that each $E|_{U_i}$ is a product. Let $\{V_i\}$ be the open cover of $P(E)$ defined by $V_i = \pi^{-1}(U_i)$. Then (5.4) means that $\{V_i\}$ is a Leray covering for the sheaves $\mathscr{F}^p$ and $\Omega^p(L(E^*))$ so that $H^*(P(E), \mathscr{F}^p)$ and $H^*(P(E), \Omega^p(L(E)^*))$ can be calculated using the open cover $\{V_i\}$. From (i) of (5.2) we obtain an isomorphism

$$(5.7) \qquad H^q(P(E), \mathscr{F}^p) \approx H^q(P(E), \Omega^p(L(E)^*))\,.$$

Composing the two isomorphisms (5.6) and (5.7) we obtain (5.1). Q.E.D.

We remark that the isomorphism (5.6) generalizes as follows. Let W be any holomorphic vector bundle over M and let S^kE^* denote the k-th symmetric tensor power of E^*. Replacing $\bigwedge^p T^*M$ by W, E^* by S^kE^*, and $L(E)^*$ by $L(E)^{*k}$ in the proofs of (ii) of (5.2) and (i) of (5.4), we obtain a natural isomorphism (see Bott [1], Kobayashi-Ochiai [1]):

$$(5.8) \qquad H^q(M, \Omega^0(W\otimes S^k(E^*)) \approx H^q(P(E), \Omega^0(\pi^*W\otimes L(E)^{*k}))\,.$$

We define negativity of a vector bundle E by negativity of the line bundle $L(E)$. Thus, we say that a holomorphic vector bundle E over M is *negative* (resp. *semi-negative of rank* $\geqq k$) if the first Chern class $c_1(L(E))$ of the line bundle $L(E)$ satisfies $c_1(L(E))<0$ (resp. $c_1(L(E))\leqq 0$ with rank $c_1(L(E))\geqq k+r-1$) in the sense of Section 3. We say that E is *positive* (resp. *semi-positive of rank* $\geqq k$) if its dual E^* is negative (resp. semi-negative of rank $\geqq k$). The extra $r-1$ in the definition of rank comes from the fact that $c_1(L(E))$ is always negative in the direction of fibres which have dimension $r-1$; this point will be clarified in the next section.

(5.9) **Theorem** *Let E be a holomorphic vector bundle of rank r over a compact Kähler manifold of dimension n. If it is semi-negative of rank $\geqq k$, then*

$$H^q(M, \Omega^p(E)) = 0 \qquad \textit{for} \quad p+q \leqq k-r\,.$$

Proof We have

$$H^q(M, \Omega^p(E)) \sim H^{n-q}(M, \Omega^{n-p}(E^*)) \approx H^{n-q}(P(E), \Omega^{n-p}(L(E)^*))$$
$$\sim H^{r+q-1}(P(E), \Omega^{r+p-1}(L(E))),$$

where $\sim$ indicates the Serre duality and $\approx$ is the isomorphism given by (5.1). Now the theorem follows from the vanishing theorem (3.4). Q.E.D.

(5.10) **Corollary** *If E is a negative holomorphic vector bundle of rank r over a compact complex manifold M of dimension n, then*

$$H^q(M; \Omega^p(E)) = 0 \qquad \textit{for} \quad p+q \leqq n-r.$$

(5.11) **Corollary** *Let E be a holomorphic vector bundle of rank r over a compact Kähler manifold M of dimension n. If it is semi-positive of rank $\geqq k$, then*

$$H^q(M; \Omega^p(E)) = 0 \qquad \textit{for} \quad p+q \geqq 2n-k+r.$$

Proof This follows from (5.9) and the Serre duality theorem (2.60). Q.E.D.

(5.12) **Corollary** *If E is a positive holomorphic vector bundle of rank r over a compact complex manifold M of dimension n, then*

$$H^q(M; \Omega^p(E)) = 0 \qquad \textit{for} \quad p+q \geqq n+r.$$

§6 Negativity and positivity of vector bundles

In Sections 3 and 5 we defined the concepts of negativity and positivity for line bundles and vector bundles. In this section we reexamine these concepts.

Let (F, h) be an Hermitian line bundle over a complex manifold M. Writing $\hat{h}(\xi)$ for $h(\xi, \xi)$, we consider $\hat{h}$ as a function on F satisfying

$$\text{(6.1)} \qquad \begin{aligned} &\hat{h}(\xi) > 0 && \text{for every nonzero} \quad \xi \in F, \\ &\hat{h}(c\xi) = |c|^2 \hat{h}(\xi) && \text{for} \quad c \in \boldsymbol{C}, \quad \xi \in F. \end{aligned}$$

Conversely, a function $\hat{h}$ satisfying (6.1) defines an Hermitian structure in F.

We recall that the first Chern class $c_1(F)$ may be represented by (cf. (II.2.22))

$$\text{(6.2)} \qquad \gamma_1 = \frac{-1}{2\pi i} d''d' \log H = -\frac{1}{2\pi i} \sum R_{\alpha\bar{\beta}} dz^\alpha \wedge d\bar{z}^\beta.$$

(Since we are in the line bundle case, H is given by $h(s, s)$ for a non-vanishing local holomorphic section s of F.)

(6.3) **Proposition** *Let (F, h) be an Hermitian line bundle over M. Then γ_1 is negative, i.e., the curvature $(R_{\alpha\bar{\beta}})$ is negative definite if and only if the function $\hat{h}$ on F is strongly plurisubharmonic outside of the zero section of F.*

We recall that a smooth real function on a complex manifold is said to be *strongly plurisubharmonic* if its complex Hessian is positive definite everywhere.

Proof We fix a point x_0 in M. To compare the curvature $(R_{\alpha\bar{\beta}})$ at x_0 and the complex Hessian of $\hat{h}$ at a point in the fibre F_{x_0}, it is most convenient to use a normal holomorphic local frame field s at x_0 (see (I.4.19) and (I.4.20)). Since we are in the line bundle case, this simply means that we choose s such that $H=1$ and $dH=0$ at x_0. Then

$$d''d' \log H = d''d'H \qquad \text{at} \quad x_0 .$$

To calculate the complex Hessian of $\hat{h}$, let U be a small neighborhood of x_0 where s is defined and let $F|_U \approx U \times \boldsymbol{C}$ be the trivialization using s. Thus, $(x, \lambda) \in U \times \boldsymbol{C}$ corresponds to $\lambda \cdot s(x) \in F_x$, and

$$\hat{h}(\lambda \cdot s(x)) = H(x)\lambda\bar{\lambda} .$$

Hence the complex Hessian of $\hat{h}$ at $\xi_0 = (x_0, \lambda_0) \in U \times \boldsymbol{C}$ is given by

$$(d'd''h)_{\xi_0} = (d'd''H)_{x_0}\lambda_0\bar{\lambda}_0 + H(x_0)(d\lambda \wedge d\bar{\lambda})_{\lambda_0} .$$

This shows that the complex Hessian of $\hat{h}$ is positive definite at ξ_0 if and only if the complex Hessian of H is positive definite at x_0. From (6.2) it is clear that the complex Hessian of H is positive definite at x_0 if and only if the curvature $(R_{\alpha\bar{\beta}})$ is negative definite at x_0. Q.E.D.

For $\varepsilon > 0$, let

$$V_\varepsilon = \{\xi \in F;\ \hat{h}(\xi) < \varepsilon\} .$$

If the curvature of h is negative, $\hat{h}$ is strongly plurisubharmonic outside of the zero section of F and V_ε is a strongly pseudoconvex neighborhood of the zero section in F.

Conversely, suppose that there exist a neighborhood V of the zero section with smooth boundary ∂V which is stable under multiplication by e^{it}, $t \in \boldsymbol{R}$, and a smooth strongly plurisubharmonic function f defined in a neighborhood W of ∂V in F such that $V \cap W = \{\xi \in W;\ f(\xi) < 1\}$. Replacing f by the function

$$\frac{1}{2\pi}\int_0^{2\pi} f(e^{it}\xi)dt$$

which is also strongly plurisubharmonic in a neighborhood of ∂V, we may assume that $f(e^{it}\xi)=f(\xi)$. Write each $\xi \in F$ as $\xi=\lambda\eta$, where $\lambda \in \boldsymbol{C}$ and $\eta \in \partial V$. (It is clear that $|\lambda|$ is uniquely determined and η is unique up to a multiplicative factor e^{it}, $t \in \boldsymbol{R}$.) Then define a function $\hat{h}$ on F by setting

$$\hat{h}(\xi)=f(\eta)\lambda\bar{\lambda}.$$

Then $\hat{h}$ satisfies (6.1) and comes from an Hermitian structure h in F. As easily seen, $\hat{h}$ is strongly plurisubharmonic outside of the zero section.

We have established

(6.4) **Proposition** *Let F be a holomorphic line bundle over a compact complex manifold M. Then $c_1(F)$ is negative if and only if there exist a neighborhood V of the zero section in F with smooth boundary ∂V which is stable under multiplication by e^{it}, $t \in \boldsymbol{R}$, and a smooth strongly plurisubharmonic function f defined in a neighborhood W of ∂V such that*

$$V \cap W=\{\xi \in W;\, f(\xi)<1\}.$$

This means that $c_1(F)$ is negative if and only if the zero section of F has a strongly pseudoconvex neighborhood in F. It follows furthermore that $c_1(F)$ is negative if and only if the zero section of F can be collapsed (i.e., blown down) to a point to yield a complex analytic variety, the point coming from the zero section being possibly an isolated singular point, see Grauert [1].

We shall now consider a holomorphic vector bundle E over M. We recall (see Section 4) that the tautological line bundle $L(E)$ over $P(E)$ is a subbundle of the pull-back bundle p^*E, where $p\colon P(E)\to M$. The natural map $\tilde{p}\colon p^*E\to E$ restricted to $L(E)$ gives a map

$$\tilde{p}\colon L(E)\longrightarrow E$$

which is biholomorphic outside of the zero sections of $L(E)$ and E and collapses the zero section of $L(E)$ (identified with $P(E)$) to the zero section of E (identified with M). In summary, we have the following diagram:

$$\begin{array}{ccc} L(E)\subset p^*E & \xrightarrow{\tilde{p}} & E \\ \downarrow & & \downarrow \\ P(E) & \xrightarrow{p} & M. \end{array}$$

In Section 5 we defined E to be negative if the line bundle $L(E)$ is negative, i.e., $c_1(L(E))$ is negative.

Since $\tilde{p}\colon L(E)\to E$ is biholomorphic outside of the zero sections, (6.4) generalizes immediately to E. Applied to $L(E)$, (6.4) yields the following

(6.5) **Proposition** *Let E be a holomorphic vector bundle over a compact complex manifold M. Then E is negative if and only if there exist a neighborhood V of the zero section in E with smooth boundary ∂V which is stable under the multiplication by e^{it}, $t\in \boldsymbol{R}$, and a smooth strongly plurisubharmonic function f defined in a neighborhood W of ∂V such that*

$$V\cap W=\{\xi\in W; f(\xi)<1\}.$$

Grauert [1] uses the existence of a strongly pseudoconvex neighborhood of the zero section in E as the definition of negativity of E.

The map $\tilde{p}\colon L(E)\to E$ leads us to the following definition. A *Finsler structure* f in E is a smooth positive function defined outside of the zero section of E such that

$$f(\lambda\xi)=f(\xi)\lambda\bar{\lambda} \qquad \text{for nonzero} \quad \lambda\in \boldsymbol{C}, \quad \xi\in E. \tag{6.6}$$

If we set $\hat{h}=f\circ\tilde{p}$, then $\hat{h}$ is a function on $L(E)$ satisfying (6.1). Conversely, every function $\hat{h}$ on $L(E)$ satisfying (6.1) arises from a Finsler structure f in E. Thus, there is a natural one-to-one correspondence between the Finsler structures in E and the Hermitian structures in $L(E)$.

From (6.3) we obtain

(6.7) **Proposition** *A holomorphic vector bundle E is negative if and only if it admits a Finsler structure f which is strongly plurisubharmonic on E outside of the zero section.*

In Kobayashi [4] it is shown that if f is strongly plurisubharmonic, then the Finsler connection can be defined and its curvature is negative. Although there is a natural correspondence between the Finsler structures in E with negative curvature and the Hermitian structures in $L(E)$ with negative curvature, it is much harder to deal with Finsler structures than with Hermitian structures.

In order to discuss semi-negative bundles, we need the concept of q-pseudoconvexity. A real smooth function f on an N-dimensional complex manifold is said to be *strongly q-pseudoconvex* if its complex Hessian has at least $N-q+1$ positive eigenvalues at every point. An N-dimensional complex manifold V is said to be *strongly q-pseudoconvex* if there is a real smooth function f such that

(1) f is strongly q-pseudoconvex outside of a compact subset K of V,

(2) $\{x \in V; f(x) < c\}$ is relatively compact for every c.

If $(R_{\alpha\bar{\beta}})$ in (6.3) is negative semi-definite of rank $\geqq k$, then the function $\hat{h}$ on F is strongly $(n+1-k)$-pseudoconvex outside of its zero section, and vice-versa. Hence a holomorphic vector bundle E is semi-negative of rank $\geqq k$ if and only if it admits a Finsler structure f which is strongly $(n+1-k)$-pseudoconvex outside of its zero section. Thus,

(6.8) **Proposition** *If a holomorphic vector bundle E over a compact complex manifold M is semi-negative of rank $\geqq k$, then E is a strongly $(n+1-k)$-pseudoconvex complex manifold.*

Andreotti and Grauert [1] proved the following finiteness theorem.

(6.9) **Theorem** *If V is a strongly q-pseudoconvex manifold and $\mathscr{F}$ is a coherent analytic sheaf over V, then*

$$\dim_{\boldsymbol{C}} H^i(V, \mathscr{F}) < \infty \qquad \text{for} \quad i \geqq q .$$

In particular,

(6.10) **Corollary** *If a holomorphic vector bundle E over a compact complex manifold M is semi-negative of rank $\geqq k$, then, for any coherent sheaf $\mathscr{F}$ on E,*

$$\dim_{\boldsymbol{C}} H^i(E, \mathscr{F}) < \infty \qquad \text{for} \quad i \geqq n+1-k .$$

From this they derived the following vanishing theorem.

(6.11) **Corollary** *Under the same assumption as in* (6.10), *for any holomorphic vector bundle W on M, we have*

$$H^i(M, \Omega^0(S^m E \otimes W)) = 0 \qquad \text{for} \quad i \leqq k-1 \quad \text{and} \quad m \geqq m_0 ,$$

where $S^m E$ denotes the m-th symmetric tensor power of E and m_0 is a positive integer *which depends on E and W.*

The case $k=n$ will be proved later (see (6.26)).

Now, going back to an Hermitian structure h in E, we shall relate the curvature of h to the curvature of the naturally induced Hermitian structure in the tautological line bundle $L(E)$. We fix a point x_0 in M and choose a normal holomorphic local frame field $s=(s_1, \cdots, s_r)$ as defined in (I.4.19). With respect to a local coordinate system $z^1, \cdots, z^n$ around x_0, the curvature of (E, h) is then given by (see (I.4.16))

(6.12) $$R_{j\bar{k}\alpha\bar{\beta}} = -\partial_\alpha \partial_{\bar{\beta}} h_{j\bar{k}} \qquad \text{at} \quad x_0 .$$

Let ξ_0 be a unit vector in the fibre E_{x_0} and $X_0=[\xi_0]$ the point of $P(E)$ represented by ξ_0. In order to calculate the curvature of $(L(E), h)$ at X_0, we apply a suitable unitary transformation to $s=(s_1, \cdots, s_r)$ so that $\xi_0=s_r(x_0)$. Expressing a variable point ξ of E as $\xi=\sum \xi^j s_j$, we take $(z^1, \cdots, z^n; \xi^1, \cdots, \xi^r)$ as a local coordinate system in E. Since $L(E)$ coincides with E outside of their zero sections, $(z^1, \cdots, z^n; \xi^1, \cdots, \xi^r)$ may be used also as a local coordinate system for $L(E)$ outside of its zero section. From the way we constructed s, we have

$$\xi^1 = \cdots = \xi^{r-1} = 0 \quad \text{and} \quad \xi^r = 1 \qquad \text{at} \quad \xi_0 .$$

Setting

(6.13) $$u^a = \xi^a/\xi^r , \qquad a=1, \cdots, r-1 ,$$

we take $(z^1, \cdots, z^n; u^1, \cdots, u^{r-1})$ as a local coordinate system around $X_0=[\xi_0]$ in $P(E)$. Then the map $t: P(E) \to L(E)$ given by

(6.14) $$t: (z^1, \cdots, z^n; u^1, \cdots, u^{r-1}) \longrightarrow (z^1, \cdots, z^n; u^1, \cdots, u^{r-1}, 1)$$

is a normal holomorphic local frame field of $L(E)$ at X_0. With respect to t, the Hermitian structure of $L(E)$ is given by the following function:

(6.15) $$H = \sum h_{a\bar{b}}(z) u^a \bar{u}^b + \sum h_{a\bar{r}}(z) u^a + \sum h_{r\bar{b}}(z) \bar{u}^b + h_{r\bar{r}}(z) .$$

Since s and t are normal holomorphic local frame fields, the curvature components of H at X_0 are given by the following Hermitian matrix of order $n+r-1$.

$$\begin{pmatrix} -\dfrac{\partial^2 \log H}{\partial z^\alpha \partial \bar{z}^\beta} & -\dfrac{\partial^2 \log H}{\partial z^\alpha \partial \bar{u}^b} \\ -\dfrac{\partial^2 \log H}{\partial u^a \partial \bar{z}^\beta} & -\dfrac{\partial^2 \log H}{\partial u^a \partial \bar{u}^b} \end{pmatrix}_{x_0} = \begin{pmatrix} -\dfrac{\partial^2 H}{\partial z^\alpha \partial \bar{z}^\beta} & -\dfrac{\partial^2 H}{\partial z^\alpha \partial \bar{u}^b} \\ -\dfrac{\partial^2 H}{\partial u^a \partial \bar{z}^\beta} & -\dfrac{\partial^2 H}{\partial u^a \partial \bar{u}^b} \end{pmatrix}_{x_0}$$

$$= \begin{pmatrix} -\dfrac{\partial^2 h_{rr}}{\partial z^\alpha \partial \bar{z}^\beta} & 0 \\ 0 & -\delta_{ab} \end{pmatrix}_{x_0}$$

We can write this more invariantly, i.e., without singling out the last coordinate ξ^r; in fact, we have

$$\left(-\frac{\partial^2 h_{rr}}{\partial z^\alpha \partial \bar{z}^\beta}\right)_{x_0} = \left(\sum R_{i\bar{j}\alpha\bar{\beta}} \xi^i \bar{\xi}^j\right)_{x_0}.$$

Hence, we set

$$(6.16) \qquad R(\xi, Z) = \sum R_{i\bar{j}\alpha\bar{\beta}} \xi^i \bar{\xi}^j Z^\alpha \bar{Z}^\beta \qquad \text{for} \quad \xi = \sum \xi^i s_i, \quad Z = \sum Z^\alpha \frac{\partial}{\partial z^\alpha},$$

and we say that the curvature R of (E, h) is *semi-negative of rank* $\geqq k$ if, for each fixed $\xi \neq 0$, $R(\xi, Z)$ is negative semi-definite of rank $\geqq k$ as an Hermitian form in Z. Then the following is clear.

(6.17) **Theorem** *Let (E, h) be an Hermitian vector bundle of rank r over M.*

(1) *The curvature of (E, h) is semi-negative of rank $\geqq k$ if and only if the curvature of the corresponding Hermitian structure in $L(E)$ is semi-negative of rank $\geqq k+r-1$.*

(2) *If the curvature of (E, h) is semi-negative of rank $\geqq k$, then E is semi-negative of rank $\geqq k$.*

(3) *If the curvature of (E, h) is semi-negative of rank $\geqq k$, then*

$$H^q(M, \Omega^p(E)) = 0 \qquad \textit{for} \quad p+q \leqq k-r,$$

provided that M is compact Kähler.

We note that the extra $r-1$ for the rank of the curvature of $L(E)$ in (1) comes from $(-\delta_{ab})$ in the expression for the curvature given above. This explains the definition of "semi-negative of rank $\geqq k$" given in Section 5, and (2) is immediate from that definition. Finally, (3) follows from (5.9).

We cannot claim the converse of (2); we can show only that if E is semi-negative of rank $\geqq k$, then it admits a Finsler structure whose curvature is semi-negative of rank $\geqq k$.

Let (E, h) be an Hermitian vector bundle and (E^*, h^*) the dual Hermitian vector bundle. Fixing a point x of M, let $s=(s_1, \cdots, s_r)$ be a normal holomorphic local frame field of E at x. Let $t=(t^1, \cdots, t^r)$ be the dual frame field for E^*. Then t is also normal. If h is given by $(h_{i\bar{j}})$ with respect to s, then h^* is given by the inverse matrix $(h^{i\bar{j}})$ with respect to t, i.e., $\sum h^{k\bar{j}} h_{i\bar{j}} = \delta^k_i$. Hence,

$$\left(\frac{\partial^2 h^{k\bar{i}}}{\partial z^\alpha \partial \bar{z}^\beta}\right)_x + \left(\frac{\partial^2 h_{i\bar{k}}}{\partial z^\alpha \partial \bar{z}^\beta}\right)_x = 0. \tag{6.18}$$

We say that the curvature R of (E, h) is *semi-positive of rank* $\geqq k$ if, for each fixed $\xi \neq 0$, $R(\xi, Z)$ is positive semi-definite of rank $\geqq k$ as an Hermitian form in Z. We have

(6.19) **Theorem** *Let (E, h) be an Hermitian vector bundle of rank r over M.*

(1) *The curvature of (E, h) is semi-positive of rank $\geqq k$ if and only if the curvature of the dual Hermitian vector bundle (E^*, h^*) is semi-negative of rank $\geqq k$.*

(2) *If the curvature of (E, h) is semi-positive of rank $\geqq k$, then E is semi-positive of rank $\geqq k$.*

(3) *If the curvature of (E, h) is semi-positive of rank $\geqq k$, then*

$$H^q(M, \Omega^p(E)) = 0 \qquad \textit{for} \quad p+q \geqq 2n-k+r.$$

provided that M is compact Kähler.

Proof (1) follows from (6.18); (2) from (1); and (3) from (2) and (5.11).

Q.E.D.

Before we prove more vanishing theorems and discuss the algebraic notion of ampleness, we have to relate the canonical line bundle $K_{P(E)}$ of $P(E)$ to the canonical line bundle K_M of M.

(6.20) **Proposition** *Let E be a holomorphic vector bundle of rank r over M. Then*

$$K_{P(E)} = L(E)^r \cdot \pi^*(K_M \det E^*),$$

where $\det E^* = \bigwedge^r E^*$ *and* $\pi: P(E) \to M$ *is the projection.*

Proof Let $T_P = T(P(E))$, $T_M = \pi^* TM$ and T_F be the subbundle of T_P consisting of vectors tangent to fibres of the fibration $\pi: P(E) \to M$. Then we have an obvious exact sequence:

$$0 \longrightarrow T_F \longrightarrow T_P \longrightarrow T_M \longrightarrow 0 .$$

Hence,

$$\det T_P = (\det T_F)(\det T_M) .$$

So the problem is reduced to showing

$$(\det T_F)^{-1} = L(E)^r \cdot \pi^*(\det E) .$$

In other words, we have to construct a non-degenerate pairing

$$\mu : \det T_F \times L(E)^r \longrightarrow \pi^*(\det E) .$$

Let $u \in P(E)$ and $\zeta = Z_2 \wedge \cdots \wedge Z_r \in \bigwedge^{r-1} T_F = \det T_F$, where $Z_i \in (T_F)_u$. Let $x = \pi(u)$ and represent u by a nonzero element $e_1 \in E_x$. Since E_x is a vector space, we identify $T_{e_1}(E_x)$ with E_x in a natural manner. Let $e_2, \cdots, e_r \in E_x$ be elements which, considered as elements in $T_{e_1}(E_x)$, are mapped onto $Z_2, \cdots, Z_r \in (T_F)_u$ by the projection $E_x - \{0\} \to P(E)_u$. Let $\varphi \in L(E)^r_u$ and write

$$\varphi = a e_1 \otimes \cdots \otimes e_1 , \qquad (e_1 : r \text{ times}) , \qquad a \in \boldsymbol{C} .$$

We define

$$\mu(\zeta, \varphi) = a e_1 \wedge \cdots \wedge e_r .$$

It is easy to verify that μ is well-defined, independently of all the choices involved, and that it is non-degenerate. Q.E.D.

Before we start the proof of the next vanishing theorem, we point out the following trivial fact.

(6.21) **Proposition** *If E is a holomorphic vector bundle and F is a holomorphic line bundle over M, then*

$$P(F^m \otimes E) = P(E) \quad \text{and} \quad L(F^m \otimes E) = \pi^* F^m \otimes L(E) \qquad \text{for} \quad m \in \boldsymbol{Z} .$$

We called a holomorphic vector bundle E *positive* or *negative* if it is semi-positive or semi-negative of maximal rank n. This means that a holomorphic line bundle F is negative if its Chern class $c_1(F)$ is represented by a negative definite closed real $(1, 1)$-form and that a holomorphic vector bundle E is negative if

$L(E)$ is negative.

(6.22) **Theorem** *Given a positive holomorphic line bundle F and an arbitrary holomorphic vector bundle W over a compact complex manifold M, there is a positive integer m_0 such that*

$$H^q(M, \Omega^0(F^m \otimes W)) = 0 \qquad \textit{for} \quad q \geqq 1 \quad \textit{and} \quad m \geqq m_0 .$$

Proof For the sake of convenience, we set $E = W^*$ and $N = n + r - 1 = \dim P(E)$. Then

$$\begin{aligned} H^q(M, \Omega^0(F^m \otimes W)) &= H^q(M, \Omega^0((F^{-m} \otimes E)^*)) \\ &= H^q(P(E), \Omega^0(\pi^* F^m \cdot L(E)^{-1})) \\ &= H^q(P(E), \Omega^N(K_{P(E)}^{-1} \cdot \pi^* F^m \cdot L(E)^{-1})) \\ &= H^q(P(E), \Omega^N(\pi^*(K_M^{-1} \cdot (\det E) \cdot F^m) \cdot L(E)^{-r-1})) , \end{aligned} \tag{6.23}$$

where the second isomorphism is by (5.1) and (6.21), the third makes use of $\Omega^N_{P(E)} = \Omega^0(K_{P(E)})$ and the fourth is by (6.20).

We calculate the curvature of the line bundle $\pi^*(K_M^{-1} \cdot (\det E) \cdot F^m) \cdot L(E)^{-r-1}$ as in the proof of (6.17), (see calculation following (6.15)), and we claim that it is positive definite if m is sufficiently large. In fact,

$$\text{curvature of } \pi^*(K_M^{-1} \cdot (\det E)) = \begin{pmatrix} * & 0 \\ 0 & 0 \end{pmatrix}$$

$$\text{curvature of } \pi^*(F^m) = \begin{pmatrix} mR_{\alpha\bar\beta} & 0 \\ 0 & 0 \end{pmatrix}$$

$$\text{curvature of } L(E)^{-r-1} = \begin{pmatrix} * & 0 \\ 0 & (r+1)I \end{pmatrix},$$

where $*$ stands for an $(n \times n)$-matrix, I is the identity matrix of order $r-1$ and $(R_{\alpha\bar\beta})$ is the curvature matrix for F. Now we see our assertion by adding these three matrices.

Applying (3.8) to the last term of (6.23), we see that our cohomology vanishes for $q + N \geqq 2N - N + 1$, i.e., for $q \geqq 1$. Q.E.D.

(6.24) **Corollary** *Given a negative holomorphic line bundle F and an arbitrary holomorphic vector bundle W over a compact complex manifold M, there is a positive integer m_0 such that*

$$H^q(M, \Omega^0(F^m \otimes W))=0 \qquad \text{for} \quad q \leqq n-1 \quad \text{and} \quad m \geqq m_0 .$$

Proof By the Serre duality theorem, we have

$$\begin{aligned} H^q(M, \Omega^0(F^m \otimes W)) &\sim H^{n-q}(M, \Omega^n(F^{-m} \otimes W^*)) \\ &= H^{n-q}(M, \Omega^0(F^{-m} \otimes (K_M \otimes W^*))) . \end{aligned}$$

Apply (6.22) to the last term. Q.E.D.

Now we extend (6.22) to vector bundles.

(6.25) **Theorem** *Given a positive holomorphic vector bundle E and an arbitrary holomorphic vector bundle W over a compact complex manifold M, there is a positive integer m_0 such that*

$$H^q(M, \Omega^0(S^m E \otimes W))=0 \qquad \text{for} \quad q \geqq 1 \quad \text{and} \quad m \geqq m_0 .$$

Proof By (5.8), we have

$$H^q(M, \Omega^0(S^m E \otimes W))=H^q(P(E^*), \Omega^0(L(E^*)^{-m} \otimes \pi^* W)) .$$

Since E is positive, $L(E^*)^{-1}$ is positive. Apply (6.22) to the right hand side. Q.E.D.

(6.26) **Corollary** *Given a negative holomorphic vector bundle E and an arbitrary holomorphic vector bundle W over a compact complex manifold M, there is a positive integer m_0 such that*

$$H^q(M, \Omega^0(S^m E \otimes W))=0 \qquad \text{for} \quad q \leqq n-1 \quad \text{and} \quad m \geqq m_0 .$$

The proof is exactly the same as that of (6.24).

We shall now explain the fact that a holomorphic vector bundle is positive if and only if it is ample in the sense of algebraic geometry. Let F be a holomorphic line bundle over a compact complex manifold M. Let $V=\Gamma(M, F)$ be the space of holomorphic sections of F. To each point x of M, we assign the subspace $V(x)$ of V consisting of sections vanishing at x. Since F is a line bundle, either $V(x)=V$ or $V(x)$ is a hyperplane of V. The latter occurs if and only if there is a section which does not vanish at x. Assuming the latter case for every x, we define a mapping

$$\varphi : M \longrightarrow P(V^*) ,$$

where $P(V^*)$ is the projective space of hyperplanes in V, i.e., lines in the dual space V^*, by setting

$$\varphi(x)=V(x)\in P(V^*)\,, \qquad x\in M\,.$$

Let $s_0, s_1, \cdots, s_N$ be a basis for V. Since the fibre F_x is 1-dimensional, the ratio $(s_0(x):s_1(x):\cdots:s_N(x))$ is well defined as a point of $P_N\boldsymbol{C}$. Then φ is given also by

$$\varphi(x)=(s_0(x):s_1(x):\cdots:s_N(x))\in P_N\boldsymbol{C}\,.$$

If this mapping φ gives an imbedding of M into $P(V^*)$, we say that F is *very ample*. A line bundle F is said to be *ample* if there is a positive integer m such that F^m is very ample.

Let $L=L(V^*)$ be the tautological line bundle over $P(V^*)$; it is a line subbundle of the product vector bundle $P(V^*)\times V^*$. If the mapping $\varphi: M\to P(V^*)$ is defined, then

$$\varphi^*L^{-1}=F\,. \tag{6.27}$$

To prove (6.27), we start with the dual pairing $\langle\ ,\ \rangle$:

$$V^*\times V\longrightarrow \boldsymbol{C}\,.$$

Since $L_x\subset V^*$, restricting the pairing to $L_x\times V$, we obtain a bilinear mapping

$$(\xi,\sigma)\in L_x\times V\longrightarrow\langle\xi,\sigma\rangle\in\boldsymbol{C}\,.$$

We claim that $\langle\xi,\sigma\rangle$ depends only on ξ and $\sigma(x)$. In other words, if $\sigma'\in V$ is another section of F such that $\sigma'(x)=\sigma(x)$, then $\langle\xi,\sigma'\rangle=\langle\xi,\sigma\rangle$. In fact, since $\sigma'-\sigma\in V(x)$ and $L_x\subset V^*$ is, by definition, the annihilator of $V(x)$, we obtain $\langle\xi,\sigma'-\sigma\rangle=0$. Thus we have a dual pairing

$$\mu: L_x\times F_x\longrightarrow\boldsymbol{C}$$

given by

$$\mu(\xi,\sigma(x))=\langle\xi,\sigma\rangle\,.$$

This proves (6.27).

We know (see (4.3)) that the line bundle L over $P(V^*)$ has negative curvature. Hence, L^{-1} has positive curvature. Its pull-back $F=\varphi^*L^{-1}$ has also positive curvature if φ is an imbedding. We have shown that if F is very ample, then F is positive. If F is ample, then F^m is very ample and hence positive for some $m>0$. Then F itself is positive. This shows that every ample line bundle is positive.

(6.28) **Theorem** *A holomorphic line bundle F over a compact complex manifold M is positive if and only if it is ample.*

The non-trivial part of (6.28), which states that every positive line bundle is ample, is known as Kodaira's imbedding theorem. For the proof, see, for example, Kodaira-Morrow [1] or Griffiths-Harris [1].

A holomorphic vector bundle E over M is said to be *ample* if the line bundle $L(E^*)^{-1}$ over $P(E^*)$ is ample. From the definition of positivity for E and from (6.28) we see immediately the following

(6.29) **Corollary** *A holomorphic vector bundle E over a compact complex manifold M is positive if and only if it is ample.*

We conclude this section by strengthening (6.25) as follows:

(6.30) **Theorem** *Let E be a positive holomorphic vector bundle over a compact complex manifold M. Then, given a coherent analytic sheaf $\mathscr{F}$ over M, there is a positive integer m_0 such that*

$$H^q(M, \Omega^0(S^m E)\otimes\mathscr{F})=0 \qquad \textit{for} \quad q\geqq 1 \quad \textit{and} \quad m\geqq m_0 .$$

Proof Without starting a discussion on coherent analytic sheaves (which will be studied in Chapter V), we shall only outline the way we derive (6.30) from (6.25). Since E is positive, the line bundle $\det(E)=\bigwedge^r E$ is positive. By the imbedding theorem of Kodaira, M is projective algebraic. Given a coherent analytic sheaf $\mathscr{F}$, there is an exact sequence

$$(6.31) \qquad 0\longrightarrow\mathscr{E}_s\xrightarrow{f_s}\cdots\longrightarrow\mathscr{E}_1\xrightarrow{f_1}\mathscr{E}_0\xrightarrow{f_0}\mathscr{F}\longrightarrow 0,$$

where $\mathscr{E}_i$ are all locally free coherent analytic sheaves, i.e., they are of the form $\Omega^0(E_i)$ for some holomorphic vector bundles E_i, (see Borel-Serre [1]); the local version of this will be proved in Chapter V. Set

$$\mathscr{F}_i=\text{Image}\, f_i=\text{Kernel}\, f_{i-1}$$

so that

$$(6.32) \qquad 0\longrightarrow\mathscr{F}_i\longrightarrow\mathscr{E}_i\longrightarrow\mathscr{F}_{i-1}\longrightarrow 0, \qquad i=1,\cdots,s,$$

are all exact, where $\mathscr{F}_0=\mathscr{F}$ and $\mathscr{F}_s=\mathscr{E}_s$. Since E is positive,

$$H^q(M, \Omega^0(S^m E)\otimes\mathscr{E}_i))=0 \qquad \text{for} \quad q\geqq 1 \quad \text{and} \quad m\geqq m_0$$

by (6.25). Hence, the long cohomology exact sequence derived from (6.32) gives isomorphisms

$$H^q(M, \Omega^0(S^m E)\otimes\mathscr{F}_{i-1}) \approx H^{q+1}(M, \Omega^0(S^m E)\otimes\mathscr{F}_i), \qquad i=1, \cdots, s.$$

Hence,

$$H^q(M, \Omega^0(S^m E)\otimes\mathscr{F}) = H^{q+1}(M, \Omega^0(S^m E)\otimes\mathscr{F}_1) = \\ \cdots = H^{q+s}(M, \Omega^0(S^m E)\otimes\mathscr{E}_s) = 0. \qquad \text{Q.E.D.}$$

There are several equivalent definitions of ampleness by Grothendieck, see Hartshorne [1], [2].

Chapter IV

Einstein-Hermitian vector bundles

Given an Hermitian vector bundle (E, h) over a compact Kähler manifold (M, g), we have a field of endomorphisms K of E whose components are given by $K^i_j = \sum g^{\alpha\bar{\beta}} R^i_{j\alpha\bar{\beta}}$. This field, which we call the mean curvature, played an important role in vanishing theorems for holomorphic sections, (see Section 1 of Chapter III). In this chapter we consider the Einstein condition, i.e., the condition that K be a scalar multiple of the identity endomorphism of E. When the Einstein condition is satisfied, E is called an Einstein-Hermitian vector bundle. In Section 1 we prove some basic properties of Einstein-Hermitian vector bundles. The reader who is familiar with stable vector bundles will recognize in many of the propositions proven in Section 1 a close parallel between the Einstein condition and the stability condition. Much of the results in Section 2 on infinitesimal deformations of Einstein-Hermitian structures will be superceded in Chapters VI and VII.

In Section 3 we shall see that an Einstein-Hermitian structure arises as the minimum of a certain functional. The reader familiar with Yang-Mills theory will immediately see a close relationship between Einstein-Hermitian structure and Yang-Mills connections. These close ties with stable bundles and Yang-Mills connections will become more apparent in subsequent chapters.

In Section 4 we prove Lübke's inequality for Einstein-Hermitian vector bundles. This generalizes Bogomolov's inequality for semi-stable bundles over algebraic surfaces and is analogous to the inequality of Chen-Ogiue for Einstein-Kähler manifolds.

In Section 5 we consider the concept of approximate Einstein-Hermitian structure, (already implicit in Donaldson's work but pointed out explicitly by Mabuchi). As we will see in Chapter VI, this is a differential geometric counterpart of the notion of semistability in algebraic geometry.

Sections 6 and 7 give some examples of Einstein-Hermitian vector bundles. The results in these two sections will not be used in later chapters.

§1 Einstein condition

Throughout this section, (E, h) will denote a holomorphic Hermitian vector bundle of rank r over an Hermitian manifold (M, g) of dimension n. We recall (see (I.4.9)) that the Hermitian connection D is a unique connection in E such that

$$Dh=0 \quad \text{and} \quad D''=d''.$$

Its curvature $R=D\circ D=D'\circ D''+D''\circ D'$ is a $(1,1)$-form with values in the bundle End(E). If $s=(s_1, \cdots, s_r)$ is a local frame field for E, the curvature form $\Omega=(\Omega^i_j)$ with respect to s is given by (see (I.1.11) and (I.4.15))

$$R(s_j)=\sum \Omega^i_j s_i, \qquad \Omega^i_j=\sum R^i_{j\alpha\bar\beta} dz^\alpha \wedge d\bar z^\beta$$

in terms of a local coordinate system $(z^1, \cdots, z^n)$ of M. As before, we write

$$h_{i\bar j}=h(s_i, s_j),$$
$$g=\sum g_{\alpha\bar\beta} dz^\alpha d\bar z^\beta.$$

As in (III.1.6) we define the mean curvature K of (E, h) by

$$K^i_j=\sum g^{\alpha\bar\beta} R^i_{j\alpha\bar\beta}, \qquad K_{j\bar k}=\sum h_{i\bar k} K^i_j. \tag{1.1}$$

Then (K^i_j) (resp. $(K_{j\bar k})$) defines an endomorphism K (resp. an Hermitian form $\hat K$) in E. By means of the operator Λ defined in (III.2.14) we can define the mean curvature K by

$$K=i\Lambda R, \tag{1.2}$$

or equivalently by (see (III.1.18))

$$K\Phi^n=inR\wedge\Phi^{n-1}. \tag{1.3}$$

As in (III.1.37), we say that (E, h) satisfies the *weak Einstein condition* (with factor φ) if

$$K=\varphi I_E, \quad \text{i.e.,} \quad K^i_j=\varphi\delta^i_j,$$

where φ is a real function defined on M. If φ is a constant, we say that (E, h) satisfies the *Einstein condition*. Then we say that (E, h) is an *Einstein-Hermitian vector bundle* over (M, g).

We list some simple consequences of the Einstein condition.

(1.4) **Proposition** (1) *Every Hermitian line bundle (E, h) over a complex manifold M satisfies the weak Einstein condition (with respect to any Hermitian metric g on M).*

(2) *If (E, h) over (M, g) satisfies the (weak) Einstein condition with factor φ,*

then the dual bundle (E^*, h^*) *satisfies (weak) Einstein condition with factor* $-\varphi$.

(3) *If* (E_1, h_1) *and* (E_2, h_2) *over* (M, g) *satisfy the (weak) Einstein condition with factor* φ_1 *and* φ_2, *respectively, then their tensor product* $(E_1 \otimes E_2, h_1 \otimes h_2)$ *satisfies the (weak) Einstein condition with factor* $\varphi_1 + \varphi_2$.

(4) *The Whitney sum* $(E_1 \oplus E_2, h_1 \oplus h_2)$ *satisfies the (weak) Einstein condition with factor* φ *if and only if both summands* (E_1, h_1) *and* (E_2, h_2) *satisfy the (weak) Einstein condition with the same factor* φ.

Proof (1) Trivial.

(2) This follows from (I.5.5) and (I.5.6).

(3) By (I.5.13), the mean curvature K of $E_1 \otimes E_2$ can be expressed in terms of the mean curvatures K_1 and K_2 of E_1 and E_2 as follows:

$$K = K_1 \otimes I_2 + I_1 \otimes K_2 ,$$

where I_1 and I_2 denote the identity endomorphisms of E_1 and E_2, respectively.

(4) By (I.5.12), the mean curvature K of $E_1 \oplus E_2$ is given by

$$K = K_1 \oplus K_2 .$$

This implies (4). Q.E.D.

(1.5) **Proposition** *Let* (E, h) *over* (M, g) *satisfy the (weak) Einstein condition with factor* φ. *Let* $\rho: GL(r; \boldsymbol{C}) \to GL(N; \boldsymbol{C})$ *be a representation such that the induced Lie algebra representation* $\rho': \mathfrak{gl}(r; \boldsymbol{C}) \to \mathfrak{gl}(N; \boldsymbol{C})$ *sends the identity* $I_r \in \mathfrak{gl}(r; \boldsymbol{C})$ *to* $cI_N \in \mathfrak{gl}(N; \boldsymbol{C})$ *for some* $c \in \boldsymbol{C}$. *Let* $E^{(\rho)}$ *be the vector bundle of rank N induced by E and* ρ, *and let* $h^{(\rho)}$ *be the induced Hermitian structure in* $E^{(\rho)}$. *Then* $(E^{(\rho)}, h^{(\rho)})$ *satisfies the (weak) Einstein condition with factor* $c\varphi$.

Proof The Lie algebra homomorphism ρ' induces a bundle homomorphism $\rho': \mathrm{End}(E) \to \mathrm{End}(E^{(\rho)})$ in a natural way. The curvature $R \in A^{1,1}(\mathrm{End}(E))$ of E and the curvature $R^{(\rho)} \in A^{1,1}(\mathrm{End}(E^{(\rho)}))$ of $E^{(\rho)}$ are related by

$$R^{(\rho)} = \rho'(R) .$$

Their mean curvatures $K \in A^0(\mathrm{End}(E))$ and $K^{(\rho)} \in A^0(\mathrm{End}(E^{(\rho)}))$ are related by

$$K^{(\rho)} = \rho'(K) .$$

If $K = \varphi I_E$, then $\rho'(K) = c\varphi I_E(\rho)$. Q.E.D.

The following corollary may be derived from (1.4) as well as from (1.5).

(1.6) **Corollary** *If (E, h) over (M, g) satisfies the (weak) Einstein condition with factor φ, then*

(1) *the tensor bundle $E^{\otimes p} \otimes E^{*\otimes q}$ with the induced Hermitian structure satisfies the (weak) Einstein condition with factor $(p-q)\varphi$;*

(2) *the symmetric tensor product S^pE with the induced Hermitian structures satisfies the (weak) Einstein condition with factor $p\varphi$;*

(3) *the exterior power $\bigwedge^p E$ with the induced Hermitian structure satisfies the (weak) Einstein condition with factor $p\varphi$.*

Let E_1 and E_2 be holomorphic vector bundles over M. We denote a sheaf homomorphism $f: \Omega^0(E_1) \to \Omega^0(E_2)$, i.e., a holomorphic cross section of $\mathrm{Hom}(E_1, E_2)$ simply by $f: E_1 \to E_2$. A bundle homomorphism $f: E_1 \to E_2$ is a sheaf homomorphism which has a constant rank.

(1.7) **Proposition** *Let (E_1, h_1) and (E_2, h_2) be Hermitian vector bundles over a compact Hermitian manifold (M, g) satisfying the (weak) Einstein condition with factor φ_1 and φ_2, respectively.*

(1) *If $\varphi_2 < \varphi_1$, then there is no nonzero sheaf homomorphism $f: E_1 \to E_2$.*

(2) *Assume $\varphi_2 \leqq \varphi_1$. Then every sheaf homomorphism $f: E_1 \to E_2$ is a bundle homomorphism. If we set $E_1' = \mathrm{Ker} f$ and $E_2' = \mathrm{Im} f$, then we have direct sum decompositions*

$$E_1 = E_1' \oplus E_1'' \quad \text{and} \quad E_2 = E_2' \oplus E_2''$$

as Hermitian vector bundles. The bundle isomorphism $f: E_1'' \to E_2'$ sends the Hermitian connection of E_1'' to the Hermitian connection of E_2'.

Proof Let $f: E_1 \to E_2$ be a sheaf homomorphism. It is a holomorphic section of the Hermitian vector bundle $\mathrm{Hom}(E_1, E_2) = E_1^* \otimes E_2$, which satisfies the weak Einstein condition with factor $\varphi_2 - \varphi_1$, (see (1.4)). If $\varphi_2 - \varphi_1 < 0$, then $\mathrm{Hom}(E_1, E_2)$ has no nonzero holomorphic sections by (III.1.9).

Assume $\varphi_2 - \varphi_1 \leqq 0$. Then every holomorphic section f of $\mathrm{Hom}(E_1, E_2)$ is parallel with respect to the Hermitian connection of $\mathrm{Hom}(E_1, E_2)$ by (III.1.9). Since f is parallel, the rank of f is constant and both E_1' and E_2' are subbundles of E_1 and E_2, respectively, invariant under the parallelism defined by the Hermitian connections. By (I.4.18) we obtain the holomorphic orthogonal decompositions above. The last assertion follows also from the fact that f is parallel. Q.E.D.

(1.8) **Proposition** *Let (E, h) over (M, g) satisfy the (weak) Einstein condition* *with factor* φ. *If* $p: \tilde{M} \to M$ *is a unramified covering, then the pull-back bundle* (p^*E, p^*h) *over* $(\tilde{M}, p^*g)$ *satisfies the* (*weak*) *Einstein condition with factor* $p^*\varphi$.

The proof is trivial.

(1.9) **Proposition** *Let* $p: \tilde{M} \to M$ *be a finite unramified covering of an Hermitian manifold* (M, g) *and* φ *a real function on* M. *If an Hermitian vector bundle* $(\tilde{E}, \tilde{h})$ *over* $(\tilde{M}, p^*g)$ *satisfies the* (*weak*) *Einstein condition with factor* $p^*\varphi$, *then its direct image bundle* $(p_*\tilde{E}, p_*\tilde{h})$ *over* (M, g) *satisfies the* (*weak*) *Einstein condition with factor* φ.

Proof We recall that the direct image sheaf p_*E is the sheaf over M defined by the presheaf

$$U \longmapsto H^0(p^{-1}U, E) \qquad \text{for open sets} \quad U \subset M .$$

In our case where p is a covering projection, p_*E is a vector bundle of rank kr if $r =$ rank E and if $p: \tilde{M} \to M$ is a k-fold covering. For $x \in M$, let $p^{-1}(x) = \{x_1, \cdots, x_k\}$. Then we have a natural isomorphism

$$(p_*E)_x = E_{x_1} \oplus \cdots \oplus E_{x_k} .$$

The Hermitian structure $\tilde{h}$ in $\tilde{E}$ induces an Hermitian inner product in $(p_*E)_x$. Applying (4) of (1.4) locally, we obtain (1.9). Q.E.D.

(1.10) **Proposition** *Let* $\rho: \pi_1(M) \to U(r)$ *be a representation of the fundamental group into the unitary group* $U(r)$. *Let* $\tilde{M}$ *be the universal covering space of* M. *Then the natural Hermitian structure in* $E = \tilde{M} \times_\rho \mathbf{C}^r$ *is flat and hence satisfies the Einstein condition with factor* 0 (*with respect to any Hermitian metric* g *on* M).

This is immediate from (I.4.21). More generally, we consider projectively flat structures, (see (I.4.22)).

(1.11) **Proposition** *Let* (E, h) *be a projectively flat vector bundle over* M. *Then, for any Hermitian metric* g *on* M, (E, h) *satisfies the weak Einstein condition.*

Proof This follows from the fact (see (I.2.8)) that (E, h) is projectively flat if and only if $R = \alpha I_E$ for some $(1, 1)$-form α. Q.E.D.

§2 Conformal invariance and deformations

Let (E, h) be an Hermitian vector bundle of rank r over an n-dimensional Hermitian manifold (M, g). We denote the Kähler form of (M, g) by Φ. In terms of a unitary frame field $\theta^1, \cdots, \theta^n$ for T^*M, Φ is given by

$$\Phi = i \sum \theta^\alpha \wedge \bar{\theta}^\alpha .$$

(2.1) **Proposition** *If (E, h) over (M, g) satisfies the (weak) Einstein condition with factor φ and if M is compact, then*

$$\int_M c_1(E, h) \wedge \Phi^{n-1} = \frac{r}{2n\pi} \int_M \varphi \Phi^n ,$$

where

$$c_1(E, h) = \frac{i}{2\pi} \operatorname{tr}(R)$$

is the first Chern form of (E, h).

Proof By (1.3),

$$in \cdot R \wedge \Phi^{n-1} = K\Phi^n = \varphi I_E \Phi^n \in A^{1,1}(\operatorname{End}(E)) .$$

Taking the trace, we obtain

$$in \cdot \operatorname{tr}(R) \wedge \Phi^{n-1} = r\varphi\Phi^n . \qquad \text{Q.E.D.}$$

We remark that if (M, g) is a Kähler manifold, the integral on the left hand side of (2.1) depends only on the cohomology classes of Φ and $c_1(E)$. Hence, in this case, the average of φ over M is determined by the cohomology classes of Φ and $c_1(E)$.

Let a be a real positive function on M, and consider a new Hermitian structure $h' = ah$ in E. Under the conformal change $h \to h' = ah$ of the Hermitian structure, the mean curvature K changes to K' by the following formula (see (III.1.29))

$$K' = K + \square(\log a) I_E . \tag{2.2}$$

Hence,

(2.3) **Proposition** *If (E, h) over (M, g) satisfies the weak Einstein condition with factor φ, then (E, h') with $h' = ah$ satisfies the weak Einstein condition with* *factor* $\varphi' = \varphi + \square(\log a)$.

Making use of (2.3) we prove

(2.4) **Proposition** *If an Hermitian vector bundle (E, h) over a compact Kähler manifold (M, g) satisfies the weak Einstein condition with factor φ, there is a conformal change $h \to h' = ah$ such that (E, h') satisfies the Einstein condition with a constant factor c. Such a conformal change is unique up to a homothety.*

Proof Let c be the constant determined by

$$c \int_M \Phi^n = \int_M \varphi \Phi^n ,$$

the average of φ over M. By (2.3), the problem is reduced to showing that given a function $f\,(=c-\varphi)$ on M such that

$$\int_M f\Phi^n = 0 , \tag{2.5}$$

there is a function u satisfying

$$f = \square u . \tag{2.6}$$

(Then $a = e^u$ is the desired function). We know that (2.6) has a solution if and only if f is orthogonal to all $\square$-harmonic functions. Since M is compact, a function is $\square$-harmonic if and only if it is constant. Since (2.5) expresses exactly the condition that f is orthogonal to the constant functions, (2.6) has a solution. The last assertion about the uniqueness follows from the fact that a $\square$-harmonic function is constant. Q.E.D.

As we have already remarked above, the constant factor c in (2.4) is determined by

$$\int_M c_1(E) \wedge \Phi^{n-1} = \frac{cr}{2n\pi} \int_M \Phi^n \tag{2.7}$$

and depends only on the cohomology classes of Φ and $c_1(E)$.

We shall now consider deformations of an Hermitian structure h on a *fixed holomorphic vector bundle E*. Let h_t be a 1-parameter family of Hermitian structures on E such that $h = h_0$. We consider the infinitesimal deformation induced by h_t:

(2.8) $$v = \partial_t h_t|_{t=0} .$$

We do our calculation using a fixed local holomorphic frame field $s = (s_1, \cdots, s_r)$. Let $\omega_t = (\omega^i_{tj})$ be the connection form h_t. We set $\omega = \omega_0$. Then

(2.9) $$\sum h_{ti\bar{k}} \omega^i_{tj} = d' h_{tj\bar{k}} .$$

Differentiating (2.9) with respect to t at $t=0$, we obtain

$$\sum v_{i\bar{k}} \omega^i_j + \sum h_{i\bar{k}} \partial_t \omega^i_{tj}|_{t=0} = d' v_{j\bar{k}} ,$$

which can be rewritten as

(2.10) $$\sum h_{i\bar{k}} \partial_t \omega^i_{tj}|_{t=0} = D' v_{j\bar{k}} .$$

Setting

(2.11) $$v^i_j = \sum h^{i\bar{k}} v_{j\bar{k}} ,$$

we can state (2.10) as follows:

(2.12) $$\partial_t \omega^i_{tj}|_{t=0} = D' v^i_j .$$

We note that $D' v^i_j$ should be regarded as (the components of) a section of $A^{1,0}(\mathrm{End}(E))$. Applying $D'' = d''$ to (2.12), we obtain

(2.13) $$\partial_t \Omega^i_{tj}|_{t=0} = D'' D' v^i_j .$$

With the usual tensor notation, we may write

(2.14) $$D'' D' v^i_j = -\sum v^i_{j\alpha\bar{\beta}} dz^\alpha \wedge d\bar{z}^\beta , \qquad \text{where} \quad v^i_{j\alpha\bar{\beta}} = \nabla_{\bar{\beta}} \nabla_\alpha v^i_j .$$

Then (2.13) can be written as follows:

(2.15) $$\partial_t R^i_{tj\alpha\bar{\beta}}|_{t=0} = -v^i_{j\alpha\bar{\beta}} .$$

Taking the trace of (2.15) with respect to g, we obtain

(2.16) $$\partial_t K^i_{tj}|_{t=0} = -\sum g^{\alpha\bar{\beta}} v^i_{j\alpha\bar{\beta}} .$$

We are now in a position to prove

(2.17) **Theorem** *Let E be a holomorphic vector bundle over a compact Kähler manifold (M, g). Let h_t be a 1-parameter family of Hermitian structures in E satisfying the Einstein condition (with factor c). Then*

(1) *the infinitesimal deformation $v = \partial_t h_t|_{t=0}$ of $h = h_0$ is parallel with respect to the Hermitian connection D defined by h;*

(2) *the infinitesimal deformation* $\partial_t D_t|_{t=0}$ *of* D *is zero.*

Proof Since $K_t = cI_E$, we obtain $\partial_t K_t|_{t=0} = 0$. From (2.16) we obtain

$$\sum g^{\alpha\bar{\beta}} v^i_{j\alpha\bar{\beta}} = 0 . \tag{2.18}$$

Set

$$f_\alpha = \sum v^i_{j\alpha} v^j_i , \qquad \text{where} \quad v^i_{j\alpha} = \nabla_\alpha v^i_j$$

and use the following general formula (cf. (III.2.22)):

$$\delta'(\sum f_\alpha dz^\alpha) = -\sum g^{\alpha\bar{\beta}} f_{\alpha\bar{\beta}} , \qquad \text{where} \quad f_{\alpha\bar{\beta}} = \nabla_{\bar{\beta}} f_\alpha .$$

Then we obtain

$$\begin{aligned}
\delta'(\sum v^i_{j\alpha} v^j_i dz^\alpha) &= -\sum g^{\alpha\bar{\beta}}(v^i_{j\alpha\bar{\beta}} v^j_i + v^i_{j\alpha} v^j_{i\bar{\beta}}) \\
&= -\sum g^{\alpha\bar{\beta}} v^i_{j\alpha} v^j_{i\bar{\beta}} \\
&= -\sum g^{\alpha\bar{\beta}} h^{i\bar{k}} h^{j\bar{m}} v_{j\bar{k}\alpha} v_{i\bar{m}\bar{\beta}} \\
&= -\sum g^{\alpha\bar{\beta}} h^{i\bar{k}} h^{j\bar{m}} v_{j\bar{k}\alpha} \bar{v}_{m\bar{i}\beta} \\
&= -\|D'v\|^2 .
\end{aligned}$$

Integrating this over M, we obtain

$$\int_M \|D'v\|^2 \Phi^n = 0 . \tag{2.19}$$

Hence, $D'v = 0$. Since $D''v_{i\bar{j}} = \overline{D'v_{j\bar{i}}} = 0$, we obtain also $D''v = 0$. The second assertion follows from (2.12). Q.E.D.

According to (2.17), the space V_h of infinitesimal variations v of an Einstein-Hermitian structure h within the space of Einstein-Hermitian structures consists of Hermitian forms $v = (v_{i\bar{j}})$ on E which are parallel with respect to the Hermitian connection D of h, i.e.,

$$V_h = \{v = (v_{i\bar{j}});\ Dv = 0 \text{ and } v_{j\bar{i}} = \bar{v}_{i\bar{j}}\} . \tag{2.20}$$

We shall show that every infinitesimal variations $v \in V_h$ of an Einstein-Hermitian structure h generates a 1-parameter family of Einstein-Hermitian structures h_t. Set

$$h_t = h + tv . \tag{2.21}$$

If t is sufficiently close to 0, h_t remains positive definite and defines an Hermitian structure in E. Let D be the Hermitian connection defined by h. It is characterized by the two properties:

$$Dh=0 \quad \text{and} \quad D''=d''.$$

Since $Dv=0$, we have $Dh_t=0$. This means that D is also the Hermitian connection of h_t. It follows that the curvature R_t of h_t is independent of t since $R_t=D\circ D$. Hence the mean curvature K_t is also independent of t and $K_t=cI_E$. (This is, of course, consistent with (2) of (2.17).)

Whether an Hermitian structure h satisfies the Einstein condition or not, we can still define the space V_h of parallel Hermitian forms v by (2.20). Let $\Psi(x)$ denote the holonomy group of (E, h) with reference point $x\in M$. Each parallel form $v\in V_h$ defines a $\Psi(x)$-invariant Hermitian form on the fibre E_x, and conversely, every $\Psi(x)$-invariant Hermitian form on E_x extends by parallel displacement to a unique element v of V_h. It follows that if the holonomy group $\Psi(x)$ is irreducible, i.e., leaves no proper subspace of E_x invariant, then V_h is 1-dimensional and is spanned by h. More generally, let

$$E_x=E_x^{(0)}+E_x^{(1)}+\cdots+E_x^{(k)} \tag{2.22}$$

be the orthogonal decomposition of E_x such that $\Psi(x)$ is trivial on $E_x^{(0)}$ and irreducible on each of $E_x^{(1)}, \cdots, E_x^{(k)}$. (Of course, $E_x^{(0)}$ may be trivial.) By parallel displacement of (2.22) we obtain an orthogonal decomposition

$$E=E^{(0)}+E^{(1)}+\cdots+E^{(k)}. \tag{2.23}$$

By (I.4.18), the decomposition above is not only orthogonal but also holomorphic. It is clear that $E^{(0)}$ is a product bundle as an Hermitian vector bundle. Corresponding to the decomposition (2.23), we obtain the decomposition of V_h:

$$V_h=V_h^{(0)}+V_h^{(1)}+\cdots+V_h^{(k)}, \tag{2.24}$$

where

$$\dim V_h^{(0)}=r_0^2, \qquad \dim V_h^{(1)}=\cdots=\dim V_h^{(k)}=1$$

with $r_0=\operatorname{rank}(E^{(0)})$.

In (2.17) we showed that an Einstein-Hermitian connection is rigid. In Chapter VI we shall see a stronger result. Namely, no holomorphic vector bundle over a compact Kähler manifold admits more than one Einstein-Hermitian connection.

§ 3 Critical Hermitian structures

We fix a holomorphic vector bundle E of rank r over a compact Kähler manifold (M, g). Given an Hermitian structure h in E, we consider the function

$$\|K\|^2 = \sum h^{i\bar{k}} h^{j\bar{m}} K_{i\bar{m}} \bar{K}_{kj} = \sum K^i_j K^j_i = \mathrm{tr}(K \circ K) \tag{3.1}$$

where K is the mean curvature of (E, h), (see (1.1)). Let Φ denote the Kähler 2-form of g. We consider the integral

$$J(h) = \frac{1}{2} \int_M \|K\|^2 \Phi^n \tag{3.2}$$

as a functional on the space of Hermitian structures h in E. We shall study critical points of this functional.

Let σ be the scalar curvature of h defined by

$$\sigma = \sum K^i_i = \sum g^{\alpha\bar{\beta}} R^i_{i\alpha\bar{\beta}} = \sum g^{\alpha\bar{\beta}} R_{\alpha\bar{\beta}}\,. \tag{3.3}$$

Let c be a constant determined by

$$rc \int_M \Phi^n = \int_M \sigma \Phi^n\,, \qquad (r = \mathrm{rank}\, E)\,. \tag{3.4}$$

Then

$$0 \leqq \|K - cI_E\|^2 = \|K\|^2 + rc^2 - 2c\sigma\,. \tag{3.5}$$

Integrating (3.5) and using (3.4), we obtain

$$\int_M \|K\|^2 \Phi^n \geqq rc^2 \int_M \Phi^n = \left(\int_M \sigma \Phi^n\right)^2 \Big/ \left(r \int_M \Phi^n\right). \tag{3.6}$$

On the other hand, taking the trace of (1.3) we obtain

$$\int_M \sigma \Phi^n = 2n\pi \int_M c_1(E) \wedge \Phi^{n-1}\,. \tag{3.7}$$

Hence,

$$J(h) = \frac{1}{2} \int_M \|K\|^2 \Phi^n \geqq 2\left(n\pi \int_M c_1(E) \wedge \Phi^{n-1}\right)^2 \Big/ \left(r \int_M \Phi^n\right). \tag{3.8}$$

where the equality holds if and only if we have the equality in (3.5), i.e., we have $K = cI_E$. In summary, we have

(3.9) **Theorem** *The function $J(h)$ is bounded below as in* (3.8) *by a constant which depends only on $c_1(E)$ and the cohomology class of Φ. Moreover, this lower bound is attained by h if and only if h satisfies the Einstein condition with constant factor c.*

In order to study other critical points of $J(h)$, let h_t be a 1-parameter family of Hermitian structures in E such that $h=h_0$. Let $D_t=D'_t+d''$ be the Hermitian connection defined by h_t. We write

$$D'_t=D'+A_t\,, \qquad \text{where} \quad A_t\in A^{1,0}(\mathrm{End}(E))\,. \tag{3.10}$$

In terms of connection forms with respect to a holomorphic local frame field $s=(s_1,\cdots,s_r)$, (3.10) may be written

$$\omega_{tj}^{i}=\omega_{j}^{i}+a_{tj}^{i}\,. \tag{3.11}$$

Applying d'' to (3.11), we obtain

$$\Omega_{tj}^{i}=\Omega_{j}^{i}+d''a_{tj}^{i}\,. \tag{3.12}$$

Let v be the infinitesimal variation of h defined by h_t as in (2.8). Comparing (3.11) with (2.12), we have

$$\partial_t a_{tj}^{i}|_{t=0}=D'v_{j}^{i}\,. \tag{3.13}$$

We define b_j^i by

$$\partial_t^2 a_{tj}^{i}|_{t=0}=b_j^i \qquad \text{with} \quad b_j^i=\sum b_{j\alpha}^{i}dz^{\alpha}\,. \tag{3.14}$$

Differentiating (3.12) with respect to t at $t=0$, we obtain (see (2.15))

$$\partial_t R_{tj\alpha\bar\beta}^{i}|_{t=0}=-v_{j\alpha\bar\beta}^{i}\,, \qquad \text{where} \quad v_{j\alpha\bar\beta}^{i}=\nabla_{\bar\beta}\nabla_{\alpha}v_{j}^{i} \tag{3.15}$$

and (see (2.16))

$$\partial_t K_{tj}^{i}|_{t=0}=-\sum g^{\alpha\bar\beta}v_{j\alpha\bar\beta}^{i}\,. \tag{3.16}$$

Taking the second derivative of (3.12) with respect to t at $t=0$, we obtain

$$\partial_t^2 R_{tj\alpha\bar\beta}^{i}|_{t=0}=-b_{j\alpha\bar\beta}^{i} \qquad \text{with} \quad b_{j\alpha\bar\beta}^{i}=\nabla_{\bar\beta}b_{j\alpha}^{i} \tag{3.17}$$

and

$$\partial_t^2 K_{tj}^{i}|_{t=0}=-\sum g^{\alpha\bar\beta}b_{j\alpha\bar\beta}^{i}\,. \tag{3.18}$$

We shall calculate the first variation of $J(h_t)$. Using the general formula (see (III.2.22))

$$\delta'(\sum f_\alpha dz^\alpha) = -\sum g^{\alpha\bar{\beta}} f_{\alpha\bar{\beta}}, \qquad \text{where} \quad f_{\alpha\bar{\beta}} = \nabla_{\bar{\beta}} f_\alpha$$

and (3.16), we obtain

$$\partial_t \|K_t\|^2|_{t=0} = \partial_t(\mathrm{tr}(K_t \circ K_t))|_{t=0} = -2\sum g^{\alpha\bar{\beta}} v^i_{j\alpha\bar{\beta}} K^j_i$$
$$= 2\delta'(\sum v^i_{j\alpha} K^j_i dz^\alpha) + 2\sum g^{\alpha\bar{\beta}} v^i_{j\alpha} K^j_{i\bar{\beta}}. \tag{3.19}$$

Integrating (3.19) over M, we obtain

$$\partial_t J(h_t)|_{t=0} = \int_M \sum g^{\alpha\bar{\beta}} v^i_{j\alpha} K^j_{i\bar{\beta}} \Phi^n = (D'v, D'K). \tag{3.20}$$

(3.21) **Theorem** *For a fixed holomorphic vector bundle E over a compact Kähler manifold (M, g), an Hermitian structure h is a critical point of the functional J if and only if K is parallel with respect to the Hermitian connection D defined by h.*

Proof From (3.20) it is clear that if $D'K=0$, then h is a critical point of J. Conversely, if h is a critical point of J, consider $h_{ti\bar{j}} = h_{i\bar{j}} + tK_{i\bar{j}}$ so that $v_{i\bar{j}} = K_{i\bar{j}}$. Then (3.20) implies $(D'K, D'K)=0$. Hence, $D'K=0$. Since K is Hermitian, $D''K=0$.
Q.E.D.

Next, we shall calculate the second variation of $J(h_t)$. Using (3.16) and (3.18) we obtain

$$\partial_t^2 J(h_t)|_{t=0} = (\delta'D'v, \delta'D'v) + (\delta'b, K)$$
$$= (\delta'D'v, \delta'D'v) + (b, D'K). \tag{3.22}$$

If h is critical, then $D'K=0$ and the second variation is given by

$$\partial_t^2 J(h_t)|_{t=0} = (\delta'D'v, \delta'D'v). \tag{3.23}$$

which is clearly non-negative. If the second variation above is zero, then $\delta'D'v=0$ and

$$(D'v, D'v) = (\delta'D'v, v) = 0, \tag{3.24}$$

which implies $D'v=0$. Since v is Hermitian, $D'v=0$ implies $D''v=0$ and $Dv=0$. We have shown

(3.25) **Theorem** *Let E be a holomorphic vector bundle over a compact Kähler manifold (M, g). If h is a critical point of the functional J, then*

$$\begin{aligned}&\text{index}(h)=0\,,\\&\text{nullity}(h)=\dim V_h\,,\end{aligned}$$

where V_h *is the space of parallel Hermitian forms* $v=(v_{i\bar{j}})$ *in* E.

If h is critical so that K is parallel, we can decompose E according to the eigenvalues of K. Decomposing E further using the holonomy group $\Psi(x)$ as in Section 2, we obtain a holomorphic orthogonal decomposition (see (2.23)):

$$E=E^{(0)}+E^{(1)}+\cdots+E^{(k)}\,, \tag{3.26}$$

where $\Psi(x)$ acts trivially on $E_x^{(0)}$ and irreducibly on $E_x^{(1)}, \cdots, E_x^{(k)}$. Hence,

(3.27) **Theorem** *Let E be a holomorphic vector bundle over a compact Kähler manifold (M, g). Assume that h is a critical point of the functional J. Let*

$$E=E^{(0)}+E^{(1)}+\cdots+E^{(k)}$$

be the holomorphic and orthogonal decomposition of (E, h) obtained in (3.26). *Let $h_0, h_1, \cdots, h_k$ be the restrictions of h to $E^{(0)}, E^{(1)}, \cdots, E^{(k)}$, respectively. Then $(E^{(0)}, h_0)$, $(E^{(1)}, h_1), \cdots, (E^{(k)}, h_k)$ are all Einstein-Hermitian vector bundles with constant factors, say* $0, c_1, \cdots, c_k$.

(3.28) *Remark* Although we can consider another functional

$$I(h)=\frac{1}{2}\int_M \|R\|^2\Phi^n\,,$$

this differs from $J(h)$ by a constant. In fact, we have

$$I(h)=J(h)+2\pi^2 n(n-1)\int_M (2c_2(E)-c_1(E)^2)\wedge\Phi^{n-2}\,. \tag{3.29}$$

This follows easily from the formula

$$n(n-1)\sum \Omega^i_j\wedge\Omega^j_i\wedge\Phi^{n-2}=(\|R\|^2-\|K\|^2)\Phi^n$$

which will be proved in (4.5) of the next section and from the formulas for $c_1(E)$ and $c_2(E)$ given in (II.2.14) and (II.2.15).

Considering J as a function on the space of Hermitian structures h in E, we want to calculate its gradient field. Integrating (3.20) by parts, we obtain

$$(3.30)\qquad \partial_t J(h_t)|_{t=0} = -\int_M \sum g^{\alpha\bar\beta} v^i_j K^j_{i\bar\beta\alpha} \Phi^n$$

We define an Hermitian form $\square\hat K$ on E by setting

$$(3.31)\qquad (\square\hat K)_{i\bar j} = -\sum v^{\alpha\bar\beta} \nabla_{\bar\beta}\nabla_\alpha K_{i\bar j} \left(= -\sum g^{\alpha\bar\beta} K_{i\bar j\alpha\bar\beta}\right).$$

Then, (3.30) may be written as

$$(3.32)\qquad \partial_t J(h_t)|_{t=0} = (\square\hat K, v).$$

Since V can be considered as a tangent vector to the space of Hermitian structures at $h=h_0$, (3.30) can be written also as

$$(3.33)\qquad dJ(v) = (\square\hat K, v).$$

Thus, $\square\hat K$ can be considered as the gradient vector field of J.

§4 Chern classes of Einstein-Hermitian vector bundles

Let (E, h) be an Hermitian vector bundle of rank r over a compact Hermitian manifold (M, g) of dimension n. The main purpose here is to derive Lübke's integral inequality involving $c_1(E)$ and $c_2(E)$. We shall first establish a few general formulas without assuming the Einstein condition.

Throughout this section, we shall use local unitary frame fields $s=(s_1, \cdots, s_r)$ for (E, h) and $\theta^1, \cdots, \theta^n$ for the cotangent bundle T^*M rather than holomorphic ones. We write

$$\Phi = \sqrt{-1}\sum \theta^\alpha \wedge \bar\theta^\alpha,$$

$$\|R\|^2 = \sum |R^i_{j\alpha\bar\beta}|^2 = \sum |R_{\bar j i\alpha\bar\beta}|^2,$$

$$\|K\|^2 = \sum |K^i_j|^2 = \sum |K_{\bar j i}|^2 = \sum |R^i_{j\alpha\bar\alpha}|^2,$$

$$\|\rho\|^2 = \sum |R_{\alpha\bar\beta}|^2 = \sum |R^i_{i\alpha\bar\beta}|^2,$$

$$\sigma = \sum R^i_{i\alpha\bar\alpha} = \sum K^i_i = \sum R_{\alpha\bar\alpha}.$$

We shall use the following formula repeatedly:

$$(4.1)\qquad n(n-1)\theta^\alpha\wedge\bar\theta^\beta\wedge\theta^\gamma\wedge\bar\theta^\delta\wedge\Phi^{n-2} = \begin{cases} -\Phi^n & \text{if } \alpha=\beta\neq\gamma=\delta, \\ \Phi^n & \text{if } \alpha=\delta\neq\beta=\gamma, \\ 0 & \text{otherwise}. \end{cases}$$

This follows from the fact that θ^λ and $\bar\theta^\lambda$ appear in pair in Φ^{n-2}.

In (II.2.13) we introduced the closed $2k$-form $\gamma_k = c_k(E, h)$ representing the k-th Chern class $c_k(E)$ of E. We shall now prove the following two formulas.

$$(4.2) \qquad c_1(E,h)^2 \wedge \Phi^{n-2} = \frac{1}{4\pi^2 n(n-1)}(\sigma^2 - \|\rho\|^2)\Phi^n,$$

$$(4.3) \qquad c_2(E,h) \wedge \Phi^{n-2} = \frac{1}{8\pi^2 n(n-1)}(\sigma^2 - \|\rho\|^2 - \|K\|^2 + \|R\|^2)\Phi^n.$$

Explicit expressions for $c_1(E,h) = \gamma_1$ and $c_2(E,h) = \gamma_2$ in terms of the curvature form $\Omega = (\Omega^i_j)$ are given in (II.2.14) and (II.2.15). Therefore it suffices to prove the following two formulas.

$$(4.4) \qquad n(n-1)\sum \Omega^i_i \wedge \Omega^j_j \wedge \Phi^{n-2} = -(\sigma^2 - \|\rho\|^2)\Phi^n,$$

$$(4.5) \qquad n(n-1)\sum \Omega^i_j \wedge \Omega^j_i \wedge \Phi^{n-2} = -(\|K\|^2 - \|R\|^2)\Phi^n.$$

Both (4.4) and (4.5) can be easily obtained by direct calculation using (4.1). In fact,

$$\begin{aligned} n(n-1)\sum \Omega^i_i \wedge \Omega^j_j \wedge \Phi^{n-2} &= n(n-1)\sum R_{\alpha\bar\beta}R_{\gamma\bar\delta}\theta^\alpha \wedge \bar\theta^\beta \wedge \theta^\gamma \wedge \bar\theta^\delta \wedge \Phi^{n-2} \\ &= -\sum (R_{\alpha\bar\alpha}R_{\gamma\bar\gamma} - R_{\alpha\bar\gamma}R_{\gamma\bar\alpha})\Phi^n \\ &= -(\sigma^2 - \|\rho\|^2)\Phi^n, \end{aligned}$$

and

$$\begin{aligned} n(n-1)\sum \Omega^i_j \wedge \Omega^j_i \wedge \Phi^{n-2} &= n(n-1)\sum R_{i\bar j\alpha\bar\beta}R_{j\bar i\gamma\bar\delta}\theta^\alpha \wedge \bar\theta^\beta \wedge \theta^\gamma \wedge \bar\theta^\delta \wedge \Phi^{n-2} \\ &= -\sum (R_{i\bar j\alpha\bar\alpha}R_{j\bar i\gamma\bar\gamma} - R_{i\bar j\alpha\bar\gamma}R_{j\bar i\gamma\bar\alpha})\Phi^n \\ &= -(\|K\|^2 - \|R\|^2)\Phi^n. \end{aligned}$$

This completes the proofs of (4.2) and (4.3).

We need also the following inequality.

$$(4.6) \qquad r\|R\|^2 - \|\rho\|^2 \geqq 0, \quad \text{and the equality holds if and only if}$$

$$rR^i_{j\alpha\bar\beta} = \delta^i_j R_{\alpha\bar\beta}.$$

To prove (4.6), we set

$$T^i_{j\alpha\bar\beta} = R^i_{j\alpha\bar\beta} - \frac{1}{r}\delta^i_j R_{\alpha\bar\beta}.$$

Then

$$
\begin{aligned}
0 \leqq \|T\|^2 &= \sum | R^i_{j\alpha\bar\beta} - \frac{1}{r}\delta^i_j R_{\alpha\bar\beta}|^2 \\
&= \sum | R^i_{j\alpha\bar\beta}|^2 + \frac{1}{r}\sum | R_{\alpha\bar\beta}|^2 - \frac{2}{r}\sum | R_{\alpha\bar\beta}|^2 \\
&= \|R\|^2 - \frac{1}{r}\|\rho\|^2 .
\end{aligned}
$$

This establishes (4.6).

We are now in a position to prove the following inequality of Lübke, [2].

(4.7) **Theorem** *Let* (E, h) *be an Hermitian vector bundle of rank* r *over a compact Hermitian manifold* (M, g) *of dimension* n *with Kähler form* Φ. *If* (E, h) *satisfies the weak Einstein condition, then*

$$
\int_M \{(r-1)c_1(E, h)^2 - 2rc_2(E, h)\} \wedge \Phi^{n-2} \leqq 0 ,
$$

and the equality holds if and only if (E, h) *is projectively flat (i.e.,* $R^i_{j\alpha\bar\beta} = (1/r)\delta^i_j R_{\alpha\bar\beta}$).

If (M, g) is Kähler so that Φ is closed, then the left hand side of the inequality above depends only on the cohomology classes $[\Phi]$, $c_1(E)$ and $c_2(E)$. Thus, together with (II.1.14), the inequality in (4.7) means

$$
(4.8) \qquad c_2(\mathrm{End}(E)) \cup [\Phi]^{n-2} = \{2r \cdot c_2(E) - (r-1)c_1(E)^2\} \cup [\Phi]^{n-2} \geqq 0 .
$$

If M is a compact complex surface (Kähler or non-Kähler), we obtain

$$
(4.9) \qquad c_2(\mathrm{End}(E)) = 2r \cdot c_2(E) - (r-1)c_1(E)^2 \geqq 0 .
$$

According to Bogomolov [1] (see also Gieseker [1]), (4.9) holds for any holomorphic semi-stable vector bundle over an algebraic surface. In Chapter V we shall see relationship between the Einstein condition and stability.

Proof of (4.7). If (E, h) satisfies the weak Einstein condition with factor φ, then $K^i_j = \varphi\delta^i_j$ so that

$$
\|K\|^2 = r\varphi^2 , \qquad \sigma = r\varphi .
$$

Hence,

$$
(4.10) \qquad r\|K\|^2 = \sigma^2 .
$$

Using (4.2), (4.3) and (4.6), we obtain

$$\{(r-1)c_1(E,h)^2 - 2rc_2(E,h)\} \wedge \Phi^{n-2} = \frac{1}{4\pi^2 n(n-1)}(\|\rho\|^2 - r\|R\|^2)\Phi^n \leqq 0\,.$$

Integrating this inequality over M, we obtain the desired inequality. The equality holds if and only if $\|\rho\|^2 - r\|R\|^2 = 0$. Hence, the second assertion of (4.7) follows from (4.6). Q.E.D.

Using calculation done to prove (4.7), we shall prove the following

(4.11) **Theorem** *Let (E, h) be an Hermitian vector bundle over a compact Kähler manifold (M, g) of dimension n. If it satisfies the Einstein condition and if $c_1(E)=0$ in $H^2(M; \boldsymbol{R})$, then*

$$\int_M c_2(E) \wedge \Phi^{n-2} \geqq 0\,,$$

and the equality holds if and only if (E, h) is flat.

Although (4.11) is a special case of (4.7), it is possible to derive (4.7) by applying (4.11) to End(E), (cf. (II.1.14)).

Proof While the first part of (4.11) is immediate from (4.7), the second half follows from the following

(4.12) **Lemma** *Let (E, h) be an Hermitian vector bundle over a compact Hermitian manifold (M, g) of dimension n. Assume that it satisfies the Einstein condition with constant factor c and that*

$$\int_M c_1(E,h) \wedge \Phi^{n-1} = 0 \quad \text{and} \quad \int_M c_1(E,h)^2 \wedge \Phi^{n-2} = 0\,.$$

Then

$$\int_M c_2(E,h) \wedge \Phi^{n-2} \geqq 0\,,$$

and the equality holds if and only if (E, h) is flat.

Proof We have only to show that the equality above implies the curvature $R=0$. By (4.7) we know already that

$$R^i_{j\alpha\bar{\beta}} = \frac{1}{r}\delta^i_j R_{\alpha\bar{\beta}}\,.$$

The problem is therefore reduced to showing $\rho = 0$. By (4.2), it is further reduced to showing $\sigma = 0$. Since h satisfies the Einstein condition, we have $\sigma = rc$. Hence, σ is constant. On the other hand,

$$0 = \int_M c_1(E, h) \wedge \Phi^{n-1} = \frac{1}{2n\pi} \int_M \sigma \Phi^n\,.$$

Hence, $\sigma = 0$. Q.E.D.

(4.13) **Corollary** *If an Hermitian vector bundle (E, h) over a compact Kähler manifold (M, g) satisfies the Einstein condition with constant factor c and if $c_1(E) = 0$ and $c_2(E) = 0$ in $H^*(M; \boldsymbol{R})$, then (E, h) is flat.*

Specializing this to the tangent bundle, we obtain the following result of Apte [1]. (See also Lascoux-Berger [1].)

(4.14) **Corollary** *A compact Einstein-Kähler manifold M satisfying $c_1(M) = 0$ and $c_2(M) = 0$ in $H^*(M; \boldsymbol{R})$ is flat.*

The classical theorem of Bieberbach states that a compact flat Riemannian manifold is covered by a Euclidean torus (see, for example, Kobayashi-Nomizu [1]). Hence, a compact flat Kähler manifold is covered by a complex Euclidean torus.

We obtain from (4.14) and from Yau's theorem (that every compact Kähler manifold with $c_1 = 0$ admits a Ricci-flat Kähler metric, Yau [1]) the following

(4.15) **Corollary** *A compact Kähler manifold M satisfying $c_1(M) = 0$ and $c_2(M) = 0$ in $H^*(M; \boldsymbol{R})$ is covered by a complex Euclidean torus.*

In the special case where (E, h) is the tangent bundle of a compact Kähler manifold, we know the following theorem of Chen-Ogiue [1] which is sharper than (4.7). Actually, the proof of (4.7) above is modeled on the proof of (4.16) below.

(4.16) **Theorem** *Let (M, g) be a compact Einstein-Kähler manifold of dimension n. Then*

$$\int_M \{(n \cdot c_1(M)^2 - 2(n+1)c_2(M)\} \wedge \Phi^{n-2} \leqq 0\,,$$

and the equality holds if and only if (M, g) *is of constant holomorphic sectional curvature.*

Proof We denote $c_i(TM, g)$ by $c_i(M, g)$. We can apply formulas (4.2) and (4.3) to $c_i(M, g)$ by setting $E = TM$ and $h = g$. In the Kähler case, we have

$$R_{i\bar{j}k\bar{m}} = R_{k\bar{m}i\bar{j}}$$

so that

$$K_{i\bar{j}} = R_{i\bar{j}}\,.$$

In particular,

$$\|K\|^2 = \|\rho\|^2\,.$$

In the Einstein-Kähler case, we have moreover

$$K^i_j = \frac{\sigma}{n}\delta^i_j\,,$$

and hence

$$\sigma^2 = n\|K\|^2\,.$$

From (4.2) and (4.3) we obtain

$$c_1(M, g)^2 \wedge \Phi^{n-2} = \frac{1}{4\pi^2 n}\|K\|^2 \Phi^n\,, \tag{4.17}$$

$$c_2(M, g) \wedge \Phi^{n-2} = \frac{1}{8\pi^2 n(n-1)}((n-2)\|K\|^2 + \|R\|^2)\Phi^n\,, \tag{4.18}$$

$$\begin{aligned}(2(n+1)c_2(M, g) - nc_1(M, g)^2) \wedge \Phi^{n-2} \\ = \frac{1}{4\pi^2 n(n-1)}((n+1)\|R\|^2 - 2\|K\|^2)\Phi^n\,.\end{aligned} \tag{4.19}$$

Since $\sigma^2 = n\|K\|^2$, we have

$$(n+1)\|R\|^2 - 2\|K\|^2 = (n+1)\|R\|^2 - \frac{2}{n}\sigma^2\,.$$

Hence, (4.16) follows from the following algebraic lemma which is valid for all Kähler manifolds.

(4.20) **Lemma** *For any Kähler manifold M, we have always*

$$(n+1)\|R\|^2 \geqq \frac{2}{n}\sigma^2,$$

and the equality holds if and only if M is a space of constant holomorphic sectional curvature.

Proof Set

$$T_{ij k\bar{m}} = R_{ij k\bar{m}} - \frac{\sigma}{n(n+1)}(\delta_{ij}\delta_{km} + \delta_{im}\delta_{kj}).$$

Then

$$\|T\|^2 = \|R\|^2 - \frac{2\sigma^2}{n(n+1)}.$$

Since the condition that the holomorphic sectional curvature be constant is expressed by $T=0$, (4.20) follows from the equality above. Q.E.D.

§5 Approximate Einstein-Hermitian structures

For simplicity we shall assume throughout this section that (M, g) is a compact Kähler manifold with Kähler form Φ. Let (E, h) be a holomorphic Hermitian vector bundle over (M, g). Let $K=K(h)$ be the mean curvature of (E, h). Let c be the (real) constant determined by (2.7); it is also the average of $(1/r)\operatorname{tr}(K)$ over M. We know from (2.7) that c depends only on $c_1(E)$ and the cohomology class of Φ, (and not on h). Since K is an Hermitian endomorphism of (E, h), we can define the length $|K-cI_E|$ by

$$|K-cI_E|^2 = \operatorname{tr}((K-cI_E)\circ(K-cI_E)). \tag{5.1}$$

We say that a holomorphic vector bundle E admits an *approximate Einstein-Hermitian structure* if for every positive ε, there is an Hermitian structure h such that

$$\operatorname*{Max}_M |K(h)-cI_E| < \varepsilon. \tag{5.2}$$

We call h satisfying (5.2) an ε-*Einstein-Hermitian structure.*

The proof of the following proposition is similar to those of (1.4) and (1.6).

(5.3) **Proposition** (1) *If E admits an approximate Einstein-Hermitian structure, so do the dual bundle E^*, the tensor bundle $E^{\otimes p} \otimes E^{*\otimes q}$, the symmetric tensor product $S^p E$ and the exterior product $\bigwedge^p E$.*

(2) *If E_1 and E_2 admit approximate Einstein-Hermitian structures, so does $E_1 \otimes E_2$.*

(3) *Assume that*

$$\frac{\deg(E_1)}{\operatorname{rank}(E_1)} = \frac{\deg(E_2)}{\operatorname{rank}(E_2)}$$

If E_1 and E_2 admit approximate Einstein-Hermitian structures, so does their Whitney sum $E_1 \oplus E_2$.

The assumption on the degree/rank ratio in (3) is equivalent to the assumption that E_1 and E_2 have the same constant c in (5.2), (see (2.7)).

The proof of the following proposition is also similar to those of (1.8) and (1.9).

(5.4) **Proposition** *Let $(\tilde{M}, p^*g)$ be a finite unramified covering of a compact Kähler manifold (M, g) with projection $p: \tilde{M} \to M$.*

(1) *If a holomorphic vector bundle E over (M, g) admits an approximate Einstein-Hermitian structure, so does p^*E over $(\tilde{M}, p^*g)$.*

(2) *If a holomorphic vector bundle $\tilde{E}$ over $(\tilde{M}, p^*g)$ admits an approximate Einstein-Hermitian structure, so does its direct image bundle $p_*\tilde{E}$ over (M, g).*

The following is a partial generalization of (1.7).

(5.5) **Proposition** *Let E and E' be holomorphic vector bundles over a compact Kähler manifold (M, g) such that*

$$\frac{\deg(E')}{\operatorname{rank}(E')} > \frac{\deg(E)}{\operatorname{rank}(E)}.$$

If both E and E' admit approximate Einstein-Hermitian structures, then there is no nonzero sheaf homomorphism $E' \to E$.

This follows from the following proposition applied to $\operatorname{Hom}(E', E)$.

(5.6) **Proposition** *Let E be a holomorphic vector bundle over a compact Kähler manifold (M, g) such that* $\deg(E)<0$. *If E admits an approximate Einstein-Hermitian structure, it admits no nonzero holomorphic sections.*

Proof This is immediate from (III.1.9). Q.E.D.

Now we generalize Lübke's inequality (4.7).

(5.7) **Theorem** *Let E be a holomorphic vector bundle of rank r over a compact Kähler manifold (M, g). If it admits an approximate Einstein-Hermitian structure, then*

$$\int_M \{(r-1)c_1(E)^2 - 2r \cdot c_2(E)\} \wedge \Phi^{n-2} \leqq 0\,.$$

Proof Since formulas (4.1) through (4.6) are valid for general Hermitian vector bundles and since (4.10) holds within ε for any positive ε if E admits an approximate Einstein-Hermitian structure, the proof of (4.7) yields also (5.7). Q.E.D.

§6 Homogeneous Einstein-Hermitian bundles

Let $M=G/G_o$ be a homogeneous Kähler manifold of a compact Lie group G. Let g be an invariant Kähler metric. Let E be a holomorphic homogeneous vector bundle over M; G acts on E compatibly with its action on M. Let $o \in M$ be the origin, the point corresponding to the coset G_o. Thus, G_o is the isotropy subgroup of G at o. Then G_o acts linearly on the fibre E_o at o.

(6.1) **Proposition** *Let E be a homogeneous holomorphic vector bundle over a compact homogeneous Kähler manifold $M=G/G_o$ as above. If the action of G_o on the fibre E_o is irreducible, then an invariant Hermitian structure h, unique up to homothety, exists and is Einstein-Hermitian.*

Proof Since G is assumed to be compact, every Hermitian structure in E gives rise to a G-invariant Hermitian structure h by averaging on G. Since G_o is irreducible on E_o, h is unique up to homothety. The mean curvature K of (E, h) is invariant by G. Since G_o acts irreducibly on E_o, $K=cI_E$ for some constant c. Q.E.D.

Even though the action of G_o is irreducible on E_o, the Hermitian vector bundle (E, h) may not be irreducible, that is, its holonomy group need not be irreducible. Without discussing the general case (for which the reader is referred to Kobayashi [7]), we shall consider here special examples.

Consider G as a principal G_o-bundle over $M = G/G_o$. Then every homogeneous vector bundle E is associated to this principal bundle. In other words, there is a representation $\rho: G_o \to GL(r; \mathbf{C})$ such that

$$E = G \times_\rho \mathbf{C}^r = (G \times \mathbf{C}^r)/G_o \,.$$

Here, the action of G_o on $G \times \mathbf{C}^r$ is given by

$$a: (u, \xi) \longmapsto (ua, \rho(a^{-1})\xi) \quad \text{for} \quad a \in G_o \,, \qquad (u, \xi) \in G \times \mathbf{C}^r \,.$$

Assume that $M = G/G_o$ is a compact irreducible Hermitian symmetric space. Then we have a natural invariant connection in the principal G_o-bundle G over M, and its holonomy group coincides with G_o. The Hermitian connection of (E, h) comes from the invariant connection of the principal G_o-bundle G, and its holonomy group is given by $\rho(G_o) \subset GL(r; \mathbf{C})$. Since G_o acts irreducibly on E_o, i.e., $\rho(G_o)$ is irreducible, (E, h) is an irreducible Einstein-Hermitian vector bundle. In summary, we have

(6.2) **Proposition** *Let E be a homogeneous holomorphic vector bundle over a compact irreducible Hermitian symmetric space $M = G/G_o$ such that G_o acts irreducibly on E_o. Then E admits a G-invariant Hermitian structure, unique up to homothety, and (E, h) is an irreducible Einstein-Hermitian vector bundle.*

(6.3) *Examples* The following homogeneous vector bundles satisfy the condition of (6.2).

(a) The tangent and cotangent bundles of a compact irreducible Hermitian symmetric space.

(b) The symmetric tensor power $S^p(TP_n)$ of the tangent bundle of the complex projective space P_n.

(c) The exterior power $\bigwedge^p(TP_n)$ of the tangent bundle of the complex projective space P_n.

Without proof we state the following result (Kobayashi [7]).

(6.4) **Theorem** *Let $M = G/G_o$ be a compact homogeneous Kähler manifold, where G is a connected, compact semisimple Lie group and $G_o = C(T)$ is the* *centralizer of a toral subgroup T of G. Let E be a homogeneous holomorphic vector bundle over M such that G_o acts irreducibly on E_o. Then with respect to an invariant Hermitian structure h (which exists and is unique up to a homothety), (E, h) is an irreducible Einstein-Hermitian vector bundle.*

We note that what we have not proved here is the fact that the holonomy group of (E, h) in (6.4) is irreducible.

We mention one example to which (6.4) applies.

(6.5) *Example* Null correlation bundles. Let $(z^0, z^1, \cdots, z^{2n+1})$ be a homogeneous coordinate system for the complex projective space P_{2n+1}. Let E be the subbundle of the tangent bundle TP_{2n+1} defined by a 1-form α:

$$\alpha = z^0 dz^1 - z^1 dz^0 + \cdots + z^{2n} dz^{2n+1} - z^{2n+1} dz^{2n} = 0 .$$

The form α is defined on $\boldsymbol{C}^{2n+2} - \{0\}$. Although it is not globally well defined on P_{2n+1}, the equation $\alpha = 0$ is well defined on P_{2n+1}; if s is a local holomorphic section of the fibering $\boldsymbol{C}^{2n+2} - \{0\} \to P_{2n+1}$, then

$$E = \{X \in TP_{2n+1};\ s^*\alpha(X) = 0\}$$

defines a subbundle of rank $2n$ independently of the choice of s. In order to view this bundle E as a homogeneous vector bundle, we consider the symplectic form

$$d\alpha = 2(dz^0 \wedge dz^1 + \cdots + dz^{2n} \wedge dz^{2n+1}) .$$

The symplectic group $Sp(n+1)$ is defined as the subgroup of $U(2n+2)$ acting on $\boldsymbol{C}^{2n+2}$ and leaving $d\alpha$ invariant. Then α itself is invariant by $Sp(n+1)$. We consider P_{2n+1} as a homogeneous space of $Sp(n+1)$ rather than $SU(2n+2)$. Thus,

$$P_{2n+1} = Sp(n+1)/Sp(n) \times T^1 .$$

Since α is invariant by $Sp(n+1)$, the subbundle $E \subset TP_{2n+1}$ is invariant by $Sp(n+1)$. The isotropy subgroup $Sp(n) \times T^1$ is the centralizer of T^1 in $Sp(n+1)$ and acts irreducibly on the fibre E_o. By (6.4), with respect to an $Sp(n+1)$-invariant Hermitian structure h, (E, h) is an irreducible Einstein-Hermitian vector bundle. When $n = 1$, this rank 2 bundle E over P_3 is known as a *null correlation bundle*. In this particular case, Lübke [1] has shown by explicit calculation that (E, h) is Einstein-Hermitian. The 1-form α defines a complex contact structure on P_{2n+1}. Null correlation bundles and their generalizations are explained in Kobayashi [7] from this view point of complex contact structures.

§ 7 Projectively flat bundles over tori

Let M be a complex manifold such that its universal covering space $\tilde{M}$ is a topologically trivial Stein manifold. (By "topologically trivial" we mean "contractible to a point"). The examples for $\tilde{M}$ we have in mind are C^n and symmetric bounded domains. Let $p: \tilde{M} \to M$ be the covering projection and Γ the covering transformation group acting on $\tilde{M}$ so that $M = \Gamma \backslash \tilde{M}$.

Let E be a holomorphic vector bundle of rank r over M. Then its pull-back $\tilde{E} = p^*E$ is a holomorphic vector bundle of the same rank over $\tilde{M}$. Since $\tilde{M}$ is topologically trivial, $\tilde{E}$ is topologically a product bundle. Since $\tilde{M}$ is Stein, by Oka's principle (Grauert [2]) $\tilde{E}$ is holomorphically a product bundle:

$$\tilde{E} = \tilde{M} \times C^r .$$

Having fixed this isomorphism, we define a holomorphic mapping

$$j: \Gamma \times \tilde{M} \longrightarrow GL(r; C) \tag{7.1}$$

by the following commutative diagram:

$$\begin{array}{ccc} C^r = \tilde{E}_x & \searrow & \\ {\scriptstyle j(\gamma, x)} \downarrow & & E_{p(x)} \qquad \text{for} \quad x \in \tilde{M}, \quad \gamma \in \Gamma . \\ C^r = \tilde{E}_{\gamma(x)} & \nearrow & \end{array}$$

Then

$$j(\gamma'\gamma, x) = j(\gamma', \gamma x) \circ j(\gamma, x) \qquad \text{for} \quad x \in \tilde{M}, \quad \gamma, \gamma' \in \Gamma . \tag{7.2}$$

The mapping j is called the *factor of automorphy* for the bundle E.

Conversely, given a holomorphic mapping $j: \Gamma \times \tilde{M} \to GL(r; C)$ satisfying (7.2), we obtain a holomorphic vector bundle

$$E = \tilde{M} \times_j C^r = \Gamma \backslash (\tilde{M} \times C^r) \tag{7.3}$$

by factoring $\tilde{M} \times C^r$ by the action of Γ:

$$\gamma(x, \zeta) = (\gamma(x), j(\gamma, x)\zeta), \qquad \gamma \in \Gamma, \quad (x, \zeta) \in \tilde{M} \times C^r . \tag{7.4}$$

In the special case where $j(\gamma, x)$ does not depend on x, i.e., j is a representation $j: \Gamma \to GL(r; C)$, the corresponding bundle E is defined by the representation j (in the sense of (I.2.4)) and hence is flat.

Given two isomorphic vector bundles E and E' over M and an isomorphism $\phi: E \to E'$, we have an isomorphism

$$\tilde{\phi}\colon \tilde{E}=\tilde{M}\times \boldsymbol{C}^r \longrightarrow \tilde{E}'=\tilde{M}\times \boldsymbol{C}^r ,$$

$$\tilde{\phi}(x,\zeta)=(x,u(x)\zeta) \qquad \text{for} \quad (x,\zeta)\in \tilde{M}\times \boldsymbol{C}^r ,$$

where $u\colon \tilde{M}\to GL(r;\boldsymbol{C})$ is holomorphic. Let j' denote the automorphic factor for E'. Since $\tilde{\phi}$ must commutes with the action of Γ, i.e., $\gamma\tilde{\phi}(x,\zeta)=\tilde{\phi}(\gamma(x,\zeta))$, we obtain

$$(\gamma x, j'(\gamma,x)u(x)\zeta)=(\gamma x, u(\gamma x)j(\gamma,x)\zeta) .$$

Hence,

$$j'(\gamma,x)=u(\gamma x)j(\gamma,x)u(x)^{-1} \qquad \text{for} \quad x\in\tilde{M}, \quad \gamma\in\Gamma . \tag{7.5}$$

Conversely, two factors of automorphy j and j' give rise to isomorphic vector bundles E and E' if they are related as in (7.5) by a holomorphic map $u\colon \tilde{M}\to GL(r;\boldsymbol{C})$. We say therefore that two such factors of automorphy j and j' are *equivalent*.

The concept of automorphic factor extends to any complex Lie group G and any holomorphic principal G-bundle P. Thus, a holomorphic map

$$j\colon \Gamma\times\tilde{M}\longrightarrow G$$

satisfying (7.2) is called a factor of automorphy. Then the equivalence classes of factors of automorphy are in one-to-one correspondence with the isomorphism classes of principal G-bundles over M.

For further details on factors of automorphy, see Gunning [1].

Let $T^n=\Gamma\backslash\boldsymbol{C}^n$ be an n-dimensional complex torus, where Γ is a lattice. Let F be a holomorphic line bundle over T^n and

$$j\colon \Gamma\times\boldsymbol{C}^n\longrightarrow \boldsymbol{C}^*=GL(1;\boldsymbol{C})$$

the corresponding factor of automorphy, (see (7.1)).

Let h be an Hermitian structure in the line bundle F. Using the projection $p\colon \boldsymbol{C}^n\to T^n$, we pull back the bundle F to obtain a line bundle $\tilde{F}$ over $\boldsymbol{C}^n$. We pull back also the Hermitian structure h to $\tilde{F}$ to obtain an Hermitian structure $\tilde{h}$ in $\tilde{F}$. Because of the isomorphism $\tilde{F}=\boldsymbol{C}^n\times\boldsymbol{C}$ we fixed to define j, we may consider $\tilde{h}$ as a positive function on $\boldsymbol{C}^n$ invariant under the action of Γ given by (7.4). The invariance condition reads as follows:

$$\tilde{h}(z)=\tilde{h}(z+\gamma)\,|j(\gamma,z)|^2 \qquad \text{for} \quad z\in\boldsymbol{C}^n, \quad \gamma\in\Gamma . \tag{7.6}$$

The connection form $\tilde{\omega}=d'\log\tilde{h}$ and the curvature form $\tilde{\Omega}=d''\tilde{\omega}=d''d'\log\tilde{h}$

satisfy

$$\tilde{\omega}(z)=\tilde{\omega}(z+\gamma)+d'\log j(\gamma, z)\,, \tag{7.7}$$

$$\tilde{\Omega}(z)=\tilde{\Omega}(z+\gamma)\,. \tag{7.8}$$

The last condition (7.8) reflects the fact that the curvature of the line bundle F is an ordinary 2-form on T^n.

If we multiply h by a positive function e^{φ} on T^n, its curvature form Ω changes by $d''d'\varphi$. It follows that by multiplying h by a suitable function e^{φ} we may assume that the curvature form Ω is a harmonic 2-form on T^n. A harmonic form on T^n has constant coefficients with respect to the natural coordinate system $z^1, \cdots, z^n$ of $\boldsymbol{C}^n$. In particular,

$$\Omega=\sum R_{j\bar{k}}dz^j\wedge d\bar{z}^k\,, \tag{7.9}$$

where $R_{j\bar{k}}$ are constant functions. Since $\Omega=-\bar{\Omega}$, we have

$$R_{j\bar{k}}=\bar{R}_{k\bar{j}}\,. \tag{7.10}$$

From such a curvature form we can recover the Hermitian structure h.

(7.11) **Proposition** *Let F be a line bundle over $T^n=\Gamma\backslash\boldsymbol{C}^n$. If h is an Hermitian structure in F such that its curvature form $\Omega=\sum R_{j\bar{k}}dz^j\wedge d\bar{z}^k$ has constant coefficients, then*

$$\log\tilde{h}=-\sum R_{j\bar{k}}z^j\bar{z}^k+f(z)+\overline{f(z)}\,,$$

where $f(z)$ is a holomorphic function defined on $\boldsymbol{C}^n$.

Proof It is clear that the Hermitian structure h_o in $\tilde{F}$ given by

$$\log h_o=-\sum R_{j\bar{k}}z^j\bar{z}^k$$

has the prescribed curvature $\Omega=\sum R_{j\bar{k}}dz^j\wedge d\bar{z}^k$. Set $\varphi=\log h-\log h_o$. Then φ is a real function on $\boldsymbol{C}^n$ satisfying the pluriharmonic condition

$$d''d'\varphi=0\,.$$

It is well known that such a function φ is the real part of a holomorphic function. In fact, from

$$d(d'\varphi-d''\varphi)=d''d'\varphi-d'd''\varphi=0$$

and from the Poincaré lemma,

$$d'\varphi - d''\varphi = d\psi = d'\psi + d''\psi$$

for some function ψ. Then $d'\varphi = d'\psi$ and $d''\varphi = -d''\psi$. It follows that $d''(\varphi+\psi)=0$ and $d(\psi+\bar{\psi})=0$. By adding a suitable constant to ψ, we may assume that $\psi+\bar{\psi}=0$. Then $\varphi+\psi$ is a holomorphic function with real part φ and imaginary part ψ. Q.E.D.

Since the Hermitian structure h in (7.11) is easily seen to be unique up to a constant multiplicative factor, it follows that the holomorphic function $f(z)$ is unique up to a constant additive factor.

We write

$$R(z, w) = \sum R_{j\bar{k}} z^j \bar{w}^k . \tag{7.12}$$

Then R is an Hermitian bilinear form on $\boldsymbol{C}^n$. Let S and A be the real and imaginary parts of R so that

$$R = S + \sqrt{-1}A , \tag{7.13}$$

where S is symmetric and A is anti-symmetric.

Then $\Omega = iA(dz, d\bar{z})$. Since $(i/2\pi)\Omega$ represents the Chern class of F, which is an integral cohomology class of T^n, it follows that $(-1/2\pi)A$ is half-integer valued on $\Gamma\times\Gamma$. (In fact, integrating $(i/2\pi)\Omega$ over the 2-cycle of T^n spanned by $\gamma, \gamma'\in\Gamma$, we obtain the value $(-1/\pi)A(\gamma, \gamma')$.) Thus,

$$A(\gamma, \gamma') \equiv 0 \pmod{\pi\boldsymbol{Z}} \qquad \text{for} \quad \gamma, \gamma'\in\Gamma . \tag{7.14}$$

From (7.6) and (7.11) we obtain

$$\begin{aligned} -R(z, z) + f(z) + \overline{f(z)} \equiv & -R(z+\gamma, z+\gamma) + f(z+\gamma) + \overline{f(z+\gamma)} \\ & + \log j(\gamma, z) + \log \overline{j(\gamma, z)} , \quad (\operatorname{mod} 2\pi i\boldsymbol{Z}) . \end{aligned}$$

It follows that

$$j(\gamma, z) = \chi(\gamma)\cdot\exp\left[R(z, \gamma) + \frac{1}{2}R(\gamma, \gamma) + f(z) - f(z+\gamma)\right] \tag{7.15}$$

with $|\chi(\gamma)| = 1$. Using (7.2) and (7.14) we see easily that χ is a semicharacter of Γ, i.e.,

$$\chi(\gamma+\gamma') = \chi(\gamma)\chi(\gamma')\cdot\exp[iA(\gamma, \gamma')] \qquad \text{for} \quad \gamma, \gamma'\in\Gamma . \tag{7.16}$$

(By (7.14), $\exp iA(\gamma, \gamma') = \pm 1$.)

From the equivalence relation defined by (7.5), it is clear that the *theta factor of automorphy* j given by (7.15) is equivalent to the following

$$(7.17) \qquad j(\gamma, z) = \chi(\gamma)\cdot\exp\left[R(z,\gamma)+\frac{1}{2}R(\gamma,\gamma)\right] \qquad (\gamma, z)\in\Gamma\times \boldsymbol{C}^n .$$

Conversely, given an Hermitian form R on $\boldsymbol{C}^n$ with imaginary part A satisfying (7.14) and a semicharacter χ satisfying (7.16), we obtain a factor of automorphy $j\colon \Gamma\times\boldsymbol{C}^n\to\boldsymbol{C}^*$ by (7.17). The fact that any line bundle F over $T^n=\Gamma\backslash\boldsymbol{C}^n$ is defined by a unique factor of automorphy of the form (7.17) is known as *Theorem of Appell-Humbert*.

Let E be a holomorphic vector bundle of rank r over a complex torus $T^n=\Gamma\backslash\boldsymbol{C}^n$ with the corresponding factor of automorphy $j\colon\Gamma\times\boldsymbol{C}^n\to GL(r;\boldsymbol{C})$. Let

$$\lambda\colon GL(r;\boldsymbol{C})\longrightarrow PGL(r;\boldsymbol{C})$$

be the natural projection to the projective general linear group and define

$$\tilde{j}=\lambda\circ j\colon \Gamma\times\boldsymbol{C}^n\longrightarrow PGL(r;\boldsymbol{C}) .$$

Then $\tilde{j}$ is a factor of automorphy for the projective bundle $P(E)$ over T^n.

Assume that E is projectively flat, i.e., $P(E)$ is flat. Then $P(E)$ is defined by a representation of Γ in $PGL(r;\boldsymbol{C})$. Replacing j by an equivalent factor of automorphy, we may assume that $\tilde{j}(\gamma, z)$ is independent of z, i.e., $\tilde{j}$ is a representation of Γ into $PGL(r;\boldsymbol{C})$. Since $\tilde{j}(\gamma, z)$ does not depend on z, we can write

$$(7.18) \qquad j(\gamma, z)=a(\gamma, z)j(\gamma, 0) \qquad \text{for} \quad (\gamma, z)\in\Gamma\times\boldsymbol{C}^n$$

where $a\colon\Gamma\times\boldsymbol{C}^n\to\boldsymbol{C}^*$ is scalar-valued. From (7.2) we obtain

$$(7.19) \qquad a(\gamma,\gamma')a(\gamma+\gamma', z)=a(\gamma, z+\gamma')a(\gamma', z) \qquad \text{for} \quad \gamma,\gamma'\in\Gamma, \quad z\in\boldsymbol{C}^n,$$

$$(7.20) \qquad a(\gamma, 0)=a(0, z)=1, \qquad \text{for} \quad \gamma\in\Gamma, \ , z\in\boldsymbol{C}^n,$$

$$(7.21) \qquad a(\gamma, -\gamma)=a(-\gamma,\gamma) \qquad \text{for} \quad \gamma\in\Gamma .$$

These properties of $a(\gamma, z)$ allow us to define the following group structure in the set $\Gamma\times\boldsymbol{C}^*$:

$$(7.22) \qquad (\gamma, c)(\gamma', c')=(\gamma+\gamma', a(\gamma', \gamma)cc') \qquad \text{for} \quad (\gamma, c), \quad (\gamma', c')\in\Gamma\times\boldsymbol{C}^* .$$

The associativity law follows from (7.19). The identity element is given by $(0, 1)$. The inverse of (γ, c) is $(-\gamma, a(\gamma, -\gamma)^{-1}c^{-1})$. We denote this group by $G(\Gamma, a)$. There is an obvious exact sequence:

$$(7.23)\qquad 1 \longrightarrow \boldsymbol{C}^* \longrightarrow G(\Gamma, a) \longrightarrow \Gamma \longrightarrow 0 .$$

We have also a natural representation

$$(7.24)\qquad \rho\colon G(\Gamma, a) \longrightarrow GL(r; \boldsymbol{C})$$

given by

$$(7.25)\qquad \rho(\gamma, c) = j(\gamma, 0)c^{-1} .$$

The group $G(\Gamma, a)$ acts freely on $\boldsymbol{C}^n \times \boldsymbol{C}^*$ by

$$(7.26)\qquad (\gamma, c)(z, d) = (z+\gamma, a(\gamma, z)cd), \quad (\gamma, c) \in G(\Gamma, a), \quad (z, d) \in \boldsymbol{C}^n \times \boldsymbol{C}^* .$$

The mapping $\boldsymbol{C}^n \to \boldsymbol{C}^n \times \boldsymbol{C}^*$ sending z to $(z, 1)$ induces an isomorphism

$$(7.27)\qquad T^n = \Gamma \backslash \boldsymbol{C}^n \approx G(\Gamma, a) \backslash \boldsymbol{C}^n \times \boldsymbol{C}^* .$$

The action (7.26) of $G(\Gamma, a)$ on $\boldsymbol{C}^n \times \boldsymbol{C}^*$ and the representation ρ of $G(\Gamma, a)$ given by (7.25) define an action of $G(\Gamma, a)$ on $(\boldsymbol{C}^n \times \boldsymbol{C}^*) \times \boldsymbol{C}^r$:

$$(7.28)\qquad \begin{aligned} &(\gamma, c)((z, d), \zeta) = ((\gamma, c)(z, d), \rho(\gamma, c)\zeta), \\ &\quad \text{for} \quad (\gamma, c) \in G(\Gamma, a), \quad (z, d) \in \boldsymbol{C}^n \times \boldsymbol{C}^*, \quad \zeta \in \boldsymbol{C}^r . \end{aligned}$$

The quotient $G(\Gamma, a) \backslash (\boldsymbol{C}^n \times \boldsymbol{C}^*) \times \boldsymbol{C}^r$ by this action is a vector bundle of rank r over $T^n = G(\Gamma, a) \backslash \boldsymbol{C}^n \times \boldsymbol{C}^*$. Again, the mapping $\boldsymbol{C}^n \times \boldsymbol{C}^r \to (\boldsymbol{C}^n \times \boldsymbol{C}^*) \times \boldsymbol{C}^r$ which sends (z, ζ) to $((z, 1), \zeta)$ induces an isomorphism

$$E = \Gamma \backslash \boldsymbol{C}^n \times \boldsymbol{C}^r \approx G(\Gamma, a) \backslash (\boldsymbol{C}^n \times \boldsymbol{C}^*) \times \boldsymbol{C}^r .$$

To summarize, we have shown that a projectively flat vector bundle E over T^n can be obtained by the representation ρ of $G(\Gamma, a)$.

If j is a factor of automorphy for E, then $\det j$ is a factor of automorphy for the determinant line bundle $\det E$. Applying (7.17) to the line bundle $\det E$, we obtain

$$(7.30)\qquad \det j(\gamma, z) = \chi(\gamma) \exp\left[R(z, \gamma) + \frac{1}{2} R(\gamma, \gamma) \right], \qquad (\gamma, z) \in \Gamma \times \boldsymbol{C}^n .$$

Comparing this with (7.18), we obtain

$$(7.31)\qquad \begin{aligned} &\det j(\gamma, 0) = \chi(\gamma) \exp\left(\frac{1}{2} R(\gamma, \gamma) \right), \\ &a(\gamma, z)^r = \exp(R(z, \gamma)) . \end{aligned}$$

Taking the r-th root of (7.31) and using (7.20), we obtain

(7.32) $$a(\gamma, z)=\exp\left(\frac{1}{r}R(z, \gamma)\right).$$

We shall now assume not only that E is projectively flat but also that it admits a projectively flat Hermitian structure. In other words, the projective bundle $P(E)$ is defined by a representation of Γ in the projective unitary group $PU(r)\subset PGL(r; \boldsymbol{C})$, (see (I.4.22)). We may therefore assume that $\tilde{j}$ sends Γ into $PU(r)$. Define the conformal unitary group $CU(r)$ by

(7.33) $$CU(r)=\{cU;\ c\in \boldsymbol{C}^* \text{ and } U\in U(r)\},$$

or $CU(r)=\lambda^{-1}(PU(r))$, where $\lambda: GL(r; \boldsymbol{C})\rightarrow PGL(r; \boldsymbol{C})$ is the natural projection. Since $\tilde{j}=\lambda\circ j$, we see that

(7.34) $$j(\gamma, z)\in CU(r) \qquad \text{for} \quad (\gamma, z)\in \Gamma\times \boldsymbol{C}^n.$$

Let h be a projectively flat Hermitian structure in E and $\tilde{h}$ be the induced Hermitian structure in the induced vector bundle $\tilde{E}=p^*E=\boldsymbol{C}^n\times \boldsymbol{C}^r$ as in $\tilde{F}$ above. We may consider $\tilde{h}$ as a function on $\boldsymbol{C}^n$ with values in the space of positive definite $r\times r$ Hermitian matrices. With matrix notation, the invariance condition (7.6) in the vector bundle case reads as follows:

(7.35) $$\tilde{h}(z)=j(\gamma, z)^*\tilde{h}(z+\gamma)j(\gamma, z),$$

where the asterisk * denotes the transpose-conjugate. The connection form $\tilde{\omega}=\tilde{h}^{-1}d'\tilde{h}$ and the curvature form $\tilde{\Omega}=d''\tilde{\omega}$ satisfy

(7.36) $$\tilde{\omega}(z)=j(\gamma, z)^{-1}\tilde{\omega}(z+\gamma)j(\gamma, z)+j(\gamma, z)^{-1}d'j(\gamma, z),$$

(7.37) $$\tilde{\Omega}(z)=j(\gamma, z)^{-1}\tilde{\Omega}(z+\gamma)j(\gamma, z).$$

Since the connection is projectively flat, the curvature must be of the following form, (see (I.2.8)):

(7.38) $$\Omega=\alpha I_r,$$

where α is a 2-form. Since its trace is the curvature of the line bundle $\det E$, we have

(7.39) $$\alpha=\frac{1}{r}\sum R_{j\bar{k}}dz^j\wedge d\bar{z}^k.$$

with constant coefficients $R_{j\bar{k}}$. From $\tilde{\Omega}=d''\tilde{\omega}$, we obtain

(7.40) $$\tilde{\omega}(z)=-\frac{1}{r}R(dz, z)I_r+\Theta(z),$$

where $R(dz, z)=\sum R_{j\bar{k}}\bar{z}^k dz^j$ and Θ is a holomorphic 1-form with values in the Lie algebra of $CU(r)$. Since

$$\Theta+\Theta^*=\psi I_r \qquad \text{(with a 1-form } \psi),$$

and Θ is holomorphic, it follows that

$$\Theta=\theta I_r, \tag{7.41}$$

where θ is a holomorphic 1-form. On the other hand, from (7.36) and from

$$\begin{aligned} j(\gamma, z)^{-1}d'j(\gamma, z) &= (j(\gamma, 0)a(\gamma, z))^{-1}d'(j(\gamma, 0)a(\gamma, z)) = (d'\log a(\gamma, z))I_r \\ &= \frac{1}{r}R(dz, \gamma)I_r \end{aligned} \tag{7.42}$$

we obtain

$$\Theta(z+\gamma)=\Theta(z),$$

i.e., θ is actually a holomorphic 1-form on the torus T^n. Hence,

$$\theta=\sum b_j dz^j \tag{7.43}$$

with constant coefficients b_j. Solving

$$\tilde{h}^{-1}d'\tilde{h}=\tilde{\omega}(z)=-\frac{1}{r}\sum R_{j\bar{k}}\bar{z}^k dz^j+\sum b_j dz^j,$$

we obtain

$$\tilde{h}(z)=\tilde{h}(0)\exp\left[-\frac{1}{r}R(z, z)+b(z)+\overline{b(z)}\right], \tag{7.44}$$

where $b(z)=\sum b_j z^j$.

In order to simplify the formula, we change the initial identification $\tilde{E}=\boldsymbol{C}^n\times\boldsymbol{C}^r$ by composing it with the isomorphism

$$(z, \zeta)\in\boldsymbol{C}^n\times\boldsymbol{C}^r \longrightarrow (z, e^{b(z)}\zeta)\in\boldsymbol{C}^n\times\boldsymbol{C}^r.$$

With this new identification $\tilde{E}=\boldsymbol{C}^n\times\boldsymbol{C}^r$, we have

$$b(z)=0.$$

By a linear change of coordinates in $\boldsymbol{C}^r$, we may assume also

$$\tilde{h}(0)=I_r.$$

Thus,

$$\tilde{h}(z)=\exp\left[-\frac{1}{r}R(z,z)\right]I_r. \tag{7.45}$$

Substituting in (7.35) the expression obtained from (7.45), we obtain

$$j(\gamma,z)=U(\gamma)\cdot\exp\left[\frac{1}{r}R(z,\gamma)+\frac{1}{2r}R(\gamma,\gamma)\right], \tag{7.46}$$

where $U(\gamma)$ is a unitary matrix. From (7.46) we obtain

$$U(\gamma)=j(\gamma,0)\exp\left[-\frac{1}{2r}R(\gamma,\gamma)\right]. \tag{7.47}$$

Using (7.46) and (7.13) we obtain

$$U(\gamma+\gamma')=U(\gamma)U(\gamma')\cdot\exp\left[\frac{i}{r}A(\gamma',\gamma)\right] \quad \text{for} \quad \gamma,\gamma'\in\Gamma. \tag{7.48}$$

Although $U\colon\Gamma\to U(r)$ is not a representation, we can "extend" it to a representation as follows. First, we introduce a group structure in the set $\Gamma\times\boldsymbol{C}^*$ by defining its multiplication rule by

$$(\gamma,c)(\gamma',c')=\left(\gamma+\gamma',\ cc'\exp\frac{i}{r}A(\gamma',\gamma)\right). \tag{7.49}$$

We denote this group by $G(\Gamma, A/r)$. Then there is an exact sequence

$$1\longrightarrow\boldsymbol{C}^*\longrightarrow G(\Gamma,A/r)\longrightarrow\Gamma\longrightarrow 0. \tag{7.50}$$

If we set

$$\rho_{A/r}(\gamma,c)=U(\gamma)c^{-1} \qquad (\gamma,c)\in G(\gamma,A/r), \tag{7.51}$$

then $\rho_{A/r}\colon G(\Gamma,A/r)\to CU(r)\subset GL(r;\boldsymbol{C})$ is a representation.

Moreover, $(G(\Gamma,A/r),\rho_{A/r})$ is isomorphic to $(G(\Gamma,a),\rho)$. Namely, we have an isomorphism

$$\begin{aligned} &f\colon G(\Gamma,a)\longrightarrow G(\Gamma,A/r),\\ &f(\gamma,c)=\left(\gamma,\ c\cdot\exp\left[-\frac{1}{2r}R(\gamma,\gamma)\right]\right) \end{aligned} \tag{7.52}$$

such that

$$\rho=\rho_{A/r}\circ f. \tag{7.53}$$

Conversely, given an Hermitian form R on $\boldsymbol{C}^n$ with imaginary part A satisfying (7.14) and a "semi-representation" $U: \Gamma \to U(r)$ satisfying (7.48) we obtain a factor of automorphy $j: \Gamma \times \boldsymbol{C}^n \to CU(r) \subset GL(r; \boldsymbol{C})$ by (7.46). The corresponding vector bundle E admits a projectively flat Hermitian structure; it is given by the natural inner product in $\boldsymbol{C}^r$.

In summary, we have

(7.54) **Theorem** *Let E be a holomorphic vector bundle of rank r over a complex torus $T^n = \Gamma \backslash \boldsymbol{C}^n$. If E admits a projectively flat Hermitian structure h, then its factor of automorphy j can be written as follows*:

$$j(\gamma, z) = U(\gamma) \cdot \exp\left[\frac{1}{r} R(z, \gamma) + \frac{1}{2r} R(\gamma, \gamma)\right] \qquad (\gamma, z) \in \Gamma \times \boldsymbol{C}^n ,$$

where

(i) *R is an Hermitian form on $\boldsymbol{C}^n$ and its imaginary part A satisfies*

$$\frac{1}{\pi} A(\gamma, \gamma') \in \boldsymbol{Z} \qquad \text{for} \quad \gamma, \gamma' \in \Gamma ;$$

(ii) *$U: \Gamma \to U(r)$ is a semi-representation in the sense that it satisfies*

$$U(\gamma + \gamma') = U(\gamma) U(\gamma') \exp \frac{i}{r} A(\gamma', \gamma) \qquad \text{for} \quad \gamma, \gamma' \in \Gamma .$$

Conversely, given an Hermitian form R on $\boldsymbol{C}^n$ with property (i) *and a semi-representation $U: \Gamma \to U(r)$, we can define a factor of automorphy $j: \Gamma \times \boldsymbol{C}^n \to CU(r)$ as above. The corresponding vector bundle E over $T^n = \Gamma \backslash \boldsymbol{C}^n$ admits a projectively flat Hermitian structure (given by* (7.45)).

We recall (Section 3 of Chapter II) that every projectively flat vector bundle E of rank r satisfies the following identity:

$$c(E) = \left(1 + \frac{c_1(E)}{r}\right)^r, \quad \text{i.e.,} \qquad c_k(E) = \binom{r}{k} \frac{1}{r^k} c_1(E)^k .$$

Materials for this section are taken largely from Matsushima [2]. See also Hano [1], Morikawa [1], Yang [1].

Chapter V

Stable vector bundles

In this chapter we shall prove the theorem that every irreducible Einstein-Hermitian vector bundle over a compact Kähler manifold is stable. In Sections 1 and 2 we consider the special case where the base space is a compact Riemann surface. For in this case, the definition of stability (due to Mumford, see Mumford and Fogarty [1]) can be given without involving coherent sheaves and the theorem can be proven as a simple application of Gauss' equation for subbundles.

However, the definition of stable vector bundle over a higher dimensional base space necessiates the introduction of coherent sheaves. Sections 3 through 6 are devoted to homological algebraic aspects of coherent sheaves and Section 7 to basic properties of stable vector bundles. Our basic references for all these are the books of Okonek, Schneider and Spindler [1] and Matsumura [1]. The former serves as an excellent introduction to stable vector bundles, especially, for differential geometers. Also very readable is the original paper of Takemoto [1] who extended the concept of stability to the case of higher dimensional algebraic manifolds. The reader may find Bănică and Stănăşilă [1] also useful as a reference on algebraic aspects of coherent sheaves. See also Siu-Trautmann [1]. The reader who wishes more details on reflexive sheaves is referred to Hartshorne [3].

In Section 8 the main theorem stated above is proved for an arbitrary dimension. This theorem is an evidence that Takemoto's stability fits our differential geometric frame perfectly. In the last two sections we study the relationship between different concepts of stability and the Einstein condition.

§ 1 Stable vector bundles over Riemann surfaces

Throughout this section, M will denote a compact complex manifold of dimension 1, i.e., a compact Riemann surface. Let E be a holomorphic vector bundle of rank r over M. The first Chern class $c_1(E)$, integrated over M, is an integer. This integer will be denoted by $c_1(E)$. It is called also the *degree* of E and is sometimes denoted by $\deg(E)$. Thus,

$$\deg(E) = c_1(E) = \int_M c_1(E). \tag{1.1}$$

We associate the following rational number to E, called the *degree/rank ratio*.

$$\mu(E)=\frac{c_1(E)}{\text{rank}(E)}\,. \tag{1.2}$$

If E is a direct sum of r line bundles $L_1, \cdots, L_r$, then $\mu(E)$ is the average of $c_1(L_1), \cdots, c_1(L_r)$.

Following Mumford, we say that E is *stable* (resp. *semi-stable*) if, for every proper subbundle E' of E, $0<\text{rank}(E')<\text{rank}(E)$, we have

$$\mu(E')<\mu(E) \qquad (\text{resp. } \mu(E')\leqq\mu(E))\,. \tag{1.3}$$

We shall give another definition of stability due to Bogomolov [1]. A *weighted flag* of E is a sequence of pairs $\mathscr{F}=\{(E_i, n_i);\ i=1, 2, \cdots, k\}$ consisting of subbundles

$$E_1\subset E_2\subset\cdots\subset E_k\subset E$$

with

$$0<\text{rank}(E_1)<\text{rank}(E_2)<\cdots<\text{rank}(E_k)<\text{rank}(E)$$

and positive integers $n_1, n_2, \cdots, n_k$. We set

$$r_i=\text{rank}(E_i)\,, \qquad r=\text{rank}(E)\,.$$

To such a flag $\mathscr{F}$ we associate a line bundle $T_{\mathscr{F}}$ by setting

$$T_{\mathscr{F}}=\prod_{i=1}^{k}((\det E_i)^r(\det E)^{-r_i})^{n_i}\,. \tag{1.4}$$

It is clearly a line subbundle of $(E\otimes E^*)^{\otimes N}$, where $N=r\sum r_i n_i$.

We recall that a vector bundle is flat if it is defined by a representation of the fundamental group of the base manifold M, (see (I.2.5)). Following Bogomolov, we say that a holomorphic vector bundle E over a compact Riemann surface M is *T-stable* if, for every weighted flag $\mathscr{F}$ of E and every flat line bundle L over M, the line bundle $T_{\mathscr{F}}\otimes L$ admits no nonzero holomorphic sections, i.e.,

$$H^0(M, \Omega^0(T_{\mathscr{F}}\otimes L))=0\,. \tag{1.5}$$

We say that E is *T-semi-stable* if, for every weighted flag $\mathscr{F}$ of E and every flat line bundle L over M, the line bundle $T_{\mathscr{F}}\otimes L$ either admits no nonzero holomorphic sections or is a product bundle, (in other words, the only holomorphic sections of $T_{\mathscr{F}}\otimes L$ are nowhere vanishing sections). (The letter T in the definition of stability stands for "tensor".)

(1.6) **Proposition** *Let E be a holomorphic vector bundle over a compact Riemann surface M. Then E is stable* (*resp. semi-stable*) *if and only if it is T-stable* (*resp. T-semi-stable*).

Proof We remark first that since $H^2(M, \boldsymbol{R}) = \boldsymbol{R}$, a line bundle F over M is positive, negative or flat according as its degree $c_1(F)$ is positive, negative or zero.

Suppose that E is stable (resp. semi-stable) and let $\mathscr{F} = \{(E_i, n_i)\}$ be any weighted flag of E. For each i, we have

$$c_1((\det E_i)^r(\det E)^{-r_i}) = rc_1(E_i) - r_i c_1(E) = rr_i(\mu(E_i) - \mu(E)),$$

which is negative (resp. non-positive) by (1.3). From (1.4) it follows that $c_1(T_{\mathscr{F}})$ is negative (resp. non-positive). If L is a flat line bundle, then $T_{\mathscr{F}} \otimes L$ is negative or flat according as $T_{\mathscr{F}}$ is negative or flat. If $T_{\mathscr{F}} \otimes L$ is negative, it has no nonzero holomorphic sections by (III.1.9). If $T_{\mathscr{F}} \otimes L$ is flat, every holomorphic section of $T_{\mathscr{F}} \otimes L$ is nowhere vanishing also by (III.1.9). Hence, E is T-stable (resp. T-semi-stable).

Suppose that E is T-semi-stable, and let E' be a subbundle of rank r' with $0 < r' < r = \text{rank}(E)$. Let n be a positive integer and consider the flag $\mathscr{F} = \{(E', n)\}$. To this flag we associate a line bundle

$$T_{\mathscr{F}} = F^n, \qquad \text{where} \quad F = (\det E')^r (\det E)^{-r'}.$$

Then the Riemann-Roch theorem states

$$\dim H^0(M, F^n) - \dim H^1(M, F^n) = nc_1(F) + \frac{1}{2} c_1(M).$$

Since $\dim H^0(M, F^n) \leqq 1$ by assumption, we see by letting $n \to \infty$ that $c_1(F) \leqq 0$. Since $c_1(F) = rc_1(E') - r'c_1(E)$, this implies $\mu(E') \leqq \mu(E)$. Hence, E is semi-stable. Assume $\mu(E') = \mu(E)$. Then $c_1(F) = 0$, i.e., F is a flat bundle. If we set $L = F^{-n}$, then $T_{\mathscr{F}} \otimes L$ is a product bundle. This shows that if E is T-stable, then it is stable. Q.E.D.

We shall show that every holomorphic vector bundle E over a compact Riemann surface has a unique maximal semistable subbundle. We start with the following

(1.7) **Lemma** *Given a holomorphic vector bundle E over a compact Riemann surface M, there is a positive integer q such that if L is a line bundle over M with* $\deg(L) \geqq q$, *then* $\text{Hom}(L, E)$ *has no nonzero holomorphic sections.*

Proof Let H be an ample line bundle over M (with an Hermitian structure whose curvature is positive). We fix also an Hermitian structure in E. Then there is a positive integer p such that the mean curvature K of $H^{-m}\otimes E$ is negative for $m \geqq p$, (see (I.5.5, 5.6, and 5.13)). By (III.1.9),

(1.8) $$H^0(M, \mathrm{Hom}(H^m, E)) = H^0(M, H^{-m}\otimes E)) = 0 \quad \text{for} \quad m \geqq p.$$

By the Riemann-Roch formula (see (III.4.4)), for any line bundle L we have

(1.9) $$\dim H^0(M, H^{-p}\otimes L) \geqq \deg(L) - p\deg(H) + 1 - g,$$

where g is the genus of M. Hence,

(1.10) $$H^0(M, \mathrm{Hom}(H^p, L)) \neq 0 \quad \text{if} \quad \deg(L) \geqq g + p\deg(H).$$

From (1.8) and (1.10) we see that

(1.11) $$H^0(M, \mathrm{Hom}(L, E)) = 0 \quad \text{if} \quad \deg(L) \geqq g + p\deg(H).$$ Q.E.D.

(1.12) **Lemma** *Given a holomorphic vector bundle E over a compact Riemann surface M, there is an integer m such that*

$$\mu(F) \leqq m$$

for all holomorphic subbundles F of E.

Proof Applying (1.7) to $\bigwedge^s E$, let $q(s)$ be the integer given by (1.7). Let F be a subbundle of rank s. The injection $F \to E$ induces an injection $\bigwedge^s F \to \bigwedge^s E$. By (1.7), $\deg(\bigwedge^s F) < q(s)$. Let $m = \mathrm{Max}\{q(s)/s;\ s = 1, 2, \cdots, r\}$. Q.E.D.

(1.13) **Proposition** *Given a holomorphic vector bundle E over a compact Riemann surface M, there is a unique subbundle E_1 such that for every subbundle F of E we have*

(i) $\mu(F) \leqq \mu(E_1)$

(ii) $\mathrm{rank}(F) \leqq \mathrm{rank}(E_1)$ *if* $\mu(F) = \mu(E_1)$.

Then E_1 is semistable.

We call E_1 the *maximal semistable subbundle* of E.

Proof The existence is clear from (1.12). From the characterizing properties of E_1, it is clear that E_1 is semistable. To prove the uniqueness, let E_1' be another subbundle satisfying (i) and (ii). Let $p: E \to E/E_1'$ be the projection. Then $p(E_1) \neq 0$.

Since $p(E_1)$ itself may not be a subbundle of E/E_1', we consider the subbundle G of E/E_1' generated by $p(E_1)$. To give a little more details, let $s_1, \cdots, s_m$ be a holomorphic local frame field for E_1. Let k be the rank of $p(E_1)$. Consider

$$p(s_{i_1}) \wedge \cdots \wedge p(s_{i_k}), \qquad i_1 < \cdots < i_k .$$

These holomorphic sections of $\bigwedge^k(E/E_1')$ define a rank k subbundle of (E/E_1') except where they all vanish. Wherever they all vanish, we factor out their common zeros; this is possible since $\dim M = 1$. Then they define a rank k subbundle G of E/E_1'.

Define the subbundle F of E_1 by the "locally defined" exact sequence

$$0 \longrightarrow F \longrightarrow E_1 \longrightarrow G \longrightarrow 0 .$$

Since E_1 is semistable, $\mu(F) \leqq \mu(E_1)$. This is equivalent to $\mu(E_1) \leqq \mu(G)$.

On the other hand, we have an exact sequence

$$0 \longrightarrow E_1' \longrightarrow p^{-1}(G) \longrightarrow G \longrightarrow 0 .$$

From the properties (i) and (ii) for E_1', we have $\mu(p^{-1}(G)) < \mu(E_1')$, i.e.,

$$\frac{\deg(G) + \deg(E_1')}{\operatorname{rank}(G) + \operatorname{rank}(E_1')} < \frac{\deg(E_1')}{\operatorname{rank}(E_1')} .$$

This implies $\mu(G) < \mu(E_1') = \mu(E_1)$, a contradiction. Q.E.D.

The following *Harder-Narasimhan filtration* theorem is now an immediate consequence of (1.13).

(1.14) **Theorem** *Given a holomorphic vector bundle E over a compact Riemann surface M, there is a unique filtration by subbundles*

$$0 = E_0 \subset E_1 \subset E_2 \subset \cdots \subset E_{s-1} \subset E_s = E$$

such that, for $1 \leqq i \leqq s-1$, E_i/E_{i-1} is the maximal semistable subbundle of E/E_{i-1}.

The Jordan-Hölder theorem holds for semistable vector bundles.

(1.15) **Theorem** *Given a semistable vector bundle E over a compact Riemann surface M, there is a filtration of E by subbundles*

$$0 = E_{k+1} \subset E_k \subset \cdots \subset E_1 \subset E_0 = E$$

such that E_i/E_{i+1} are stable and $\mu(E_i/E_{i+1}) = \mu(E)$ for $i = 0, 1, \cdots, k$.

Moreover,

$$Gr(E) = (E_0/E_1) \oplus (E_1/E_2) \oplus \cdots \oplus (E_k/E_{k+1})$$

is uniquely determined by E up to an isomorphism.

The proof is standard.

For more details on vector bundles over compact Riemann surfaces, see Astérisque notes by Seshadri [1].

§2 Einstein-Hermitian vector bundles over Riemann surfaces

Let (E, h) be an Hermitian vector bundle over a Riemann surface M. Let $g = g_{1\bar{1}} dz^1 d\bar{z}^1$ be any Hermitian metric on M, (expressed in terms of a local coordinate system z^1). The mean curvature K of (E, h) is given, in terms of its components, by

$$K^i_j = g^{1\bar{1}} R^i_{j1\bar{1}} . \tag{2.1}$$

The weak Einstein condition in this case is equivalent to

$$R^i_{j1\bar{1}} = g_{1\bar{1}} \varphi \delta^i_j , \tag{2.2}$$

where φ is a function on M. Since any two Hermitian metrics on M are conformal to each other, this weak Einstein condition is independent of the choice of g. From (I.2.8) we obtain

(2.3) **Proposition** *An Hermitian vector bundle (E, h) over a Riemann surface (M, g) satisfies the weak Einstein condition if and only if it is projectively flat.*

Let E' be a holomorphic subbundle of an Einstein-Hermitian vector bundle (E, h) over a compact Riemann surface (M, g). If we denote the curvature forms of (E, h) and $(E', h|_{E'})$ by Ω and Ω', respectively, then the vector bundle analogue of the Gauss-Codazzi equation (see (I.6.11)) states

$$\Omega'^a_b = \Omega^a_b - \sum \omega^\lambda_b \wedge \omega^\lambda_a , \qquad 1 \leqq a, b \leqq p < \lambda \leqq r , \tag{2.4}$$

where $r = \operatorname{rank}(E)$ and $p = \operatorname{rank}(E')$. The first Chern classes of E and E' are represented by (II.2.14):

$$c_1(E, h) = \frac{i}{2\pi} \sum_{j=1}^{r} \Omega^j_j , \qquad c_1(E', h) = \frac{i}{2\pi} \sum_{a=1}^{p} \Omega'^a_a .$$

By (2.3) (see also (2.2)), $\Omega^i_j = \alpha \delta^i_j$ with a suitable (1, 1)-form α. Hence,

$$\deg(E)=\int_M c_1(E)=\frac{i}{2\pi}\int_M r\alpha\,,$$

$$\deg(E')=\int_M c_1(E')=\frac{i}{2\pi}\int_M (p\alpha-\sum\omega_a^\lambda\wedge\bar{\omega}_a^\lambda)\,,$$

$$\mu(E)=\frac{i}{2\pi}\int_M \alpha\,,$$

$$\mu(E')=\frac{i}{2\pi}\int_M \alpha-\frac{i}{2p\pi}\int_M \sum\omega_a^\lambda\wedge\bar{\omega}_a^\lambda\,.$$

This implies $\mu(E')\leqq\mu(E)$, and the equality holds if and only if $\omega_a^\lambda=0$, i.e., the second fundamental form vanishes. From (I.6.14) it follows that if $\mu(E)=\mu(E')$, then

$$E=E'\oplus E''\,,$$

where E'' is holomorphic and orthogonal to E'. By (IV.1.4), both (E', h) and (E'', h) satisfy the Einstein condition with the same factor as (E, h). In particular,

$$\mu(E)=\mu(E')=\mu(E'')\,.$$

We have established

(2.5) **Theorem** *If (E, h) is an Einstein-Hermitian vector bundle over a compact Riemann surface (M, g), then E is semistable and decomposes into a direct sum*

$$E=E_1\oplus\cdots\oplus E_k\,,\qquad \textit{(holomorphic and orthogonal)}$$

of stable Einstein-Hermitian vector bundles $E_1,\cdots,E_k$ such that $\mu(E)=\mu(E_1)=\cdots=\mu(E_k)$.

This will be generalized to compact Kähler manifolds (M, g) of higher dimensions in Section 8.

(2.6) *Remark* It is possible to derive quickly from (III.1.38) that every Einstein-Hermitian vector bundle E over a compact Riemann surface is T-semistable (see Section 1 for the definition). But (2.5) asserts much more.

Conversely,

(2.6) **Theorem** *If E is a stable vector bundle over a compact Riemann surface (M, g), there is an Einstein-Hermitian structure h in E.*

This follows from the following reformulation of the result of Narashimhan-Seshadri [1], (see Atiyah-Bott [1] for a reformulation which is very close to what follows).

(2.7) **Theorem** *A holomorphic vector bundle E over a compact Riemann surface M is stable if and only if the associated projective bundle $P(E)$ comes from an irreducible representation of the fundamental group $\pi_1(M)$ into the projective unitary group $PU(r)$, that is, if and only if E admits a projectively flat Hermitian structure.*

Clearly, (2.6) follows from (2.3) and (2.7). A direct proof of (2.6) has been given by Donaldson [1].

§3 Coherent sheaves——homological algebra of stalks

Throughout this section we denote the ring of germs of holomorphic functions at the origin of $\boldsymbol{C}^n$ by A and its maximal ideal by $\mathfrak{m}$. Thus, A is the ring of convergent power series in n complex variables and $\mathfrak{m}$ is the ideal consisting of power series without constant term. The ring A is an example of regular local ring. Most of the results in this section are valid for any regular local ring, and some of them hold for more general local rings.

We denote the free A-module $A\oplus\cdots\oplus A$ of rank r by A^r. An A-module M is said to be of *finite type* if it is finitely generated, i.e., if there is a surjective homomorphism $A^r \rightarrow M$ for some finite r.

(3.1) **Nakayama's lemma** *If an A-module M of finite type satisfies*

$$M=\mathfrak{m}M\,,$$

then $M=0$.

Proof Assuming $M\neq 0$, let $\{u_1, \cdots, u_r\}$ be a minimal set of generators of M. Since $u_r \in M=\mathfrak{m}M$, we can write

$$u_r = a_1u_1 + \cdots + a_{r-1}u_{r-1} + a_ru_r \qquad \text{with} \quad a_i \in \mathfrak{m}\,,$$

or

$$(1-a_r)u_r = a_1u_1 + \cdots + a_{r-1}u_{r-1}\,.$$

Since $1-a_r$ is a unit, $u_r \in Au_1 + \cdots + Au_{r-1}$, showing that $\{u_1, \cdots, u_{r-1}\}$ generates M. This contradicts the minimality of $\{u_1, \cdots, u_r\}$. Q.E.D.

We recall that an A-module M of finite type is said to be *projective* if given A-modules L and N of finite rank and homomorphisms α and β, with β surjective, in the following diagram, there is a homomorphism γ such that $\alpha = \beta \circ \gamma$.

(3.2)
$$\begin{array}{ccccc} M & & & & \\ \gamma \downarrow & \searrow^{\alpha} & & & \\ L & \xrightarrow{\beta} & N & \longrightarrow & 0 \, . \end{array}$$

(3.3) **Lemma** *An A-module M of finite type is free if and only if it is projective.*

Proof It is trivial that every free module is projective. Assume that M is projective. Choose elements $u_1, \cdots, u_r$ in M such that their images $\bar{u}_1, \cdots, \bar{u}_r$ in $M/\mathfrak{m}M$ form a basis of the vector space $M/\mathfrak{m}M$. Let L be the free A-module generated by r letters $e_1, \cdots, e_r$. Let $\beta: L \to M$ be the homomorphism defined by $\beta(e_i) = u_i$, and let Q be the cokernel of β so that

$$L \xrightarrow{\beta} M \xrightarrow{\pi} Q \longrightarrow 0$$

is exact. Then the induced sequence

$$L/\mathfrak{m}L \xrightarrow{\bar{\beta}} M/\mathfrak{m}M \xrightarrow{\bar{\pi}} Q/\mathfrak{m}Q \longrightarrow 0$$

is exact. (In fact, if $u \in M$ and $\pi(u) \in \mathfrak{m}Q$, then $\pi(u) = a \cdot \pi(u')$ for some $a \in \mathfrak{m}$ and $u' \in M$. Since $\pi(u - au') = 0$, there is an $x \in L$ such that $u - au' = \beta(x)$, i.e., $u \equiv \beta(x) \bmod \mathfrak{m}M$.) On the other hand, from the way $u_1, \cdots, u_r$ were chosen, $\bar{\beta}$ is clearly surjective. Hence, $Q/\mathfrak{m}Q = 0$. By Nakayama's lemma, $Q = 0$, which shows that β is surjective.

Since M is projective, there is a homomorphism $\gamma: M \to L$ which makes the following diagram commutative:

$$\begin{array}{ccccc} M & & & & \\ \gamma \downarrow & \searrow^{id} & & & \\ L & \xrightarrow[\beta]{} & M & \longrightarrow & 0 \end{array}$$

Then γ is injective. Let K be the kernel of β so that the sequence

$$0 \longrightarrow K \xrightarrow{\iota} L \xrightarrow{\beta} M \longrightarrow 0$$

is exact. Then, $L=\iota(K)+\gamma(M)$. Identifying $\iota(K)$ with K and $\gamma(M)$ with M, we write

$$L=K+M.$$

Then

$$L/\mathfrak{m}L=K/\mathfrak{m}K+M/\mathfrak{m}M.$$

Since $L/\mathfrak{m}L\approx M/\mathfrak{m}M$ in a natural way, we obtain $K/\mathfrak{m}K=0$. Again by Nakayama's lemma, $K=0$. Hence, $L=M$. Q.E.D.

We recall the definitions of Ext and Tor. Given an A-module M of finite type, choose a projective resolution

$$\text{(3.4)}\quad E.(M):\ \cdots\longrightarrow E_m\longrightarrow E_{m-1}\longrightarrow\cdots\longrightarrow E_0\longrightarrow M\longrightarrow 0$$

of M, where the E_m are projective A-modules. If N is another A-module of finite type, then we define

$$\text{(3.5)}\qquad \mathrm{Ext}^i_A(M,N)=H^i(\mathrm{Hom}_A(E.(M),N)),$$

$$\text{(3.6)}\qquad \mathrm{Tor}^A_i(M,N)=H_i(E.(M)\otimes_A N).$$

Then

$$\text{(3.7)}\qquad \mathrm{Ext}^0_A(M,N)=\mathrm{Hom}_A(M,N),$$

$$\text{(3.8)}\qquad \mathrm{Tor}^A_0(M,N)=M\otimes_A N.$$

The basic properties of the Ext and Tor functors are given by

(3.9) **Theorem** *Short exact sequences of A-modules*

$$0\longrightarrow M'\longrightarrow M\longrightarrow M''\longrightarrow 0,$$

$$0\longrightarrow N'\longrightarrow N\longrightarrow N''\longrightarrow 0$$

induce long exact sequences

(a) $\cdots\longrightarrow \mathrm{Ext}^i_A(M,N)\longrightarrow \mathrm{Ext}^i_A(M',N)\longrightarrow \mathrm{Ext}^{i+1}_A(M'',N)\longrightarrow\cdots,$

(b) $\cdots\longrightarrow \mathrm{Ext}^i_A(M,N)\longrightarrow \mathrm{Ext}^i_A(M,N'')\longrightarrow \mathrm{Ext}^{i+1}_A(M,N')\longrightarrow\cdots,$

(c) $\cdots\longrightarrow \mathrm{Tor}^A_i(M,N)\longrightarrow \mathrm{Tor}^A_i(M'',N)\longrightarrow \mathrm{Tor}^A_{i-1}(M',N)\longrightarrow\cdots,$

(d) $\cdots\longrightarrow \mathrm{Tor}^A_i(M,N)\longrightarrow \mathrm{Tor}^A_i(M,N'')\longrightarrow \mathrm{Tor}^A_{i-1}(M,N')\longrightarrow\cdots.$

(3.10) **Lemma** *An A-module M of finite type is free if (and only if)*

$$\mathrm{Tor}_1^A(\boldsymbol{C}, M)=0\,,$$

where $\boldsymbol{C}=A/\mathfrak{m}$ is considered as an A-module in a natural way.

Proof Let $u_1, \cdots, u_r$ be elements of M such that their images $\bar{u}_1, \cdots, \bar{u}_r$ in $M/\mathfrak{m}M$ form a basis over $\boldsymbol{C}=A/\mathfrak{m}$. Let L be the free A-module generated by r letters $e_1, \cdots, e_r$. Let $\beta\colon L\to M$ be the homomorphism defined by $\beta(e_i)=u_i$. Since the cokernel Q of β is zero as we saw in the proof of (3.3), we have the following exact sequence:

$$0\longrightarrow K\longrightarrow L\longrightarrow M\longrightarrow 0\,,$$

where $K=\mathrm{Ker}\,\beta$. This induces the following exact sequence

$$0\longrightarrow \mathrm{Tor}_1^A(\boldsymbol{C}, M)\longrightarrow \boldsymbol{C}\otimes_A K\longrightarrow \boldsymbol{C}\otimes_A L\longrightarrow \boldsymbol{C}\otimes_A M\longrightarrow 0\,,$$

which may be written as

$$0\longrightarrow \mathrm{Tor}_1^A(\boldsymbol{C}, M)\longrightarrow K/\mathfrak{m}K\longrightarrow L/\mathfrak{m}L\longrightarrow M/\mathfrak{m}M\longrightarrow 0\,.$$

Since $\bar{\beta}\colon L/\mathfrak{m}L\to M/\mathfrak{m}M$ is an isomorphism, $K/\mathfrak{m}K$ is isomorphic to $\mathrm{Tor}_1^A(\boldsymbol{C}, M)$. But, by Nakayama's lemma $K/\mathfrak{m}K=0$ if and only if $K=0$. Hence, $\mathrm{Tor}_1^A(\boldsymbol{C}, M)=0$ if and only if $L\approx M$. Q.E.D.

(3.11) **Syzygy Theorem** *If M is an A-module of finite type and if*

$$0\longrightarrow F\longrightarrow E_{n-1}\longrightarrow \cdots\longrightarrow E_1\longrightarrow E_0\longrightarrow M\longrightarrow 0$$

is an exact sequence of A-modules where E_k's are all free, then F is also free.

Proof Setting

$$R_k=\mathrm{Image}(E_k\to E_{k-1})=\mathrm{Ker}(E_{k-1}\to E_{k-2})\,,$$

we break up the sequence into short exact sequences:

$$0\longrightarrow R_{k+1}\longrightarrow E_k\longrightarrow R_k\longrightarrow 0\,,\quad k=0, \cdots, n-1\,.$$

This gives rise to a long exact sequence

$$\longrightarrow \mathrm{Tor}_{q+1}^A(\boldsymbol{C}, E_k)\longrightarrow \mathrm{Tor}_{q+1}^A(\boldsymbol{C}, R_k)\longrightarrow \mathrm{Tor}_q^A(\boldsymbol{C}, R_{k+1})$$
$$\longrightarrow \mathrm{Tor}_q^A(\boldsymbol{C}, E_k)\longrightarrow\,,$$

where $C=A/\mathfrak{m}$. Since $\mathrm{Tor}_q^A(C, E_k)=0$ for $q\geqq 1$ (for E_k is free), we have

$$\mathrm{Tor}_{q+1}^A(C, R_k)\approx \mathrm{Tor}_q^A(C, R_{k+1}) \qquad \text{for} \quad q\geqq 1 .$$

Hence,

$$\mathrm{Tor}_{d+1}^A(C, M)\approx \mathrm{Tor}_1^A(C, R_{d+1}) .$$

The proof will be complete if we show

$$(3.12) \qquad \mathrm{Tor}_{d+1}^A(C, M)=0 \qquad \text{for} \quad d\geqq n ,$$

since $\mathrm{Tor}_1^A(C, R_{d+1})=0$ implies that R_{d+1} is free by (3.10). In order to prove (3.12), we construct a free resolution of $C=A/\mathfrak{m}$ as follows. We recall that A is the ring of germs of holomorphic functions at the origin 0 of C^n, i.e., the ring of convergent power series in $z^1, \cdots, z^n$. Set

$$\begin{aligned} K_i=\Omega_0^i &= \text{the } A\text{-module of germs of holomorphic } i\text{-forms at } 0 \\ &\approx A\otimes \Lambda^i C^n \end{aligned}$$

and define an A-homomorphism $\partial\colon K_i\to K_{i-1}$ by

$$\partial(dz^{j_1}\wedge\cdots\wedge dz^{j_i})=\sum(-1)^{\alpha-1}z^{j_\alpha}dz^{j_1}\wedge\cdots\wedge \widehat{dz^{j_\alpha}}\wedge\cdots\wedge dz^{j_i} .$$

Then we obtain a free resolution of $C=A/\mathfrak{m}$ (called the Koszul complex):

$$(3.13) \quad 0\longrightarrow K_n\longrightarrow K_{n-1}\longrightarrow\cdots\longrightarrow K_1\longrightarrow K_0\longrightarrow C\longrightarrow 0 .$$

The exactness of (3.13) follows from $d\circ\partial+\partial\circ d=1$. Since $\mathrm{Tor}_*^A(C, M)$ is the homology of the complex

$$0\longrightarrow K_n\otimes_A M\longrightarrow K_{n-1}\otimes_A M\longrightarrow\cdots\longrightarrow K_1\otimes_A M\longrightarrow K_0\otimes_A M\longrightarrow 0 ,$$

it follows that $\mathrm{Tor}_{d+1}^A(C, M)=0$ for $d\geqq n$. Q.E.D.

The *homological dimension* of M, denoted by $\mathrm{dh}(M)$, is defined to be the length d of a minimal free resolution

$$(3.14) \qquad 0\longrightarrow E_d\longrightarrow\cdots\longrightarrow E_1\longrightarrow E_0\longrightarrow M\longrightarrow 0 .$$

In the course of the proof of (3.11), we prove also the following:

$$(3.15) \quad \mathrm{dh}(M)=d \Leftrightarrow \begin{cases}\mathrm{Tor}_d^A(C, M)\neq 0\\ \mathrm{Tor}_{d+1}^A(C, M)=0\end{cases} \Leftrightarrow \begin{cases}\mathrm{Tor}_d^A(C, M)\neq 0\\ \mathrm{Tor}_i^A(C, M)=0\end{cases} \quad \text{for} \quad i\geqq d+1 .$$

The Syzygy theorem states

$$\mathrm{dh}(M) \leqq n .$$

To define the homological codimension of M, we consider first the notion of M-sequence of A. A sequence $\{a_1, \cdots, a_p\}$ of elements of $\mathfrak{m}$ is called an *M-sequence* of A if, for each i, $0 \leqq i \leqq p-1$, a_{i+1} is not a zero divisor on $M_i = M/(a_1, \cdots, a_i)M$, where $(a_1, \cdots, a_i)$ denotes the ideal of A generated by $a_1, \cdots, a_i$. Then the sequence of ideals

$$(a_1) \subset (a_1, a_2) \subset \cdots \subset (a_1, \cdots, a_i) \subset \cdots$$

is strictly increasing and there is a maximal M-sequence. If $\{a_1, \cdots, a_p\}$ is a maximal M-sequence, then p is called the *homological codimension* (or the *depth*) of M and is denoted by $\mathrm{codh}(M)$. The fact that p is independent of the choice of the sequence $\{a_1, \cdots, a_p\}$ follows from the following

(3.16) **Proposition** *There exists an M-sequence $\{a_1, \cdots, a_p\}$ of length p of A if and only if*

$$\mathrm{Ext}^i_A(\boldsymbol{C}, M) = 0 \qquad \textit{for} \quad i < p .$$

Proof For each nonzero u in M, set

$$\mathrm{Ann}(u) = \{a \in A;\ au = 0\} .$$

Then $\mathrm{Ann}(u)$ is an ideal of A, and the set of zero divisors of M is given as the union of these ideals $\mathrm{Ann}(u)$, $0 \neq u \in M$. Let $\mathrm{Ann}(u_0)$ be a maximal one among all these ideals $\mathrm{Ann}(u)$. Then it is a prime ideal. In fact, if $ab \in \mathrm{Ann}(u_0)$ and $b \notin \mathrm{Ann}(u_0)$, then $bu_0 \neq 0$ but $abu_0 = 0$. Hence, $a \in \mathrm{Ann}(bu_0)$. Since the obvious inclusion $\mathrm{Ann}(u_0) \subset \mathrm{Ann}(bu_0)$ and maximality of $\mathrm{Ann}(u_0)$ imply the equality $\mathrm{Ann}(bu_0) = \mathrm{Ann}(u_0)$, we obtain $a \in \mathrm{Ann}(u_0)$, showing that $\mathrm{Ann}(u_0)$ is a prime ideal. It follows that the set of zero divisors of M is the union of ideals $\mathrm{Ann}(u_0)$ which are prime.

Considering first the case $p = 1$, assume $\mathrm{Ext}^0_A(\boldsymbol{C}, M) = 0$. We want to show that there is an M-sequence $\{a_1\}$ of A. If there is none, every element of $\mathfrak{m}$ is a zero divisor of M so that $\mathfrak{m}$ is (contained in) the union of ideals $\mathrm{Ann}(u_0)$ which are prime. Since the complex subspace $\mathfrak{m}$ cannot be a union of finitely many proper subspaces, it follows that $\mathfrak{m} = \mathrm{Ann}(u_0)$ for some u_0. Then we have a nonzero homomorphism (in fact, injection) $A/\mathfrak{m} \to M$ induced by the map $a \in A \to au_0 \in M$. Since $\mathrm{Hom}_A(A/\mathfrak{m}, M) = \mathrm{Ext}^0_A(A/\mathfrak{m}, M) = 0$, this is a contradiction. Therefore there is an element a_1 in $\mathfrak{m}$ such that $\{a_1\}$ is an M-sequence.

Assume $\mathrm{Ext}^i_A(\boldsymbol{C}, M) = 0$ for $i < p$. Set $M_1 = M/(a_1)M$. From the exact sequence

$$0 \longrightarrow M \xrightarrow{a_1} M \longrightarrow M_1 \longrightarrow 0,$$

we obtain a long exact sequence

$$\longrightarrow \mathrm{Ext}^i_A(C, M) \longrightarrow \mathrm{Ext}^i_A(C, M_1) \longrightarrow \mathrm{Ext}^{i+1}_A(C, M) \longrightarrow .$$

From our assumption we obtain $\mathrm{Ext}^i_A(C, M_1)=0$ for $i<p-1$. By induction on p, there is an M_1-sequence $\{a_2, \cdots, a_p\}$ of A. Then $\{a_1, a_2, \cdots, a_p\}$ is an M-sequence of A.

To prove the converse, assume that there is an M-sequence $\{a_1, \cdots, a_p\}$ of A. By induction on p, $\mathrm{Ext}^i_A(C, M_1)=0$ for $i<p-1$. From the long exact sequence

$$\longrightarrow \mathrm{Ext}^{i-1}_A(C, M_1) \longrightarrow \mathrm{Ext}^i_A(C, M) \longrightarrow \mathrm{Ext}^i_A(C, M) \longrightarrow,$$

we see that

$$0 \longrightarrow \mathrm{Ext}^i_A(C, M) \xrightarrow{a_1} \mathrm{Ext}^i_A(C, M)$$

is exact for $i<p$. Since $C=A/\mathfrak{m}$ and $a_1 \in \mathfrak{m}$, it follows that a_1 annihilates C. Hence, for every $f \in \mathrm{Hom}_A(C, M)$, we have $a_1 f=0$. Therefore, a_1 maps $\mathrm{Ext}^i_A(C, M)$ to zero. Hence, $\mathrm{Ext}^i_A(C, M)=0$ for $i<p$. Q.E.D.

Thus, we proved

(3.17) $$\mathrm{codh}(M)=p \Leftrightarrow \begin{cases} \mathrm{Ext}^p_A(C, M)\neq 0 \\ \mathrm{Ext}^i_A(C, M)=0 \end{cases} \quad \text{for} \quad i<p .$$

The following formula justifies the term "codimension".

(3.18) $$\mathrm{dh}(M)+\mathrm{codh}(M)=n .$$

We prove (3.18) by induction on $\mathrm{codh}(M)$. Assume $\mathrm{codh}(M)=0$, i.e., there is no M-sequence of A. Then the proof of (3.16) (specifically, the part where we considered the case $p=1$) shows that there is an injective homomorphism $C=A/\mathfrak{m}\rightarrow M$. Since $\mathrm{Tor}^A_n(C, M)$ is left exact, we have an exact sequence

$$0 \longrightarrow \mathrm{Tor}^A_n(C, C) \longrightarrow \mathrm{Tor}^A_n(C, M) .$$

But $\mathrm{Tor}^A_n(C, C)$ is nonzero. (Using the resolution (3.13) of C, we see easily that $\mathrm{Tor}^A_i(C, C)$ is isomorphic to $\bigwedge^i C^n$). Hence, $\mathrm{Tor}^A_n(C, M)\neq 0$, which implies $\mathrm{dh}(M)=n$.

Assume now that (3.18) holds for all A-modules whose homological codimension is less than $\mathrm{codh}(M)>0$. Let a_1 be an element of $\mathfrak{m}$ which is not a zero divisor in M. Set $M_1=M/(a_1)M$. Then

$$\text{(3.19)} \qquad \operatorname{codh}(M_1) = \operatorname{codh}(M) - 1 .$$

(In fact, the inequality $\operatorname{codh}(M_1) \leqq \operatorname{codh}(M) - 1$ follows directly from the definition of homological codimension. We obtain the opposite inequality from (3.16) since the condition $\operatorname{Ext}^i_A(C, M) = \operatorname{Ext}^{i+1}_A(C, M) = 0$ implies $\operatorname{Ext}^i_A(C, M_1) = 0$ by the long exact sequence for Ext.)

By the induction hypothesis, we have

$$\operatorname{dh}(M_1) + \operatorname{codh}(M_1) = n .$$

The problem is reduced to showing

$$\operatorname{dh}(M_1) = \operatorname{dh}(M) + 1 .$$

From the exact sequence

$$0 \longrightarrow M \xrightarrow{a_1} M \longrightarrow M_1 \longrightarrow 0 ,$$

we obtain a long exact sequence

$$\begin{aligned} \operatorname{Tor}^A_p(C, M) \xrightarrow{a_1} \operatorname{Tor}^A_p(C, M) \longrightarrow \operatorname{Tor}^A_p(C, M_1) \\ \longrightarrow \operatorname{Tor}^A_{p-1}(C, M) \xrightarrow{a_1} \operatorname{Tor}^A_{p-1}(C, M) . \end{aligned}$$

Since a_1 annihilates $C = A/\mathfrak{m}$, this reduces to the following short exact sequence:

$$0 \longrightarrow \operatorname{Tor}^A_p(C, M) \longrightarrow \operatorname{Tor}^A_p(C, M_1) \longrightarrow \operatorname{Tor}^A_{p-1}(C, M) \longrightarrow 0 .$$

Since $\operatorname{Tor}^A_{p-1}(C, M) = 0$ implies $\operatorname{Tor}^A_p(C, M) = 0$ (see (3.15)), $\operatorname{Tor}^A_p(C, M_1) = 0$ if and only if $\operatorname{Tor}^A_{p-1}(C, M) = 0$. Q.E.D.

In (3.15) we expressed $\operatorname{dh}(M)$ in terms of $\operatorname{Tor}^A_*(C, M)$. We can characterize $\operatorname{dh}(M)$ in terms of $\operatorname{Ext}^*_A(M, A)$ as well.

$$\text{(3.20)} \qquad \operatorname{dh}(M) = d \Leftrightarrow \begin{cases} \operatorname{Ext}^d_A(M, A) \neq 0 \\ \operatorname{Ext}^i_A(M, A) = 0 \end{cases} \quad \text{for} \quad i > d .$$

To prove (3.20), we reformulate it as follows:

$$\operatorname{dh}(M) \leqq d \Leftrightarrow \operatorname{Ext}^i_A(M, A) = 0 \qquad \text{for} \quad i > d .$$

The proof is by induction on d starting with $n+1$ and going down to 0. If $d = n+1$, the assertion is evident since we have always $\operatorname{dh}(M) \leqq n+1$ as well as $\operatorname{Ext}^i_A(M, A) = 0$ for $i > n+1$. To go from $d = q+1$ to q, assuming $\operatorname{Ext}^i_A(M, A) = 0$ for $i > q$ we shall prove $\operatorname{dh}(M) \leqq q$, (the converse being obvious). By the induction hypothesis, we have $\mathrm{dh}(M) \leqq q+1$, i.e., there is a free resolution of M of length $q+1$:

$$0 \longrightarrow E_{q+1} \xrightarrow{\partial_q} E_q \longrightarrow \cdots \longrightarrow E_1 \longrightarrow E_0 \longrightarrow M \longrightarrow 0 .$$

Since $\mathrm{Ext}_A^{q+1}(M, A)=0$, it follows that

$$\mathrm{Hom}_A(E_q, A) \xrightarrow{\partial_q^*} \mathrm{Hom}_A(E_{q+1}, A) \longrightarrow 0$$

is exact. Since E_{q+1} is free of rank, say s, summing s copies of the preceding exact sequence we obtain an exact sequence

$$\mathrm{Hom}_A(E_q, E_{q+1}) \xrightarrow{\partial_q^*} \mathrm{Hom}_A(E_{q+1}, E_{q+1}) \longrightarrow 0 .$$

In particular, there is an element $f \in \mathrm{Hom}_A(E_q, E_{q+1})$ such that

$$f \circ \partial_q = 1_{E_{q+1}} \in \mathrm{Hom}_A(E_{q+1}, E_{q+1}) ,$$

which means that $\partial_q(E_{q+1})$ is projective. Since every projective A-module is free, we have a free resolution of M of length q:

$$0 \longrightarrow \partial_q(E_{q+1}) \longrightarrow E_{q-1} \longrightarrow \cdots \longrightarrow E_1 \longrightarrow E_0 \longrightarrow M \longrightarrow 0 .$$

Hence, $\mathrm{dh}(M) \leqq q$, thus completing the proof of (3.20).

Using (3.20) we prove the following

(3.21) **Proposition** *If*

$$0 \longrightarrow M' \longrightarrow M \longrightarrow M'' \longrightarrow 0$$

is a short exact sequence of A-modules of finite type, then

$$\mathrm{dh}(M) \leqq \max(\mathrm{dh}(M'), \mathrm{dh}(M'')) ,$$

and if the strict inequality holds, then

$$\mathrm{dh}(M'') = \mathrm{dh}(M') + 1 .$$

Proof The given short exact sequence gives rise to a long exact sequence

$$\mathrm{Ext}_A^i(M'', A) \longrightarrow \mathrm{Ext}_A^i(M, A) \longrightarrow \mathrm{Ext}_A^i(M', A)$$
$$\longrightarrow \mathrm{Ext}_A^{i+1}(M'', A) \longrightarrow \mathrm{Ext}_A^{i+1}(M, A) .$$

If $\mathrm{Ext}_A^i(M', A) = \mathrm{Ext}_A^i(M'', A) = 0$, then $\mathrm{Ext}_A^i(M, A) = 0$, which implies the first assertion. If $i > d$, then $\mathrm{Ext}_A^i(M', A) \approx \mathrm{Ext}_A^{i+1}(M'', A)$. This implies the second assertion. Q.E.D.

(3.22) **Proposition** *If an A-module M of finite type is a k-th syzygy module in the sense that there is an exact sequence*

$$0 \longrightarrow M \longrightarrow E_1 \longrightarrow \cdots \longrightarrow E_k ,$$

where $E_1, \cdots, E_k$ are free A-modules of finite type, then

$$\mathrm{dh}(M) \leqq n-k .$$

Proof Let

$$M_0 = M ,$$

$$M_i = \mathrm{Image}(E_i \longrightarrow E_{i+1}) , \qquad i=1, \cdots, k-1 ,$$

$$M_k = E_k/M_{k-1} .$$

Then we have short exact sequences

$$0 \longrightarrow M_{i-1} \longrightarrow E_i \longrightarrow M_i \longrightarrow 0 , \qquad i=1, \cdots, k .$$

By (3.21) we have either

$$0 = \mathrm{dh}(E_i) = \max(\mathrm{dh}(M_i), \mathrm{dh}(M_{i-1}))$$

or

$$\mathrm{dh}(M_i) = \mathrm{dh}(M_{i-1}) + 1 .$$

Hence,

$$\mathrm{dh}(M_i) = \begin{cases} 0 & \text{if } M_i \text{ is free} \\ \mathrm{dh}(M_{i-1}) + 1 & \text{otherwise} \end{cases} \tag{3.23}$$

It follows that

$$0 \leqq \mathrm{dh}(M_0) \leqq \max(0, \mathrm{dh}(M_i) - i) .$$

Since $\mathrm{dh}(M_k) \leqq n$, we obtain $\mathrm{dh}(M) \leqq n-k$. Q.E.D.

§4 Coherent sheaves——torsion and reflexivity of stalks

As in the preceding section, let A denote the ring of germs of holomorphic functions at the origin 0 of $\boldsymbol{C}^n$. Let K be the quotient field of A; it is the field of germs of meromorphic functions at $0 \in \boldsymbol{C}^n$.

The *rank* of an A-module M of finite type is defined by

(4.1) $$\operatorname{rank} M = \dim_K(K \otimes_A M) .$$

An element u of M is called a *torsion element* if $au = 0$ for some nonzero element a of A. The set $T(M)$ of torsion elements of M is an A-submodule of M, called the *torsion submodule* of M. If M has no torsion elements, i.e., $T(M) = 0$, then M is said to be *torsion-free*. Every free A-module is obviously torsion-free. Every submodule of a torsion-free module is again torsion-free. Conversely,

(4.2) **Proposition** *If M is a torsion-free A-module of rank r, then it is a submodule of a free A-module of rank r.*

Proof Since M is torsion-free, the natural map

$$i \colon M \longrightarrow K \otimes_A M = K^r$$

is injective. Since M is of finite type, the injection i followed by multiplication by some nonzero element a of A gives an injection

$$j \colon M \longrightarrow A^r .$$

(We multiply by a suitable element a to clear the "denominators".) Q.E.D.

Given an A-module M of finite type, the *dual* of M is defined to be

(4.3) $$M^* = \operatorname{Hom}_A(M, A) .$$

It is again an A-module of finite type. There is a natural homomorphism σ_M of M into its double dual M^{**}; for each $u \in M$, $\sigma_M(u) \in \operatorname{Hom}_A(M^*, A)$ is given by

(4.4) $$(\sigma_M(u))(f) = f(u) \qquad \text{for} \quad f \in M^* .$$

Then

(4.5) $$\operatorname{Ker} \sigma_M = \bigcap_{f \in M^*} \operatorname{Ker} f .$$

(4.6) **Proposition** *If M is an A-module of finite type and $T(M)$ its torsion submodule, then*

$$T(M) = \operatorname{Ker} \sigma_M .$$

Proof If $u \in T(M)$, then $au = 0$ for some nonzero element $a \in A$, and

$$a \cdot f(u) = f(au) = f(0) = 0 \qquad \text{for all} \quad f \in M^* .$$

Hence, $f(u)=0$ for all $f\in M^*$. By (4.5), $u\in \operatorname{Ker}\sigma_M$.

Let u be an element of M not belonging to $T(M)$. We have only to prove that $f(u)\neq 0$ for some $f\in M^*$. Since $M/T(M)$ is torsion-free, it is a submodule of a free module A^r by (4.2). Thus, there is an injection

$$j: M/T(M) \longrightarrow A^r .$$

Let $p: M\to M/T(M)$ be the natural projection. Since $0\neq p(u)\in M/T(M)$, $j\circ p(u)$ is a nonzero element of A^r. Let $q: A^r\to A$ be the projection to one of the factors of A^r such that $q\circ j\circ p(u)\neq 0$. Then we set $f=q\circ j\circ p$. Q.E.D.

(4.7) **Corollary** *An A-module M of finite type is torsion-free if and only if the natural homomorphism $\sigma_M: M\to M^{**}$ is injective.*

If $M=T(M)$, M is called a *torsion module*. From (4.6) we obtain

(4.8) **Corollary** *An A-module M of finite type is a torsion module if and only if $M^*=0$.*

(4.9) **Corollary** *The natural map $M\to M/T(M)$ induces an isomorphism $(M/T(M))^*\approx M^*$.*

If $\sigma_M: M\to M^{**}$ is an isomorphism, we say that M is *reflexive*. If M is free, it is reflexive. If it is reflexive, it is torsion-free. Thus,

$$\text{free} \Rightarrow \text{reflexive} \Rightarrow \text{torsion-free} .$$

Dualizing $\sigma_M: M\to M^{**}$, we obtain a homomorphism

$$\sigma_M^*: (M^{**})^* \longrightarrow M^* .$$

On the other hand, let

$$\sigma_{M^*}: M^* \longrightarrow (M^*)^{**}$$

be the natural homomorphism for M^*. Then

$$\sigma_M^*\circ\sigma_{M^*}=1_{M^*} . \tag{4.10}$$

In fact, if $u\in M$ and $f\in M^*$, then

$$(\sigma_M^*\circ\sigma_{M^*}f)(u)=(\sigma_{M^*}f)(\sigma_M u)=(\sigma_M u)(f)=f(u) .$$

From (4.10) we see that $\sigma_{M^*}: M^*\to M^{***}$ is injective and M^* is torsion-free.

(4.11) **Proposition** *If M is an A-module of finite type, then the natural homomorphism $\sigma_{M^*}: M^* \to M^{***}$ is an isomorphism.*

Proof By (4.9) we may assume, replacing M by $M/T(M)$, that M is torsion-free. Let K be the quotient field of A and V the K-vector space of dimension r ($=\operatorname{rank} M$) defined by $V = M \otimes_A K$. Let

$$i_M: M \longrightarrow V = M \otimes_A K$$

be the natural map, which is injective since M is torsion-free. Under this imbedding i_M, M is a lattice in V in the sense that there exist two free A-submodules L_1, L_2 of V of rank r such that $L_1 \subset M \subset L_2 \subset V$. (In fact, if $x_1, \cdots, x_r \in M$ form a basis for V over K, we can take as L_1 the free A-submodule generated by $x_1, \cdots, x_r$. We can obtain L_2 from the proof of (4.2).)

We have a commutative diagram

$$\begin{array}{ccc} M & \xrightarrow{i_M} & V = M \otimes_A K \\ f \downarrow & & \downarrow f \otimes 1 \\ A & \xrightarrow{i_A} & K = A \otimes_A K \end{array} \qquad \text{for} \quad f \in M^* = \operatorname{Hom}_A(M, A)\,.$$

This diagram means that the correspondence $f \to \varphi = f \otimes 1$ gives a natural identification of M^* with the A-submodule

$$\{\varphi \in V^*;\ \varphi \circ i_M(M) \subset i_A(A)\}$$

of V^*. Since i_A and i_M are injections, we make the following identifications:

$$A = i_A(A) \subset K\,, \qquad M = i_M(M) \subset V\,.$$

Thus, omitting i_A and i_M, the correspondence $f \to \varphi = f \otimes 1$ gives a natural identification of M^* with the dual lattice of $M \subset V$, i.e.,

$$M^* \approx \{\varphi \in V^*;\ \varphi(M) \subset A\} \subset V^*\,.$$

Similarly, M^{**} can be considered as the dual lattice of the lattice $M^* \subset V^*$. Thus,

$$M \subset M^{**} \subset V (= V^{**})\,.$$

Considering the dual lattice of M and M^{**}, we obtain the inclusion

$$(M^{**})^* \subset M^* \subset V^*\,.$$

This shows that $\sigma_M^*: M^{***} \to M^*$ is also injective. Q.E.D.

(4.12) **Proposition** *If M is an A-module of finite type, then $M^{**}/\sigma_M(M)$ is a torsion module.*

Proof Dualizing the exact sequence

$$M \longrightarrow M^{**} \longrightarrow M^{**}/\sigma_M(M) \longrightarrow 0,$$

we obtain an exact sequence

$$0 \longrightarrow (M^{**}/\sigma_M(M))^* \longrightarrow M^{***} \longrightarrow M^*.$$

Since $M^{***} \to M^*$ is an isomorphism by (4.11), we have $(M^{**}/\sigma_M(M))^* = 0$. By (4.8), $M^{**}/\sigma_M(M)$ is a torsion module. Q.E.D.

(4.13) **Proposition** *Let M be an A-module of finite type. If it is reflexive, it can be included in an exact sequence*

$$(*) \qquad 0 \longrightarrow M \longrightarrow E \longrightarrow F \longrightarrow 0$$

with E free and F torsion-free. Conversely, if M is included in an exact sequence $(*)$ with E reflexive and F torsion-free, then M is reflexive.

Proof Suppose M is reflexive. Take an exact sequence

$$E_1 \longrightarrow E_0 \longrightarrow M^* \longrightarrow 0,$$

where E_0 and E_1 are free. Dualizing it, we obtain an exact sequence

$$0 \longrightarrow M^{**} \longrightarrow E_0^* \longrightarrow E_1^*.$$

Let $E = E_0^*$ and $F = \text{Image}(E_0^* \to E_1^*)$.

Conversely, suppose M is included in the exact sequence $(*)$ with the stated properties. Being a submodule of a reflexive module E, M is torsion-free. Hence, $\sigma_M : M \to M^{**}$ is injective. Dualizing the injection $j: M \to E$ twice, we obtain a homomorphism $j^{**}: M^{**} \to E^{**} = E$. We claim that j^{**} is injective. (In fact, if $x_1, \cdots, x_r \in M$ form a basis for $V = M \otimes_A K$, then $j(x_1), \cdots, j(x_r)$ are linearly independent over K in E, so that the map $j \otimes 1: M \otimes_A K \to E \otimes_A K$ is injective. Composing this injection $j \otimes 1$ with the natural inclusion $M^{**} \subset M \otimes_A K$ obtained in the proof of (4.11), we have an inclusion $M^{**} \subset E \otimes_A K$. Hence, $j^{**}: M^{**} \to E$ is also injective.) Since

$$M \subset M^{**} \subset E,$$

M^{**}/M may be considered as a submodule of $E/M = F$. Since M^{**}/M is a torsion module (see (4.12)) and F is torsion-free, we obtain $M^{**}=M$. Q.E.D.

We defined the concept of k-th syzygy module in (3.22). The following proposition is clear from (4.2), (4.13) and (3.22).

(4.14) **Proposition** *Let M be an A-module of finite type.*

(a) *It is torsion-free if and only if it is a first syzygy module. If it is torsion-free, then* $\mathrm{dh}(M) \leqq n-1$.

(b) *It is reflexive if and only if it is a second syzygy module. If it is reflexive, then* $\mathrm{dh}(M) \leqq n-2$.

The following proposition generalizes (4.11), (where $N=A$).

(4.15) **Proposition** *Let M and N be A-modules of finite type. If N is reflexive, so is* $\mathrm{Hom}_A(M, N)$.

Proof Let

$$0 \longrightarrow M_1 \longrightarrow E \longrightarrow M \longrightarrow 0$$

be an exact sequence such that E is free. Then

$$0 \longrightarrow \mathrm{Hom}_A(M, N) \longrightarrow \mathrm{Hom}_A(E, N) \longrightarrow \mathrm{Hom}_A(M_1, N)$$

is exact. Since $\mathrm{Hom}_A(E, N)=N\oplus\cdots\oplus N$, it follows that $\mathrm{Hom}_A(E, N)$ is reflexive. Since N is torsion-free, $\mathrm{Hom}_A(M_1, N)$ is torsion-free. Apply (4.13) to the exact sequence

$$0 \longrightarrow \mathrm{Hom}_A(M, N) \longrightarrow \mathrm{Hom}_A(E, N) \longrightarrow L \longrightarrow 0\,,$$

where $L=\mathrm{Image}(\mathrm{Hom}_A(E, N) \to \mathrm{Hom}_A(M_1, N))$. Q.E.D.

§5 Local properties of coherent sheaves

Let M be a complex manifold of dimension n and $\mathcal{O}=\mathcal{O}_M$ the structure sheaf of M, i.e., the sheaf of germs of holomorphic functions on M. (Thus, in the previous notation, $\mathcal{O}=\Omega^0$.) We write

$$\mathcal{O}^p=\mathcal{O}\oplus\cdots\oplus\mathcal{O}\,, \qquad (p \text{ times})\,.$$

An *analytic sheaf* over M is a sheaf of $\mathcal{O}$-modules over M. We say that an analytic sheaf $\mathcal{S}$ over M is *locally finitely generated* if, given any point x_0 of M, there exists a neighborhood U of x_0 and finitely many sections of $\mathscr{S}_U$ that generate each stalk $\mathscr{S}_x$, $x \in U$, as an $\mathcal{O}_x$-module. This means that we have an exact sequence

$$\mathcal{O}_U^p \longrightarrow \mathscr{S}_U \longrightarrow 0 . \tag{5.1}$$

In particular, each stalk $\mathscr{S}_x$ is an $\mathcal{O}_x$-module of finite type, to which the results of Sections 3 and 4 can be applied.

We say that an analytic sheaf $\mathscr{S}$ is *coherent* if, given any point x_0 of M, there exists a neighborhood U of x_0 and an exact sequence

$$\mathcal{O}_U^q \longrightarrow \mathcal{O}_U^p \longrightarrow \mathscr{S}_U \longrightarrow 0 . \tag{5.2}$$

This means that the kernel of (5.1) is also finitely generated.

For the proof of the following lemma of Oka, see for example Gunning-Rossi [1] or Hörmander [1].

(5.3) **Oka's lemma** *The kernel of any homomorphism $\mathcal{O}^q \to \mathcal{O}^p$ is locally finitely generated.*

It follows from (5.3) that the kernel $\mathscr{R}$ of $\mathcal{O}^q \to \mathcal{O}^p$ is coherent. (In fact, apply (5.3) again to the kernel of $\mathcal{O}_U^r \to \mathscr{R}_U \subset \mathcal{O}_U^p$.)

Given a coherent sheaf $\mathscr{S}$ over M and a point x_0 of M, consider the exact sequence (5.2). Applying Oka's lemma to (5.2), we obtain an exact sequence

$$\mathcal{O}_U^r \longrightarrow \mathcal{O}_U^q \longrightarrow \mathcal{O}_U^p \longrightarrow \mathscr{S}_U \longrightarrow 0 ,$$

(where U is a neighborhood of x_0 possibly smaller than the neighborhood U in the sequence (5.2). But we denote this smaller neighborhood again by U.) Repeat this process (taking smaller and smaller neighborhoods U of x_0). As we shall see, after a finite number of steps we obtain a free resolution:

$$0 \longrightarrow \mathcal{O}_U^{p_d} \longrightarrow \cdots \longrightarrow \mathcal{O}_U^{p_1} \longrightarrow \mathcal{O}_U^{p_0} \longrightarrow \mathscr{S}_U \longrightarrow 0 , \qquad (d \leqq n) . \tag{5.4}$$

This follows from Syzygy Theorem (3.11).

From the definition of homological dimension we know that $\mathscr{S}_x$ is a free $\mathcal{O}_x$-module if and only if $\mathrm{dh}(\mathscr{S}_x)=0$, or equivalently, (see (3.18)), if and only if $\mathrm{codh}(\mathscr{S}_x)=n$.

For each integer m, $0 \leqq m \leqq n$, the *m-th singularity set* of $\mathscr{S}$ is defined to be

$$S_m(\mathscr{S}) = \{x \in M;\ \mathrm{codh}(\mathscr{S}_x) \leqq m\} = \{x \in M;\ \mathrm{dh}(\mathscr{S}_x) \geqq n-m\} . \tag{5.5}$$

Evidently,

$$S_0(\mathscr{S})\subset S_1(\mathscr{S})\subset\cdots\subset S_{n-1}(\mathscr{S})\subset S_n(\mathscr{S})=M\,.$$

We call $S_{n-1}(\mathscr{S})$ the *singularity set* of $\mathscr{S}$. It is clear that

(5.6) $$S_{n-1}(\mathscr{S})=\{x\in M;\ \mathscr{S}_x \text{ is not free}\}\,.$$

The fact that $S_m(\mathscr{S})$ is a closed subset of M follows from the following

(5.7) **Lemma** *The homological dimension* $\mathrm{dh}(\mathscr{S}_x)$ *is upper semicontinuous in* x. (*Hence,* $\mathrm{codh}(\mathscr{S}_x)$ *is lower semicontinuous in* x.)

Proof Let $d=\mathrm{dh}(\mathscr{S}_{x_0})$. Then we have a free resolution of $\mathscr{S}_{x_0}$ of length d. Since $\mathscr{S}$ is coherent, we obtain a free resolution (5.4) of $\mathscr{S}_U$ of length d for some neighborhood U of x_0. This shows

$$\mathrm{dh}(\mathscr{S}_x)\leqq d \qquad \text{for} \quad x\in U\,. \qquad \text{Q.E.D.}$$

The following theorem is due to Scheja [1].

(5.8) **Theorem** *The m-th singularity set* $S_m(\mathscr{S})$ *of a coherent sheaf* $\mathscr{S}$ *is a closed analytic subset of M of dimension* $\leqq m$.

Proof The theorem is of local character. We fix a point $x\in M$ and show that, for a suitable neighborhood U of x, $S_m(\mathscr{S})\cap U$ is an analytic subset of U. Without being explicit, we shall shrink U to a smaller neighborhood whenever necessary.

Let

(5.9) $$\longrightarrow \mathscr{E}_{n-m} \xrightarrow{h} \mathscr{E}_{n-m-1} \longrightarrow \cdots \longrightarrow \mathscr{E}_0 \longrightarrow \mathscr{S}_U \longrightarrow 0$$

be a free resolution of $\mathscr{S}_U$. Fixing a local basis in each $\mathscr{E}_p$, we can represent h by a matrix $(h^i_j(y))$ of holomorphic functions on U. Let

$$r=\max_{y\in U}\mathrm{rank}(h^i_j(y))\,.$$

Then

(5.10) $$S_m(\mathscr{S})\cap U=\{y\in U;\ \mathrm{rank}(h^i_j(y))<r\}\,.$$

Hence, $S_m(\mathscr{S})\cap U$ is an analytic subset of U.

In order to prove the inequality $\dim S_m(\mathscr{S})\leqq m$, we consider first the case $m=0$. Let $\mathscr{H}^0_{\{x\}}\mathscr{S}$ denote the sheaf defined by the presheaf

$$V \longrightarrow \{s \in \Gamma(V, \mathscr{S});\ s|_{V-x}=0\}\ .$$

Then it is a coherent subsheaf of $\mathscr{S}$. Hence, the quotient sheaf $\mathscr{F}=\mathscr{S}/\mathscr{H}^0_{\{x\}}\mathscr{S}$ is also coherent. We claim

$$\operatorname{codh}(\mathscr{F}_x) \geqq 1\ .$$

If not (i.e., if $\operatorname{codh}(\mathscr{F}_x)=0$), then there exists a nonzero element f_x of $\mathscr{F}_x$ which is annihilated by every element a_x of the maximal ideal $\mathfrak{m}_x$ of $\mathcal{O}_x$. Let s be a local section of $\mathscr{S}$ in U representing f_x. Similarly, let a be a holomorphic function in U extending a_x. Since $a_x f_x=0$, it follows that $a \cdot s$ is a local section of $\mathscr{H}^0_{\{x\}}\mathscr{S}$ and hence $a \cdot s=0$ in $U-\{x\}$. Since this holds for all local holomorphic functions a vanishing at x, we have $s=0$ in $U-\{x\}$. This means that s is a section of $\mathscr{H}^0_{\{x\}}\mathscr{S}$. Since the element $f_x \in \mathscr{F}_x=(\mathscr{S}/\mathscr{H}^0_{\{x\}}\mathscr{S})_x$ is represented by s, we obtain a contradicting conclusion that $f_x=0$. This proves our claim.

Since $\operatorname{codh}(\mathscr{F}_y)$ is lower semicontinuous in y (see (5.7)), it follows that

$$\operatorname{codh}(\mathscr{F}_y) \geqq 1 \qquad \text{for} \quad y \in U\ .$$

Hence,

$$S_0(\mathscr{F}) \cap U = \varnothing\ .$$

Since $\mathscr{F} \approx \mathscr{S}$ on $U-\{x\}$, we obtain

$$S_0(\mathscr{F}) \cap (U-\{x\}) = S_0(\mathscr{S}) \cap (U-\{x\})\ ,$$

which shows that $S_0(\mathscr{S}) \cap U$ is either the singleton $\{x\}$ or the empty set. Hence, $\dim S_0(\mathscr{S}) \leqq 0$.

We prove the general case by induction on m. Suppose $\dim S_m(\mathscr{S}) > m$ and let x be a point where $S_m(\mathscr{S})$ has dimension $>m$. Since $S_0(\mathscr{S})$ is discrete, we may assume without loss of generality that $x \notin S_0(\mathscr{S})$. Then there exists an element a_x in the maximal ideal $\mathfrak{m}_x$ of $\mathcal{O}_x$ such that a_x is not a zero divisor on $\mathscr{S}_x$. If a is the local holomorphic function determined by the germ a_x, then the local sheaf-homomorphism $\alpha: \mathscr{S} \to \mathscr{S}$ defined by multiplication by a is injective in some neighborhood U of x. (In fact, let $\mathscr{G}=\alpha(\mathscr{S})$. Then $\mathscr{G}$ is a coherent sheaf. Since a_x is not a zero divisor on $\mathscr{S}_x$, the homomorphism $\alpha_x: \mathscr{S}_x \to \mathscr{G}_x$ is bijective. Let $\beta_x: \mathscr{G}_x \to \mathscr{S}_x$ be the inverse homomorphism. Since $\mathscr{G}$ is coherent, β_x extends to a local homomorphism $\beta: \mathscr{G} \to \mathscr{S}$ in some neighborhood of x. Then $\beta \circ \alpha$ is an endomorphism of $\mathscr{S}$ such that $(\beta \circ \alpha)_x$ coincides with the identity endomorphism of $\mathscr{S}_x$. Since $\mathscr{S}$ is coherent, $\beta \circ \alpha$ coincides with the identity endomorphism of $\mathscr{S}$ in some neighborhood of x. Then α is injective there.)

Now we claim

$$S_m(\mathscr{S}) \cap U \cap \{a=0\} \subset S_{m-1}(\mathscr{S}/\alpha(\mathscr{S})).$$

In fact, given a point y in $S_m(\mathscr{S}) \cap U \cap \{a=0\}$, if we choose an $\mathscr{S}_y$-sequence $\{a_{1y}, \cdots, a_{my}\}$ for $\mathscr{O}_y$ such that $a_1 = a$, then $\{a_{2y}, \cdots, a_{my}\}$ is an $(\mathscr{S}/\alpha(\mathscr{S}))_y$-sequence for $\mathscr{O}_y$. Hence, y belongs to $S_{m-1}(\mathscr{S}/\alpha(\mathscr{S}))$.

Since $S_m(\mathscr{S}) \cap U$ has dimension $>m$ at x by assumption and since a vanishes at x, $S_m(\mathscr{S}) \cap U \cap \{a=0\}$ has dimension $>m-1$ at x. This contradicts the inductive hypothesis that $\dim S_{m-1}(\mathscr{S}/\alpha(\mathscr{S})) \leqq m-1$. Q.E.D.

(5.11) **Corollary** *If a coherent sheaf $\mathscr{S}$ is a k-th syzygy sheaf in the sense that for every point x of M there is an open neighborhood U together with an exact sequence*

$$0 \longrightarrow \mathscr{S}_U \longrightarrow \mathscr{E}_1 \longrightarrow \cdots \longrightarrow \mathscr{E}_k$$

such that $\mathscr{E}_1, \cdots, \mathscr{E}_k$ are locally free coherent sheaves over U, then

$$\dim S_m(\mathscr{S}) \leqq m-k.$$

Proof Writing $\mathscr{S}$ for $\mathscr{S}_U$, let

$$\mathscr{S}_0 = \mathscr{S},$$

$$\mathscr{S}_i = \mathrm{Image}(\mathscr{E}_i \to \mathscr{E}_{i+1}), \qquad i=1, \cdots, k-1,$$

$$\mathscr{S}_k = \mathscr{E}_k / \mathscr{S}_{k-1}$$

so that we have short exact sequences

$$0 \longrightarrow \mathscr{S}_{i-1} \longrightarrow \mathscr{E}_i \longrightarrow \mathscr{S}_i \longrightarrow 0, \qquad i=1, \cdots, k.$$

Then (see the proof of (3.22))

$$\mathrm{dh}(\mathscr{S}_{i,x}) = \begin{cases} 0 & \text{if } \mathscr{S}_{i,x} \text{ is free}, \\ \mathrm{dh}(\mathscr{S}_{i-1,x})+1 & \text{otherwise}. \end{cases} \tag{5.12}$$

It follows that

$$S_m(\mathscr{S}_0) \subset S_{m-1}(\mathscr{S}_1) \subset S_{m-2}(\mathscr{S}_2) \subset \cdots \subset S_{m-k}(\mathscr{S}_k).$$

Since $\dim S_{m-k}(\mathscr{S}_k) \leqq m-k$ by (5.8), we obtain $\dim S_m(\mathscr{S}) \leqq m-k$. Q.E.D.

Since every coherent sheaf $\mathscr{S}$ is locally free outside its singularity set $S_{n-1}(\mathscr{S})$, we define

$$(5.13) \qquad \operatorname{rank} \mathscr{S} = \operatorname{rank} \mathscr{S}_x, \qquad x \in M - S_{n-1}(\mathscr{S}).$$

If every stalk of $\mathscr{S}$ is torsion-free, $\mathscr{S}$ is said to be *torsion-free*. Every locally free sheaf is obviously torsion-free. Any coherent subsheaf of a torsion-free sheaf is again torsion-free. Conversely, we have

(5.14) **Proposition** *If $\mathscr{S}$ is a torsion-free coherent sheaf of rank r, then it is locally a subsheaf of a free sheaf of rank r, i.e., for every point x of M, there exists a neighborhood U and an injective homomorphism*

$$j\colon \mathscr{S}_U \longrightarrow \mathcal{O}_U^r .$$

Proof By (4.2), there is an injection

$$j_x\colon \mathscr{S}_x \longrightarrow \mathcal{O}_x^r .$$

Since $\mathscr{S}$ is coherent, j extends to a homomorphism $j\colon \mathscr{S}_U \to \mathcal{O}_U^r$ for a small neighborhood U of x. Then j is an injection in a possibly smaller neighborhood of x. Q.E.D.

From (5.11) and (5.14) we obtain

(5.15) **Corollary** *If $\mathscr{S}$ is a torsion-free coherent sheaf, then*

$$\dim S_m(\mathscr{S}) \leqq m-1 \qquad \textit{for all} \quad m .$$

This means that a torsion-free sheaf $\mathscr{S}$ is locally free outside the set $S_{n-1}(\mathscr{S})$ of codimension at least 2. In particular, every torsion-free coherent sheaf over a Riemann surface is locally free.

The *dual* of a coherent sheaf $\mathscr{S}$ is defined to be the coherent sheaf

$$(5.16) \qquad \mathscr{S}^* = \operatorname{Hom}(\mathscr{S}, \mathcal{O}) .$$

There is a natural homomorphism σ of $\mathscr{S}$ into its double dual $\mathscr{S}^{**}$:

$$(5.17) \qquad \sigma\colon \mathscr{S} \longrightarrow \mathscr{S}^{**} .$$

Then the kernel $\operatorname{Ker}\sigma$ consists exactly of torsion elements of $\mathscr{S}$ (see (4.6)). The coherent subsheaf $\operatorname{Ker}\sigma$ of $\mathscr{S}$, denoted by $\mathscr{T}(\mathscr{S})$, is called the *torsion subsheaf* of $\mathscr{S}$. It is clear that σ is injective if and only if $\mathscr{S}$ is torsion-free.

If $\sigma\colon \mathscr{S} \to \mathscr{S}^{**}$ is bijective, $\mathscr{S}$ is said to be *reflexive*. Every locally free sheaf is obviously reflexive. Every reflexive sheaf is torsion-free. From (4.11) we obtain

(5.18) **Proposition** *The dual $\mathscr{S}^*$ of any coherent sheaf $\mathscr{S}$ is reflexive.*

The proof of the following proposition is identical to that of (4.13).

(5.19) **Proposition** *If a coherent sheaf $\mathscr{S}$ is reflexive, then every point x of M has a neighborhood U such that $\mathscr{S}_U$ can be included in an exact sequence*

$$0 \longrightarrow \mathscr{S}_U \longrightarrow \mathscr{E}_1 \longrightarrow \mathscr{E}_2$$

such that $\mathscr{E}_0$ and $\mathscr{E}_1$ are free coherent sheaves.

From (5.11) and (5.19) we obtain

(5.20) **Corollary** *If $\mathscr{S}$ is a reflexive coherent sheaf, then*

$$\dim S_m(\mathscr{S}) \leqq m-2 \qquad \text{for all} \quad m\,.$$

This means that a reflexive sheaf $\mathscr{S}$ is locally free outside the set $S_{n-1}(\mathscr{S})$ of codimension at least 3. In particular, every reflexive sheaf over a complex analytic surface (of complex dimension 2) is locally free.

We say that a coherent sheaf $\mathscr{S}$ over M is *normal* if for every open set U in M and every analytic subset $A \subset U$ of codimension at least 2, the restriction map

$$\Gamma(U, \mathscr{S}) \longrightarrow \Gamma(U-A, \mathscr{S})$$

is an isomorphism. By Hartogs' extension theorem, the structure sheaf $\mathscr{O} = \mathscr{O}_M$ is normal. We note that, by (5.14), this restriction map is injective if $\mathscr{S}$ is torsion-free.

(5.21) **Proposition** *A coherent sheaf $\mathscr{S}$ is reflexive if and only if it is torsion-free and normal.*

Proof Since $\mathscr{O}$ is normal, the dual sheaf of any coherent sheaf is normal. Hence, if $\mathscr{S}$ is reflexive, i.e., $\mathscr{S} = \mathscr{S}^{**}$, then it is normal.

Conversely, assume that $\mathscr{S}$ is torsion-free and normal. Since $\mathscr{S}$ is torsion-free, the natural map $\sigma: \mathscr{S} \to \mathscr{S}^{**}$ is injective and the singularity set $A = S_{n-1}(\mathscr{S})$ is of codimension at least 2 by (5.15). For every open set U in M, we have the following commutative diagram:

$$\begin{array}{ccc} \Gamma(U-A, \mathscr{S}) & \xrightarrow{\sigma} & \Gamma(U-A, \mathscr{S}^{**}) \\ \uparrow & & \uparrow \\ \Gamma(U, \mathscr{S}) & \xrightarrow{\sigma} & \Gamma(U, \mathscr{S}^{**}) . \end{array}$$

The vertical arrows are isomorphisms since $\mathscr{S}$ is normal and $\mathscr{S}^{**}$ is reflexive and hence normal. The top horizontal arrow is an isomorphism since $\sigma: \mathscr{S} \to \mathscr{S}^{**}$ is an isomorphism outside the singularity set $A = S_{n-1}(\mathscr{S})$. Hence, the bottom horizontal arrow is also an isomorphism. It follows that $\mathscr{S}$ is reflexive. Q.E.D.

The following proposition may be considered as the converse to (5.19).

(5.22) **Proposition** *Let*

$$0 \longrightarrow \mathscr{S}' \longrightarrow \mathscr{S} \longrightarrow \mathscr{S}'' \longrightarrow 0$$

be an exact sequence of coherent sheaves where $\mathscr{S}$ *is reflexive and* $\mathscr{S}''$ *is torsion-free. Then* $\mathscr{S}'$ *is normal and hence reflexive.*

Proof Let $U \subset M$ be open and $A \subset U$ an analytic subset of codimension at least 2. Then we have the following diagram

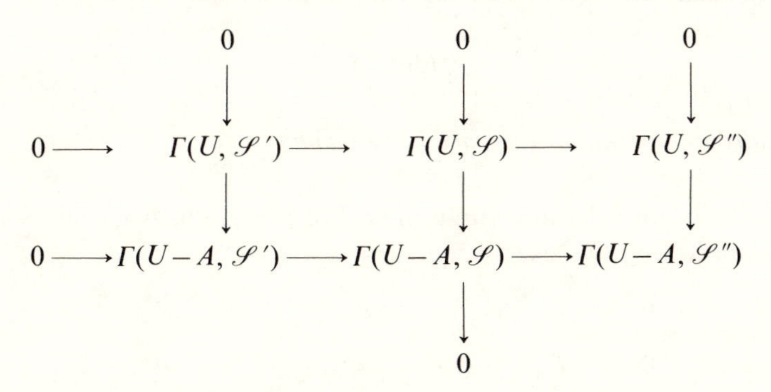

where all vertical and horizontal sequences are exact. (The left vertical sequence is exact because $\mathscr{S}'$ is torsion-free.) A simple diagram chasing shows that the restriction map $\Gamma(U, \mathscr{S}') \to \Gamma(U-A, \mathscr{S}')$ is surjective. Q.E.D.

(5.23) **Proposition** *Let* $\mathscr{F}$ *and* $\mathscr{S}$ *be coherent sheaves over* M. *If* $\mathscr{F}$ *is reflexive, so is* $\mathrm{Hom}(\mathscr{S}, \mathscr{F})$.

Proof This follows from (4.15). It can be proved also in the same way as (4.15) using (5.22) in place of (4.13). Q.E.D.

§6 Determinant bundles

Every exact sequence

$$0 \longrightarrow E_m \longrightarrow \cdots \longrightarrow E_1 \longrightarrow E_0 \longrightarrow 0 \tag{6.1}$$

of holomorphic vector bundles induces an exact sequence

$$0 \longrightarrow \mathscr{E}_m \longrightarrow \cdots \longrightarrow \mathscr{E}_1 \longrightarrow \mathscr{E}_0 \longrightarrow 0 \tag{6.2}$$

of locally free coherent sheaves, where $\mathscr{E}_i$ denotes the sheaf $\mathcal{O}(E_i)$ of germs of holomorphic sections of E_i. Conversely, every exact sequence (6.2) of locally free coherent sheaves comes from an exact sequence (6.1) of the corresponding holomorphic vector bundles. This can be easily verified by induction on m.

For a holomorphic vector bundle E of rank r, its *determinant bundle* $\det E$ is defined by

$$\det E = \bigwedge^r E\,.$$

(6.3) **Lemma** *Given an exact sequence* (6.1), *the line bundle*

$$\bigotimes_{i=0}^{m} (\det E_i)^{(-1)^i}$$

is canonically isomorphic to the trivial line bundle.

Proof The proof is by induction on m. For $m=1$, this is trivial. Now, reduce (6.1) to the following two exact sequences:

$$0 \longrightarrow E \longrightarrow E_1 \longrightarrow E_0 \longrightarrow 0\,,$$

$$0 \longrightarrow E_m \longrightarrow \cdots \longrightarrow E_2 \longrightarrow E \longrightarrow 0\,,$$

where $E = \mathrm{Ker}(E_1 \to E_0) = \mathrm{Im}(E_2 - E_1)$, and use the induction. Q.E.D.

Given a coherent sheaf $\mathscr{S}$, we shall define its *determinant bundle* $\det \mathscr{S}$. Let

$$0 \longrightarrow \mathscr{E}_n \longrightarrow \cdots \longrightarrow \mathscr{E}_1 \longrightarrow \mathscr{E}_0 \longrightarrow \mathscr{S}_U \longrightarrow 0 \tag{6.4}$$

be a resolution of $\mathscr{S}_U$ by locally free coherent sheaves, where U is a small open set in the base manifold M. Let E_i denote the vector bundle corresponding to the sheaf $\mathscr{E}_i$. We set

$$\text{(6.5)} \qquad \det \mathscr{S}_U = \bigotimes_{i=0}^{n} (\det E_i)^{(-1)^i}.$$

We need to check that $\det \mathscr{S}_U$ is independent of the choice of the resolution (6.4). Let

$$\text{(6.4)}' \qquad 0 \longrightarrow \mathscr{E}'_n \longrightarrow \cdots \longrightarrow \mathscr{E}'_1 \longrightarrow \mathscr{E}'_0 \longrightarrow \mathscr{S}_U \longrightarrow 0$$

be another locally free resolution of $\mathscr{S}_U$. Let E'_i be the vector bundle corresponding to $\mathscr{E}'_i$. We want to show that there is a canonical isomorphism

$$\text{(6.6)} \qquad \bigotimes_{i=0}^{n} (\det E_i)^{(-1)^i} \approx \bigotimes_{i=0}^{n} (\det E'_i)^{(-1)^i}.$$

To simplify the notation, we omit the subscript U from $\mathscr{S}_U$. The first step in constructing the canonical isomorphism (6.6) is to construct a third locally free resolution of $\mathscr{S}$ (the middle exact sequence in (6.7) below) which maps surjectively to both (6.4) and (6.4)$'$:

$$\text{(6.7)} \qquad \begin{array}{ccccccccccc}
0 & \longrightarrow & \mathscr{E}_n & \xrightarrow{f_n} & \cdots & \xrightarrow{f_2} & \mathscr{E}_1 & \xrightarrow{f_1} & \mathscr{E}_0 & \xrightarrow{f_0} & \mathscr{S} & \longrightarrow & 0 \\
 & & \uparrow & & & & \uparrow & & \uparrow & & \| & & \\
0 & \longrightarrow & \mathscr{E}''_n & \xrightarrow{f''_n} & \cdots & \xrightarrow{f''_2} & \mathscr{E}''_1 & \xrightarrow{f''_1} & \mathscr{E}''_0 & \xrightarrow{f''_0} & \mathscr{S} & \longrightarrow & 0 \\
 & & \downarrow & & & & \downarrow & & \downarrow & & \| & & \\
0 & \longrightarrow & \mathscr{E}'_n & \xrightarrow{f'_n} & \cdots & \xrightarrow{f'_2} & \mathscr{E}'_1 & \xrightarrow{f'_1} & \mathscr{E}'_0 & \xrightarrow{f'_0} & \mathscr{S} & \longrightarrow & 0
\end{array}$$

where the vertical arrows are all surjective. We construct first $\mathscr{E}''_0$. Let

$$\mathscr{G}_0 = \{(u, u') \in \mathscr{E}_0 \oplus \mathscr{E}'_0;\ f_0(u) = f'_0(u')\}$$

and let $\mathscr{E}''_0$ be a locally free sheaf which maps surjectively onto $\mathscr{G}_0$. Let $f''_0\colon \mathscr{E}''_0 \to \mathscr{G}_0 \to \mathscr{S}$ be the composed map. The surjective maps $\mathscr{E}''_0 \to \mathscr{E}_0$ and $\mathscr{E}''_0 \to \mathscr{E}'_0$ are obtained by composing $\mathscr{E}''_0 \to \mathscr{G}_0$ with the natural projections $\mathscr{G}_0 \to \mathscr{E}_0$ and $\mathscr{G}_0 \to \mathscr{E}'_0$. We construct next $\mathscr{E}''_1$. Let $\mathscr{S}_0$, $\mathscr{S}'_0$ and $\mathscr{S}''_0$ be the kernels of f_0, f'_0 and f''_0, respectively. The relevant part of (6.7) looks as follows:

$$\begin{array}{ccccc}
\mathscr{E}_1 & \xrightarrow{f_1} & \mathscr{S}_0 & \longrightarrow & 0 \\
 & & \uparrow p & & \\
 & & \mathscr{S}_0'' & & \\
 & & \downarrow p' & & \\
\mathscr{E}_1' & \xrightarrow{f_1'} & \mathscr{S}_0' & \longrightarrow & 0
\end{array}$$

where the vertical arrows are both surjective. Let

$$\mathscr{G}_1 = \{(u, u') \in \mathscr{E}_1 \oplus \mathscr{E}_1'; f_1(u) = p(u'') \text{ and } f_1'(u') = p'(u'') \text{ for some } u'' \in \mathscr{S}_0''\},$$

and let $\mathscr{E}_1''$ be a locally free sheaf which maps surjectively onto $\mathscr{G}_1$. The constructions of $\mathscr{E}_1'' \to \mathscr{E}_0''$, $\mathscr{E}_1'' \to \mathscr{E}_1$ and $\mathscr{E}_1'' \to \mathscr{E}_1'$ are the same as in the case of $\mathscr{E}_0''$. Inductively we can complete the construction of the diagram (6.7).

The second step is to show that the top two resolutions of $\mathscr{S}$ in (6.7) give rise to the same determinant bundle. We consider the following commutative diagram:

(6.8)

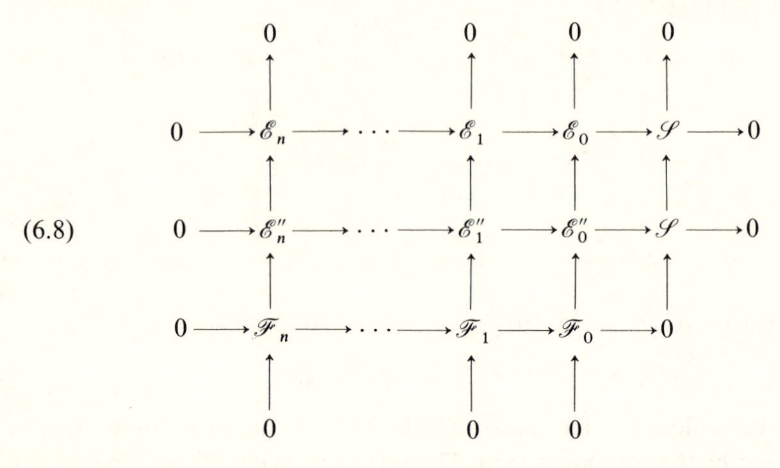

where $\mathscr{F}_i = \mathrm{Ker}(\mathscr{E}_i'' \to \mathscr{E}_i)$. Being the kernel of a surjective map between two locally free sheaves, $\mathscr{F}_i$ is also locally free. Let F_i be the vector bundle corresponding to $\mathscr{F}_i$. Then, by (6.3)

$$\det E_i'' = (\det E_i) \otimes (\det F_i), \qquad \text{(canonically)}$$

$$\bigotimes_{i=0}^{n} (\det F_i)^{(-1)^i} = \text{trivial line bundle}, \qquad \text{(canonically)}.$$

Hence, we have the desired canonical isomorphism:

$$\bigotimes_{i=0}^{n} (\det E_i'')^{(-1)^i} = \bigotimes_{i=0}^{n} ((\det E_i)\otimes(\det F_i))^{(-1)^i} = \bigotimes_{i=0}^{n} (\det E_i)^{(-1)^i}.$$

Similarly, we obtain a canonical isomorphism

$$\bigotimes_{i=0}^{n} (\det E_i'')^{(-1)^i} = \bigotimes_{i=0}^{n} (\det E_i')^{(-1)^i}.$$

This completes the proof that $\det \mathscr{S}$ is well defined.

(6.9) **Proposition** *If*

$$0 \longrightarrow \mathscr{S}' \longrightarrow \mathscr{S} \longrightarrow \mathscr{S}'' \longrightarrow 0$$

is an exact sequence of coherent sheaves, then there is a canonical isomorphism

$$\det \mathscr{S} = (\det \mathscr{S}')\otimes(\det \mathscr{S}'').$$

Proof The proof is fairly standard. All we have to do is to construct the following commutative diagram:

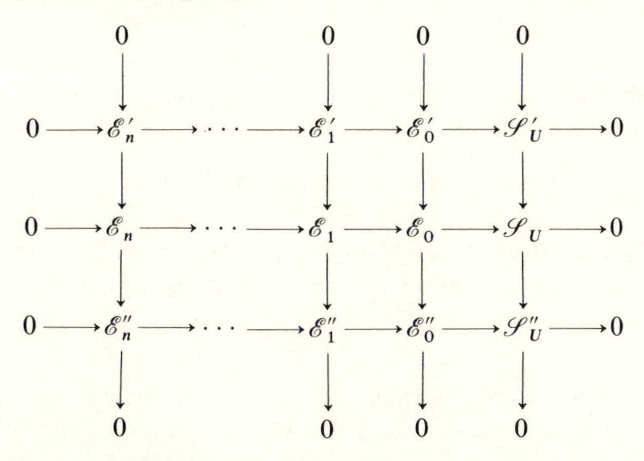

where the horizontal sequences are locally free resolutions and the vertical sequences are all exact. To construct the diagram above, we choose first $\mathscr{E}_0'$ and $\mathscr{E}_0''$. Set $\mathscr{E}_0 = \mathscr{E}_0' \oplus \mathscr{E}_0''$ and define maps $\mathscr{E}_0 \to \mathscr{S}_U$ and $\mathscr{E}_0' \to \mathscr{E}_0 \to \mathscr{E}_0''$ in an obvious manner. Then set $\mathscr{S}_1' = \mathrm{Ker}(\mathscr{E}_0' \to \mathscr{S}_U')$, $\mathscr{S}_1 = \mathrm{Ker}(\mathscr{E}_0 \to \mathscr{S}_U)$ and $\mathscr{S}_1'' = \mathrm{Ker}(\mathscr{E}_0'' \to \mathscr{S}_0'')$. Repeating this process we obtain the desired diagram. (In the process we may have to shrink the neighborhood U.) Q.E.D.

(6.10) **Proposition** *If $\mathcal{S}$ is a torsion-free coherent sheaf of rank r, then there is a canonical isomorphism*

$$\det \mathcal{S} = (\bigwedge^r \mathcal{S})^{**}.$$

Proof Let $A = S_{n-1}(\mathcal{S})$ be the singularity set of $\mathcal{S}$. Since $\mathcal{S}$ is torsion-free, A is an analytic subset of codimension at least 2 in M, (see (5.15)). Since $\mathcal{S}$ is locally free over $M-A$, there is a canonical isomorphism f: $\det \mathcal{S}_{M-A} \to (\bigwedge^r \mathcal{S})^{**}_{M-A}$. Since $(\bigwedge^r \mathcal{S})^{**}$ is reflexive (see (5.18)) and hence normal (see (5.21)), $\mathrm{Hom}(\det \mathcal{S}, (\bigwedge^r \mathcal{S})^{**})$ is also normal. Hence, f extends to a homomorphism $\tilde{f}$: $\det \mathcal{S} \to (\bigwedge^r \mathcal{S})^{**}$. Let g be the inverse map of f. Since $\det \mathcal{S}$ is normal, g extends to a homomorphism $\tilde{g}$: $(\bigwedge^r \mathcal{S})^{**} \to \det \mathcal{S}$. Since $f \circ g$ and $g \circ f$ are the identity endomorphisms of $(\bigwedge^r \mathcal{S})^{**}_{M-A}$ and $(\det \mathcal{S})_{M-A}$, respectively, $\tilde{f} \circ \tilde{g}$ and $\tilde{g} \circ \tilde{f}$ are the identity endomorphisms of $(\bigwedge^r \mathcal{S})^{**}$ and $\det \mathcal{S}$, respectively. Q.E.D.

(6.11) *Remark* Since $(\bigwedge^r \mathcal{S})^{**}$ is defined without using a locally free resolution of $\mathcal{S}$, the proof of (6.10) may be used to show quickly that when $\mathcal{S}$ is torsion-free the definition (6.5) of $\det \mathcal{S}_U$ is independent of the choice of the resolution (6.4).

(6.12) **Proposition** *If $\mathcal{S}$ is a torsion-free coherent sheaf, then there is a canonical isomorphism*

$$(\det \mathcal{S})^* = (\det \mathcal{S}^*).$$

Proof By (6.10), $(\det \mathcal{S})^* = (\bigwedge^r \mathcal{S})^*$ and $(\det \mathcal{S}^*) = (\bigwedge^r \mathcal{S}^*)^{**}$. Let $A = S_{n-1}(\mathcal{S})$. Then $(\bigwedge^r \mathcal{S})^*_{M-A} = (\bigwedge^r \mathcal{S}^*)^{**}_{M-A}$. The remainder of the proof is similar to that of (6.10). Q.E.D.

(6.13) **Proposition** *Every monomorphism $\mathcal{S}' \to \mathcal{S}$ between torsion-free coherent sheaves of the same rank induces a sheaf monomorphism* $\det \mathcal{S}' \to \det \mathcal{S}$.

Proof Let $A = S_{n-1}(\mathcal{S})$ and $A' = S_{n-1}(\mathcal{S}')$. Then outside of $A \cup A'$ the map $\mathcal{S}' \to \mathcal{S}$ and the induced map $\det \mathcal{S}' \to \det \mathcal{S}$ are isomorphisms. Hence, $\mathrm{Ker}(\det \mathcal{S}' \to \det \mathcal{S})$ is a torsion sheaf. Being also a subsheaf of a torsion-free sheaf, it must be zero. Q.E.D.

(6.14) **Proposition** *If $\mathcal{S}$ is a torsion sheaf, then* $\det \mathcal{S}$ *admits a non-trivial holomorphic section. Moreover, if* $\mathrm{supp}(\mathcal{S}) = \{x \in M;\ \mathcal{S}_x \neq 0\}$ *has codimension at* *least* 2, *then* $\det \mathscr{S}$ *is a trivial line bundle.*

Proof Let $A = S_{n-1}(\mathscr{S})$. Since $\mathscr{S}$ is locally free over $M - A$ and $\mathscr{S}$ is a torsion sheaf, it follows that $\operatorname{supp}(\mathscr{S}) \subset A$. In the locally free resolution (6.4) of $\mathscr{S}_U$, let $\mathscr{S}_1 = \operatorname{Ker}(\mathscr{E}_0 \to \mathscr{S}_U) = \operatorname{Im}(\mathscr{E}_1 \to \mathscr{E}_0)$, so that

$$0 \longrightarrow \mathscr{S}_1 \longrightarrow \mathscr{E}_0 \longrightarrow \mathscr{S}_U \longrightarrow 0$$

is exact. By (6.9),

$$\det \mathscr{S}_U = (\det \mathscr{S}_1)^* \otimes (\det \mathscr{E}_0) = \operatorname{Hom}(\det \mathscr{S}_1, \det \mathscr{E}_0)\,.$$

The injective map $\det \mathscr{S}_1 \to \det \mathscr{E}_0$ induced by the injection $\mathscr{S}_1 \to \mathscr{E}_0$ (see (6.13)) can be considered as a non-trivial section of $\det \mathscr{S}_U$. The fact that this section is independent of the choice of the resolution (6.4) and hence is globally well defined over M can be proved in the same way as $\det \mathscr{S}_U$ was shown to be independent of the resolution (6.4). The relevant part of the diagram (6.8) is given by

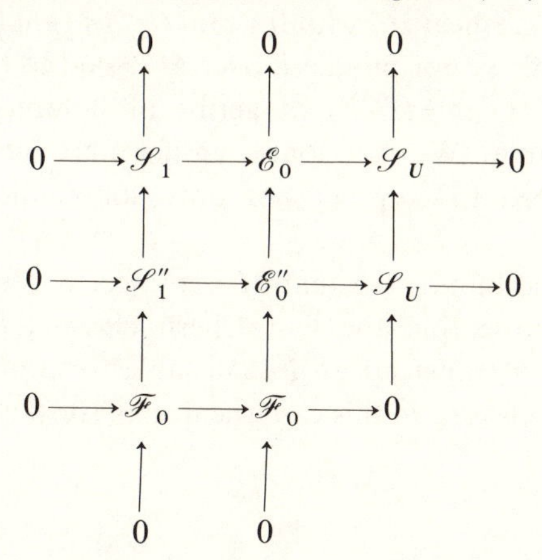

Q.E.D.

Having defined the determinant bundle $\det \mathscr{S}$ of a coherent sheaf $\mathscr{S}$, we define the first Chern class $c_1(\mathscr{S})$ by

(6.15) $$c_1(\mathscr{S}) = c_1(\det \mathscr{S})\,.$$

§7 Stable vector bundles

Let $\mathscr{S}$ be a torsion-free coherent sheaf over a compact Kähler manifold (M, g) of dimension n. Let Φ be the Kähler form of (M, g); it is a real positive closed $(1,1)$-form on M. Let $c_1(\mathscr{S})$ be the first Chern class of $\mathscr{S}$, (see (6.15)); it is represented by a real closed (1, 1)-form on M. The *degree* (or more precisely, the Φ-degree) of $\mathscr{S}$ is defined to be

$$\text{(7.1)} \qquad \deg(\mathscr{S}) = \int_M c_1(\mathscr{S}) \wedge \Phi^{n-1} .$$

The *degree/rank ratio* $\mu(\mathscr{S})$ is defined to be

$$\text{(7.2)} \qquad \mu(\mathscr{S}) = \deg(\mathscr{S})/\text{rank}(\mathscr{S}) .$$

Following Takemoto [1], we say that $\mathscr{S}$ is *Φ-semistable* if for every coherent subsheaf $\mathscr{S}'$, $0 < \text{rank } \mathscr{S}'$, we have

$$\mu(\mathscr{S}') \leqq \mu(\mathscr{S}) .$$

If moreover the strict inequality

$$\mu(\mathscr{S}') < \mu(\mathscr{S})$$

holds for all coherent subsheaf $\mathscr{S}'$ with $0 < \text{rank}(\mathscr{S}') < \text{rank}(\mathscr{S})$, we say that $\mathscr{S}$ is *Φ-stable*. A holomorphic vector bundle E over M is said to be Φ-semistable (resp. Φ-stable) if the sheaf $\mathcal{O}(E) = \Omega^0(E)$ of germs of holomorphic sections is Φ-semistable (resp. Φ-stable). We note that even if we are interested in stability of vector bundles we need to consider not only subbundles but also coherent subsheaves.

If H is an ample line bundle (so that M is projective algebraic) and if Φ is a closed (1, 1)-form representing the first Chern class $c_1(H)$, then we say *H-semistable* (resp. *H-stable*) instead of Φ-semistable (resp. Φ-stable). Since $c_1(H)$ and $c_1(\mathscr{S})$ are integral classes, in this case the degree (or the H-degree) of $\mathscr{S}$ is an integer.

(7.3) **Lemma** *If*

$$0 \longrightarrow \mathscr{S}' \longrightarrow \mathscr{S} \longrightarrow \mathscr{S}'' \longrightarrow 0$$

is an exact sequence of coherent sheaves over a compact Kähler manifold (M, g), *then*

$$r'(\mu(\mathscr{S}) - \mu(\mathscr{S}')) + r''(\mu(\mathscr{S}) - \mu(\mathscr{S}'')) = 0 ,$$

where $r' = \text{rank } \mathscr{S}'$ *and* $r'' = \text{rank } \mathscr{S}''$.

Proof By (6.9),

$$c_1(\mathscr{S}) = c_1(\mathscr{S}') + c_1(\mathscr{S}'') .$$

Hence,

$$(r'+r'')\mu(\mathscr{S})=\deg(\mathscr{S})=\deg(\mathscr{S}')+\deg(\mathscr{S}'')=r'\mu(\mathscr{S}')+r''\mu(\mathscr{S}'') .$$

Q.E.D.

From (7.3) we see that stability and semistability of $\mathscr{S}$ can be defined in terms of quotient sheaves $\mathscr{S}''$ instead of subsheaves $\mathscr{S}'$. Namely,

(7.4) **Proposition** *Let $\mathscr{S}$ be a torsion-free coherent sheaf over a compact Kähler manifold (M, g). Then*

(a) *$\mathscr{S}$ is Φ-semistable if and only if $\mu(\mathscr{S})\leqq\mu(\mathscr{S}'')$ holds for every quotient sheaf $\mathscr{S}''$ such that $0<\operatorname{rank}\mathscr{S}''$;*

(b) *$\mathscr{S}$ is Φ-stable if and only if $\mu(\mathscr{S})<\mu(\mathscr{S}'')$ holds for every quotient sheaf $\mathscr{S}''$ such that $0<\operatorname{rank}\mathscr{S}''<\operatorname{rank}\mathscr{S}$.*

In (7.4) we do not have to consider all quotient sheaves. First we prove

(7.5) **Lemma** *If $\mathscr{T}$ is a coherent torsion sheaf, then*

$$\deg(\mathscr{T})\geqq 0 .$$

Proof This follows from (6.14). If V denotes the divisor of M defined as the zeros of a holomorphic section of $\det\mathscr{T}$, then

$$\deg(\mathscr{T})=\int_V \Phi^{n-1}\geqq 0 . \qquad \text{Q.E.D.}$$

(7.6) **Proposition** *Let $\mathscr{S}$ be a torsion-free coherent sheaf over a compact Kähler manifold (M, g). Then*

(a) *$\mathscr{S}$ is Φ-semistable if and only if either one of the following conditions holds:*

(a′) *$\mu(\mathscr{S}')\leqq\mu(\mathscr{S})$ for any subsheaf $\mathscr{S}'$ such that the quotient $\mathscr{S}/\mathscr{S}'$ is torsion-free;*

(a″) *$\mu(\mathscr{S})\leqq\mu(\mathscr{S}'')$ for any torsion-free quotient sheaf $\mathscr{S}''$.*

(b) *$\mathscr{S}$ is Φ-stable if and only if either one of the following conditions holds:*

(b′) *$\mu(\mathscr{S}')<\mu(\mathscr{S})$ for any subsheaf $\mathscr{S}'$ such that the quotient $\mathscr{S}/\mathscr{S}'$ is torsion-free and $0<\operatorname{rank}\mathscr{S}'<\operatorname{rank}\mathscr{S}$;*

(b″) *$\mu(\mathscr{S})<\mu(\mathscr{S}'')$ for any torsion-free quotient sheaf $\mathscr{S}''$ such that $0<\operatorname{rank}\mathscr{S}''<\operatorname{rank}\mathscr{S}$.*

Proof Given an exact sequence

$$0 \longrightarrow \mathscr{S}' \longrightarrow \mathscr{S} \longrightarrow \mathscr{S}'' \longrightarrow 0,$$

let $\mathscr{T}''$ be the torsion subsheaf of $\mathscr{S}''$. Set $\mathscr{S}_1'' = \mathscr{S}''/\mathscr{T}''$ and define $\mathscr{S}_1'$ by the exact sequence

$$0 \longrightarrow \mathscr{S}_1' \longrightarrow \mathscr{S} \longrightarrow \mathscr{S}_1'' \longrightarrow 0.$$

Then $\mathscr{S}'$ is a subsheaf of $\mathscr{S}_1'$ and the quotient sheaf $\mathscr{S}_1'/\mathscr{S}'$ is isomorphic to the torsion sheaf $\mathscr{T}''$. From (6.9) and (7.5) we have

$$\mu(\mathscr{S}_1'') \leqq \mu(\mathscr{S}'') \quad \text{and} \quad \mu(\mathscr{S}') \leqq \mu(\mathscr{S}_1').$$

Our assertions follow from these inequalities. Q.E.D.

The following proposition should be compared with (IV.1.4).

(7.7) **Proposition** *Let $\mathscr{S}$ be a torsion-free coherent sheaf over a compact Kähler manifold (M, g). Then*

(a) *If* rank$(\mathscr{S})=1$, *then $\mathscr{S}$ is Φ-stable;*

(b) *Let $\mathscr{L}$ be (the sheaf of germs of holomorphic sections of) a line bundle over M. Then $\mathscr{S} \otimes \mathscr{L}$ is Φ-stable (resp. Φ-semistable) if and only if $\mathscr{S}$ is Φ-stable (resp. Φ-semistable);*

(c) *$\mathscr{S}$ is Φ-stable (resp. Φ-semistable) if and only if its dual $\mathscr{S}^*$ is Φ-stable (resp. Φ-semistable).*

Proof Both (a) and (b) are trivial. In order to prove (c), we need the following lemma which follows immediately from (6.12).

(7.8) **Lemma** *If $\mathscr{S}$ is a torsion-free coherent sheaf, then*

$$\mu(\mathscr{S}) = -\mu(\mathscr{S}^*).$$

Assume first that $\mathscr{S}^*$ is Φ-stable and consider an exact sequence

$$0 \longrightarrow \mathscr{S}' \longrightarrow \mathscr{S} \longrightarrow \mathscr{S}'' \longrightarrow 0$$

such that $\mathscr{S}''$ is torsion-free. Dualizing it, we have an exact sequence

$$0 \longrightarrow \mathscr{S}''^* \longrightarrow \mathscr{S}^* \longrightarrow \mathscr{S}'^*.$$

Using (7.8), we have

$$\mu(\mathscr{S}) = -\mu(\mathscr{S}^*) < -\mu(\mathscr{S}''^*) = \mu(\mathscr{S}'') .$$

By (7.6), $\mathscr{S}$ is Φ-stable.

Assume next that $\mathscr{S}$ is Φ-stable, and consider an exact sequence

$$0 \longrightarrow \mathscr{F}' \longrightarrow \mathscr{S}^* \longrightarrow \mathscr{F}'' \longrightarrow 0$$

such that $\mathscr{F}''$ is torsion-free. Dualizing it, we have an exact sequence

$$0 \longrightarrow \mathscr{F}''^* \longrightarrow \mathscr{S}^{**} \longrightarrow \mathscr{F}'^* .$$

Considering $\mathscr{S}$ as a subsheaf of $\mathscr{S}^{**}$ under the natural injection $\sigma: \mathscr{S} \to \mathscr{S}^{**}$, we define $\mathscr{S}'$ and $\mathscr{S}''$ by

$$\mathscr{S}' = \mathscr{S} \cap \mathscr{F}''^* , \qquad \mathscr{S}'' = \mathscr{S}/\mathscr{S}' .$$

Then define $\mathscr{T}''$ by the exact sequence

$$0 \longrightarrow \mathscr{F}''^*/\mathscr{S}' \longrightarrow \mathscr{S}^{**}/\mathscr{S} \longrightarrow \mathscr{T}'' \longrightarrow 0 .$$

Since $\mathscr{S}^{**}/\mathscr{S}$ is a torsion-sheaf by (4.12), so are $\mathscr{F}''^*/\mathscr{S}'$ and $\mathscr{T}''$. By (6.12), $\det(\mathscr{S}^{**}) = (\det \mathscr{S}^*)^* = (\det \mathscr{S})^{**} = \det \mathscr{S}$. By (6.9), $\det(\mathscr{S}^{**}/\mathscr{S})$ is a trivial line bundle. In general, if $\mathscr{T}$ is a torsion sheaf such that $\det \mathscr{T}$ is a trivial line bundle and if

$$0 \longrightarrow \mathscr{T}' \longrightarrow \mathscr{T} \longrightarrow \mathscr{T}'' \longrightarrow 0$$

is exact, then both $\det \mathscr{T}'$ and $\det \mathscr{T}''$ are trivial line bundles by (6.9) and (6.14). Hence, $\det(\mathscr{F}''^*/\mathscr{S}')$ is a trivial line bundle; i.e., $\det(\mathscr{F}''^*) = \det \mathscr{S}'$. In particular, $\deg(\mathscr{F}''^*) = \deg(\mathscr{S}')$. Since $\operatorname{rank}(\mathscr{F}''^*) = \operatorname{rank}(\mathscr{S}')$, we obtain

$$\mu(\mathscr{S}') = \mu(\mathscr{F}''^*) .$$

Hence,

$$\mu(\mathscr{F}'') = -\mu(\mathscr{F}''^*) = -\mu(\mathscr{S}') > -\mu(\mathscr{S}) = \mu(\mathscr{S}^*) ,$$

where the first and the last equalities are consequences of (6.12) and the inequality follows from the assumption that $\mathscr{S}$ is Φ-stable.

The proof for the semistable case is similar. Q.E.D.

(7.9) **Proposition** *Let $\mathscr{S}_1$ and $\mathscr{S}_2$ be two torsion-free coherent sheaves over a compact Kähler manifold (M, g). Then $\mathscr{S}_1 \oplus \mathscr{S}_2$ is Φ-semistable if and only if $\mathscr{S}_1$ and $\mathscr{S}_2$ are both Φ-semistable with $\mu(\mathscr{S}_1) = \mu(\mathscr{S}_2)$.*

Proof Assume that $\mathcal{S}_1$ and $\mathcal{S}_2$ are Φ-semistable with $\mu=\mu(\mathcal{S}_1)=\mu(\mathcal{S}_2)$. Then $\mu(\mathcal{S}_1\oplus\mathcal{S}_2)=\mu$, and for every subsheaf $\mathcal{F}$ of $\mathcal{S}_1\oplus\mathcal{S}_2$ we have the following commutative diagram with exact horizontal sequences:

$$\begin{array}{ccccccccc} 0 & \longrightarrow & \mathcal{S}_1 & \longrightarrow & \mathcal{S}_1\oplus\mathcal{S}_2 & \longrightarrow & \mathcal{S}_2 & \longrightarrow & 0 \\ & & \uparrow & & \uparrow & & \uparrow & & \\ 0 & \longrightarrow & \mathcal{F}_1 & \longrightarrow & \mathcal{F} & \longrightarrow & \mathcal{F}_2 & \longrightarrow & 0\,, \end{array}$$

where $\mathcal{F}_1=\mathcal{F}\cap(\mathcal{S}_1\oplus 0)$ and $\mathcal{F}_2$ is the image of $\mathcal{F}$ under the projection $\mathcal{S}_1\oplus\mathcal{S}_2\to\mathcal{S}_2$ so that the vertical arrows are all injective. Since $\mathcal{S}_i$ is Φ-semistable, we have

$$\deg(\mathcal{F}_i)\leqq\mu\cdot\operatorname{rank}(\mathcal{F}_i)\,.$$

Hence,

$$\mu(\mathcal{F})=(\deg(\mathcal{F}_1)+\deg(\mathcal{F}_2))/(\operatorname{rank}\mathcal{F}_1+\operatorname{rank}\mathcal{F}_2)\leqq\mu\,.$$

This shows that $\mathcal{S}_1\oplus\mathcal{S}_2$ is Φ-semistable.

Conversely, assume that $\mathcal{S}_1\oplus\mathcal{S}_2$ is Φ-semistable. Since $\mathcal{S}_i$ is a quotient sheaf of $\mathcal{S}_1\oplus\mathcal{S}_2$ and at the same time a subsheaf of $\mathcal{S}_1\oplus\mathcal{S}_2$, we have $\mu(\mathcal{S}_1\oplus\mathcal{S}_2)=\mu(\mathcal{S}_i)$. Any subsheaf $\mathcal{F}$ of $\mathcal{S}_i$ is a subsheaf of $\mathcal{S}_1\oplus\mathcal{S}_2$. Hence, $\mu(\mathcal{F})\leqq\mu(\mathcal{S}_1\oplus\mathcal{S}_2)=\mu(\mathcal{S}_i)$. This shows that $\mathcal{S}_i$ is Φ-semistable. Q.E.D.

(7.10) *Remark* It is clear that if $\mathcal{S}_1$ and $\mathcal{S}_2$ are nonzero, $\mathcal{S}_1\oplus\mathcal{S}_2$ can never be Φ-stable.

The following proposition should be compared with (IV.1.7) and (IV.5.5).

(7.11) **Proposition** *Let $\mathcal{S}_1$ and $\mathcal{S}_2$ be Φ-semistable sheaves over a compact Kähler manifold (M, g). Let $f\colon \mathcal{S}_1\to\mathcal{S}_2$ be a homomorphism.*

(1) *If $\mu(\mathcal{S}_1)>\mu(\mathcal{S}_2)$, then $f=0$;*

(2) *If $\mu(\mathcal{S}_1)=\mu(\mathcal{S}_2)$ and if $\mathcal{S}_1$ is Φ-stable, then* $\operatorname{rank}(\mathcal{S}_1)=\operatorname{rank}(f(\mathcal{S}_1))$ *and f is injective unless $f=0$;*

(3) *If $\mu(\mathcal{S}_1)=\mu(\mathcal{S}_2)$ and if $\mathcal{S}_2$ is Φ-stable, then* $\operatorname{rank}(\mathcal{S}_2)=\operatorname{rank}(f(\mathcal{S}_1))$ *and f is generically surjective unless $f=0$.*

Proof Assume $f\neq 0$. Set $\mathcal{F}=f(\mathcal{S}_1)$. Then $\mathcal{F}$ is a torsion-free quotient sheaf of $\mathcal{S}_1$.

(1) Since

$$\mu(\mathcal{F})\leqq\mu(\mathcal{S}_2)<\mu(\mathcal{S}_1)\leqq\mu(\mathcal{F})\,,$$

we have a contradiction.

(2) If $\mathscr{S}_1$ is Φ-stable and if $\operatorname{rank}(\mathscr{S}_1) > \operatorname{rank}(\mathscr{F})$, then

$$\mu(\mathscr{F}) \leqq \mu(\mathscr{S}_2) = \mu(\mathscr{S}_1) < \mu(\mathscr{F}),$$

which is impossible. Hence, $\operatorname{rank}(\mathscr{S}_1) = \operatorname{rank}(\mathscr{F})$.

(3) If $\mathscr{S}_2$ is Φ-stable and if $\operatorname{rank}(\mathscr{S}_2) > \operatorname{rank}(\mathscr{F})$, then

$$\mu(\mathscr{F}) < \mu(\mathscr{S}_2) = \mu(\mathscr{S}_1) \leqq \mu(\mathscr{F}),$$

which is impossible. Hence, $\operatorname{rank}(\mathscr{S}_2) = \operatorname{rank}(\mathscr{F})$. Q.E.D.

(7.12) **Corollary** *Let E_1 and E_2 be Φ-semistable vector bundles over a compact Kähler manifold (M, g) such that* $\operatorname{rank}(E_1) = \operatorname{rank}(E_2)$ *and* $\deg(E_1) = \deg(E_2)$. *If E_1 or E_2 is Φ-stable, then any nonzero sheaf homomorphism $f: E_1 \to E_2$ is an isomorphism.*

Proof By (7.11), f is an injective sheaf homomorphism. The induced homomorphism $\det(f): \det E_1 \to \det E_2$ is also nonzero. Consider $\det(f)$ as a holomorphic section of the line bundle $\operatorname{Hom}(\det E_1, \det E_2) = (\det E_1)^{-1}(\det E_2)$ and apply (III.1.24). Then we see that $\det(f)$ is an isomorphism. Hence, f is an isomorphism. Q.E.D.

(7.13) **Corollary** *If $\mathscr{S}$ is a Φ-semistable sheaf over a compact Kähler manifold M such that* $\deg(\mathscr{S}) < 0$, *then $\mathscr{S}$ admits no nonzero holomorphic section.*

Proof Let $\mathcal{O}$ be the sheaf of germs of holomorphic functions on M. Apply (7.11) to $f: \mathcal{O} \to \mathscr{S}$. Q.E.D.

The proposition above should be compared with (IV.5.6).

A holomorphic vector bundle E over a compact complex manifold M is said to be *simple* if every sheaf homomorphism $f: E \to E$ (i.e., every holomorphic section of $\operatorname{Hom}(E, E) = E^* \otimes E$) is a scalar multiple of the identity endomorphism.

(7.14) **Corollary** *Every Φ-stable vector bundle E over a compact Kähler manifold M is simple.*

Proof Given an endomorphism $f: E \to E$, let a be an eigenvalue of $f: E_x \to E_x$ at an arbitrarily chosen point $x \in M$. Applying (7.12) to $f - aI_E$, we see that $f - aI_E = 0$. Q.E.D.

We consider now the Harder-Narasimhan filtration theorem in higher dimension, (see Shatz [1] and Maruyama [6] in the algebraic case).

(7.15) **Theorem** *Given a torsion-free coherent sheaf $\mathscr{E}$ over a compact Kähler manifold (M, g), there is a unique filtration by subsheaves*

$$0=\mathscr{E}_0\subset\mathscr{E}_1\subset\mathscr{E}_2\subset\cdots\subset\mathscr{E}_{s-1}\subset\mathscr{E}_s=\mathscr{E}$$

such that, for $1\leqq i\leqq s-1$, $\mathscr{E}_i/\mathscr{E}_{i-1}$ is the maximal Φ-semistable subsheaf of $\mathscr{E}/\mathscr{E}_{i-1}$.

Proof The main step in the proof lies in the following

(7.16) **Lemma** *Given $\mathscr{E}$, there is an integer m_0 such that*

$$\mu(\mathscr{F})\leqq m_0$$

for all coherent subsheaves $\mathscr{F}$ of $\mathscr{E}$.

Proof To give the main idea of the proof, we assume that $\mathscr{E}$ is a vector bundle. In the general case, we use (5.14) to reduce the proof to the vector bundle case.

We prove first that given a holomorphic vector bundle E, there is an integer q_0 such that $\mu(L)\leqq q_0$ for all line subbundles L of E. Choose an Hermitian structure h in E. As in Section 6 of Chapter I, we choose a local unitary frame field $e_1, \cdots, e_r$ for E in such a way that e_1 spans L. Using (I.6.13), we can express the curvature $\sum S_{\alpha\bar\beta}\theta^\alpha\wedge\bar\theta^\beta$ of L in terms of the curvature $\Omega^i_j=\sum R^i_{j\alpha\bar\beta}\theta^\alpha\wedge\bar\theta^\beta$ of E:

$$S_{\alpha\bar\beta}=R^1_{1\alpha\bar\beta}-\sum_{\lambda=2}^{r} A^\lambda_{1\alpha}\bar A^\lambda_{1\beta}\,.$$

Hence,

$$\sum S_{\alpha\bar\alpha}=\sum_\alpha R^1_{1\alpha\bar\alpha}-\sum_{\alpha,\lambda}|A^\lambda_{1\alpha}|^2\leqq\sum_\alpha|R^1_{1\alpha\bar\alpha}|\leqq\sum_{i,\alpha}|R^i_{i\alpha\bar\alpha}|\leqq\left(nr\cdot\sum_{i,\alpha}|R^i_{i\alpha\bar\alpha}|^2\right)^{1/2}.$$

The right hand side is now independent of L and frames $e_1, \cdots, e_r, \theta^1, \cdots, \theta^n$. Since

$$\deg(L)=\int_M\frac{i}{2\pi}\sum S_{\alpha\bar\beta}\theta^\alpha\wedge\theta^\beta\wedge\Phi^{n-1}=\int_M\frac{1}{2n\pi}\left(\sum S_{\alpha\bar\alpha}\right)\Phi^n,$$

we see that $\deg(L)$ is bounded by a number which depends only on (E, h).

Applying the result above to $\bigwedge^p E$, we obtain a number $q_0(\bigwedge^p E)$. Let m_0 be the maximum of $q_0(\bigwedge^p E)$, $p=1, \cdots, r$. If $\mathscr{F}$ is a vector subbundle F of rank p of E, then apply the result above to the line subbundle $L = \det F \subset \Lambda^p E$. In the general case, we consider a line bundle $L = \det \mathscr{F} \otimes [D]$, where D is a certain effective divisor and use the argument given in (8.5) in the next section, (in particular, see (*) and (**) in (8.5)). Details are left to the reader.

Once we have (7.16), we can prove the following lemma in the same way as (1.13).

(7.17) **Lemma** *Given $\mathscr{E}$, there is a unique subsheaf $\mathscr{E}_1$ with torsion-free quotient $\mathscr{E}/\mathscr{E}_1$ such that, for every subsheaf $\mathscr{F}$ of $\mathscr{E}$,*

(i) $\mu(\mathscr{F}) \leqq \mu(\mathscr{E}_1)$,

(ii) $\operatorname{rank}(\mathscr{F}) \leqq \operatorname{rank}(\mathscr{E}_1)$ *if* $\mu(\mathscr{F}) = \mu(\mathscr{E}_1)$.

Then $\mathscr{E}_1$ is Φ-semistable.

Proof The existence is clear from (7.16). From (i) it is clear that $\mathscr{E}_1$ is Φ-semistable. To prove the uniqueness, let $\mathscr{E}'_1$ be another subsheaf satisfying (i) and (ii). Let $p: \mathscr{E} \to \mathscr{E}/\mathscr{E}'_1$ be the projection. Then $p(\mathscr{E}_1) \neq 0$. Define a subsheaf $\mathscr{F}$ of $\mathscr{E}_1$ by the exact sequence

$$0 \longrightarrow \mathscr{F} \longrightarrow \mathscr{E}_1 \longrightarrow p(\mathscr{E}_1) \longrightarrow 0 .$$

Since $\mathscr{E}_1$ is semistable, $\mu(\mathscr{F}) \leqq \mu(\mathscr{E}_1) \leqq \mu(p(\mathscr{E}_1))$.

On the other hand, we have an exact sequence

$$0 \longrightarrow \mathscr{E}'_1 \longrightarrow p^{-1}(p(\mathscr{E}_1)) \longrightarrow p(\mathscr{E}_1) \longrightarrow 0 .$$

From (i) and (ii) for $\mathscr{E}'_1$, we have $\mu(p^{-1}p(\mathscr{E}_1)) < \mu(\mathscr{E}'_1)$, i.e.,

$$\frac{\deg(p(\mathscr{E}_1)) + \deg(\mathscr{E}'_1)}{\operatorname{rank}(p(\mathscr{E}_1)) + \operatorname{rank}(\mathscr{E}'_1)} < \frac{\deg(\mathscr{E}'_1)}{\operatorname{rank}(\mathscr{E}'_1)} .$$

This implies $\mu(p(\mathscr{E}_1)) < \mu(\mathscr{E}'_1) = \mu(\mathscr{E}_1)$, a contradiction.

Now the theorem follows from (7.17). Q.E.D.

The proof for the following Jordan-Hölder theorem is standard.

(7.18) **Theorem** *Given a Φ-semistable sheaf $\mathscr{E}$ over a compact Kähler manifold (M, g), there is a filtration of $\mathscr{E}$ by subsheaves*

$$0 = \mathscr{E}_{k+1} \subset \mathscr{E}_k \subset \cdots \subset \mathscr{E}_1 \subset \mathscr{E}_0 = \mathscr{E}$$

such that $\mathscr{E}_i/\mathscr{E}_{i+1}$ are Φ-stable and $\mu(\mathscr{E}_i/\mathscr{E}_{i+1}) = \mu(\mathscr{E})$ for $i = 0, 1, \cdots, k$. Moreover,

$$Gr(\mathscr{E})=(\mathscr{E}_0/\mathscr{E}_1)\oplus(\mathscr{E}_1/\mathscr{E}_2)\oplus\cdots\oplus(\mathscr{E}_k/\mathscr{E}_{k+1})$$

is uniquely determined by $\mathscr{E}$ *up to an isomorphism.*

§8 Stability of Einstein-Hermitian vector bundles

In defining the concept of stability for a holomorphic vector bundle E, it was necessary to consider not only subbundles of E but also subsheaves of $\mathscr{E}=\mathcal{O}(E)$. However, as a first step in proving that every Einstein-Hermitian vector bundle is semistable and is a direct sum of stable bundles, we shall consider only subbundles. Since every coherent sheaf over M is a vector bundle outside its singularity set, it will become necessary to consider bundles over noncompact manifolds. We have to define therefore the local versions of the degree deg($\mathscr{S}$) and the degree/rank ratio $\mu(\mathscr{S})$. Given a coherent sheaf $\mathscr{S}$ over a Kähler manifold (M, g), we define

(8.1)
$$d(\mathscr{S})=c_1(\mathscr{S})\wedge\Phi^{n-1},$$

where $c_1(\mathscr{S})$ denotes the first Chern form of the determinant bundle $\det\mathscr{S}$ with respect to an Hermitian structure in $\det\mathscr{S}$. Therefore, $d(\mathscr{S})$ is an (n, n)-form on M and depends on the choice of the Hermitian structure in $\det\mathscr{S}$ and also on g. If M is compact, it can be integrated over M and gives deg($\mathscr{S}$).

(8.2) **Proposition** *Let (E, h) be an Einstein-Hermitian vector bundle over a Kähler manifold (M, g) with constant factor c. Let*

$$0\longrightarrow E'\longrightarrow E\longrightarrow E''\longrightarrow 0$$

be an exact sequence of vector bundles. Then

$$\frac{d(E')}{\operatorname{rank} E'}\leqq\frac{d(E)}{\operatorname{rank} E},$$

and if the equality holds, the exact sequence above splits and both E' and E'' are Einstein-Hermitian vector bundles with factor c (with respect to the naturally induced Hermitian structures).

Proof The proof is essentially the same as that of (2.5). We denote the curvature forms of (E, h) and $(E', h|_{E'})$ by Ω and Ω', respectively. They are related by the Gauss-Codazzi equation as follows (see (2.4) and (I.6.11)):

$$\Omega'^a_b=\Omega^a_b-\sum\omega^\lambda_b\wedge\bar{\omega}^\lambda_a,\qquad 1\leqq a, b\leqq p<\lambda\leqq r,$$

where $r = \operatorname{rank} E$ and $p = \operatorname{rank} E'$. Then

$$c_1(E, h) = \frac{i}{2\pi} \sum_{j=1}^{r} \Omega^j_j,$$

$$c_1(E', h) = \frac{i}{2\pi} \sum_{a=1}^{p} \Omega'^a_a = \frac{i}{2\pi} (\sum \Omega^a_a - \sum \omega^\lambda_a \wedge \bar{\omega}^\lambda_a)$$

From (IV.1.3) we obtain

$$d(E) = c_1(E, h) \wedge \Phi^{n-1} = \frac{1}{2n\pi} \sum K^j_j \Phi^n = \frac{rc}{2n\pi} \Phi^n,$$

$$d(E') = c_1(E', h) \wedge \Phi^{n-1} = \frac{1}{2n\pi} \sum K^a_a \Phi^n - \frac{i}{2\pi} \sum \omega^\lambda_a \wedge \bar{\omega}^\lambda_a \wedge \Phi^{n-1}$$

$$= \frac{pc}{2n\pi} \Phi^n - \frac{i}{2\pi} \sum \omega^\lambda_a \wedge \bar{\omega}^\lambda_a \wedge \Phi^{n-1}.$$

Hence,

$$\frac{d(E)}{r} - \frac{d(E')}{p} = \frac{i}{2p\pi} \sum \omega^\lambda_a \wedge \bar{\omega}^\lambda_a \wedge \Phi^{n-1}.$$

If we write

$$\omega^\lambda_a = \sum A^\lambda_{a\alpha} \theta^\alpha$$

as in (I.6.13), then using (III.1.18) we obtain

$$i \sum \omega^\lambda_a \wedge \bar{\omega}^\lambda_a \wedge \Phi^{n-1} = \frac{1}{n} \sum A^\lambda_{a\alpha} \bar{A}^\lambda_{a\alpha} \Phi^n.$$

This proves that

$$\frac{d(E)}{r} \geqq \frac{d(E')}{p}$$

and that the equality holds if and only if the second fundamental form A vanishes identically. Our proposition now follows from (I.6.14). Q.E.D.

(8.3) Theorem *Let (E, h) be an Einstein-Hermitian vector bundle over a compact Kähler manifold (M, g). Then E is Φ-semistable and (E, h) is a direct sum*

$$(E, h) = (E_1, h_1) \oplus \cdots \oplus (E_k, h_k)$$

of Φ-stable Einstein-Hermitian vector bundles $(E_1, h_1), \cdots, (E_k, h_k)$ with the same factor c as (E, h).

Proof To prove that E is Φ-semistable, let $\mathscr{F}$ be a subsheaf of $\mathscr{E}=\mathscr{O}(E)$ of rank $p<r=\operatorname{rank} E$ such that $\mathscr{E}/\mathscr{F}$ is torsion-free, (see (7.6)). The inclusion map

$$j: \mathscr{F} \longrightarrow \mathscr{E}$$

induces a homomorphism $\det(j)=(\bigwedge^p j)^{**}$:

$$\det(j): \det \mathscr{F} = (\bigwedge^p \mathscr{F})^{**} \longrightarrow (\bigwedge^p \mathscr{E})^{**} = \bigwedge^p \mathscr{E} .$$

Then $\det(j)$ is injective since it is injective outside the singularity set $S_{n-1}(\mathscr{F})$ and hence its kernel must be a torsion sheaf. Tensoring this homomorphism with $(\det \mathscr{F})^*$, we obtain a non-trivial homomorphism

$$f: \mathscr{O}_M \longrightarrow \bigwedge^p \mathscr{E} \otimes (\det \mathscr{F})^* ,$$

which may be considered as a holomorphic section of the vector bundle $\bigwedge^p E \otimes (\det \mathscr{F})^*$.

The factor c for (E, h) is given by (see (IV.2.7))

$$\text{(8.4)} \qquad c=\frac{2n\pi \cdot \mu(E)}{n! \cdot \operatorname{vol}(M)}, \qquad \text{where} \quad \operatorname{vol}(M)=\frac{1}{n!}\int_M \Phi^n .$$

Since every line bundle admits an Einstein-Hermitian structure (IV.1.4), we choose an Einstein-Hermitian structure with constant factor c' in the line bundle $\det \mathscr{F}$. Then

$$c'=\frac{2n\pi \cdot \mu(\det \mathscr{F})}{n! \cdot \operatorname{vol}(M)}=\frac{2np\pi \cdot \mu(\mathscr{F})}{n! \cdot \operatorname{vol}(M)} .$$

Then the vector bundle $\bigwedge^p E \otimes (\det \mathscr{F})^*$ is Einstein-Hermitian with factor $pc-c'$, (see (IV.1.4)). Since this bundle admits a non-trivial section f, the vanishing theorem (III.1.9) implies

$$pc-c' \geqq 0 .$$

This is nothing but the desired inequality $\mu(\mathscr{F}) \leqq \mu(\mathscr{E})$. This proves that E is Φ-semistable.

In order to prove the second assertion of the theorem, assume that the equality $\mu(\mathscr{F})=\mu(\mathscr{E})$ holds for the subsheaf $\mathscr{F}$ above. Then $pc-c'=0$. Then, by the same vanishing theorem, the section f must be parallel. In other words, the line bundle $\det \mathscr{F}$, under the injection $\det(j): \det \mathscr{F} \to \bigwedge^p E$, is a parallel line subbundle of $\bigwedge^p E$.

Let $M' = M - S_{n-1}(\mathscr{F})$, where $S_{n-1}(\mathscr{F})$ denotes the singularity set of $\mathscr{F}$. Let F be the vector bundle over M' corresponding to $\mathscr{F}$, i.e., $\mathscr{F}|_{M'} = \mathcal{O}(F)$. Then the bundle F, under the injection $j: F \to E|_{M'}$, is a parallel subbundle of E. Then (I.4.18) implies a holomorphic orthogonal decomposition

$$E|_{M'} = F \oplus G ,$$

where G is an Hermitian vector bundle over M'. If we set

$$\mathscr{G} = \mathscr{E}/\mathscr{F} ,$$

then G is clearly the bundle corresponding to $\mathscr{G}|_{M'}$, i.e., $\mathscr{G}|_{M'} = \mathcal{O}(G)$.

(It is possible to obtain the decomposition $E|_{M'} = F \oplus G$ from (8.2) as follows. From $\mu(\mathscr{F}) = \mu(\mathscr{E})$ and from the fact that c and c' are constant, we obtain $(1/p)d(F) = (1/r)d(E)$ on M'. Then the assertion follows from (8.2).)

We shall now show that the exact sequence

$$0 \longrightarrow \mathscr{F} \longrightarrow \mathscr{E} \longrightarrow \mathscr{G} \longrightarrow 0$$

splits. We know already that it splits over M'. By (5.22), $\mathscr{F}$ is reflexive. Hence, $\mathrm{Hom}(\mathscr{E}, \mathscr{F})$ and $\mathrm{Hom}(\mathscr{F}, \mathscr{F})$ are also reflexive and, in particular, normal (see (5.23)). Thus,

$$\Gamma(M, \mathrm{Hom}(\mathscr{E}, \mathscr{F})) = \Gamma(M', \mathrm{Hom}(\mathscr{E}, \mathscr{F})) ,$$

$$\Gamma(M, \mathrm{Hom}(\mathscr{F}, \mathscr{F})) = \Gamma(M', \mathrm{Hom}(\mathscr{F}, \mathscr{F})) .$$

Hence, the splitting homomorphism $p' \in \Gamma(M', \mathrm{Hom}(\mathscr{E}, \mathscr{F}))$ with $p' \circ j = id_{\mathscr{F}|_{M'}} \in \Gamma(M', \mathrm{Hom}(\mathscr{F}, \mathscr{F}))$ extends uniquely to a splitting homomorphism $p \in \Gamma(M, \mathrm{Hom}(\mathscr{E}, \mathscr{F}))$ with $p \circ j = id_{\mathscr{F}} \in \Gamma(M, \mathrm{Hom}(\mathscr{F}, \mathscr{F}))$. This proves that $\mathscr{E} = \mathscr{F} \oplus \mathscr{G}$. Since $\mathscr{E}$ is locally free, both $\mathscr{F}$ and $\mathscr{G}$ are projective and hence locally free. We have therefore a holomorphic decomposition of the bundle

$$E = F \oplus G$$

over M. Since this decomposition is orthogonal on M', it is orthogonal on M. Now the theorem follows from (IV.1.4). Q.E.D.

(8.5) *Remark* The proof of (8.3) presented here is due to Lübke [3] and simplifies greatly the proof in my lecture notes (Kobayashi [6]) which runs as follows.

To prove that E is Φ-semistable, let $\mathscr{F} \subset \mathscr{E} = \mathcal{O}(E)$ be a subsheaf of rank p and let

$$\tilde{j}\colon \det \mathscr{F}(=\Lambda^p \mathscr{F})^{**}) \longrightarrow \Lambda^p \mathscr{E}$$

be the natural sheaf homomorphism. Let α be a local holomorphic frame field (i.e., non-vanishing section) of $\det \mathscr{F}$, and let

$$B=\{x\in M;\ \tilde{j}(\alpha)(x)=0\}\ ,$$

(i.e., B is the zero set of the holomorphic section f of the bundle $\Lambda^p E\otimes(\det \mathscr{F})^*$ which corresponds to $\tilde{j}$). Let D_i, $i=1,\cdots,k$, be the irreducible components of B of codimension 1. Let V denote the union of all irreducible components of B of higher codimension so that

$$B=\bigcup D_i \cup V\,.$$

For each D_i, define its multiplicity n_i as follows. If $x\in D_i-\bigcup_{j\neq i} D_j \cup V$ and if D_i is defined by $w=0$ in a neighborhood of x, then n_i is the largest integer m such that $\tilde{j}(\alpha)/w^m$ is holomorphic. Then $\tilde{j}(\alpha)/w^{n_i}$ is a local holomorphic section of $\Lambda^p E$ not vanishing at x. We set

$$D=\sum n_i D_i$$

and let $[D]$ denote the line bundle defined by the divisor D. Let δ be the natural holomorphic section of $[D]$; δ vanishes along each D_i with multiplicity exactly n_i. Let

$$j'\colon \det \mathscr{F}\otimes[D] \longrightarrow \Lambda^p E$$

be defined by

$$j'=\tilde{j}\otimes\frac{1}{\delta}\,.$$

Then, as a mapping of the line bundle $\det \mathscr{F}\otimes[D]$ into the vector bundle $\Lambda^p E$, j' is injective over $M-V$, (for $1/\delta$ cancels the zeros of $\tilde{j}(\alpha)$ on $\bigcup D_i$).

Pull back the Hermitian structure $\Lambda^p h$ of $\Lambda^p E$ by j', and let

$$u=j'^*(\Lambda^p h)\,.$$

Then u defines an Hermitian structure in the line bundle $\det \mathscr{F}\otimes[D]_{M-V}$. (Over V, u is degenerate.) Let

$$W=S_{n-1}(\mathscr{F})\cup V\cup\left(\bigcup_i D_i\right),$$

where $S_{n-1}(\mathscr{F})$ is the singularity set of $\mathscr{F}$. Then there exists a subbundle F of

$E|_{M-W}$ such that

$$\mathscr{F}|_{M-W} = \mathcal{O}(F).$$

In order to prove that E is Φ-semistable, it suffices to show the following three inequalities:

$$(*) \qquad \int_M c_1(\mathscr{F}) \wedge \Phi^{n-1} \leqq \int_M c_1(\det \mathscr{F} \otimes [D]) \wedge \Phi^{n-1},$$

and the equality holds if and only if $D=0$;

$$(**) \qquad \int_M c_1(\det \mathscr{F} \otimes [D]) \wedge \Phi^{n-1} = \int_{M-V} c_1(\det \mathscr{F} \otimes [D], u) \wedge \Phi^{n-1};$$

$$(***) \qquad \frac{1}{p} c_1(\det \mathscr{F} \otimes [D], u) \wedge \Phi^{n-1} \leqq \frac{1}{r} c_1(E, h) \wedge \Phi^{n-1}$$

everywhere on $M-V$.

Proof of $(*)$ This is clear from

$$\int_M c_1([D]) \wedge \Phi^{n-1} = \sum n_i \int_{D_i} \Phi^{n-1} \geqq 0$$

Proof of $(**)$ Let τ be a local holomorphic frame field for $\det \mathscr{F} \otimes [D]$. Since $j': \det \mathscr{F} \otimes [D] \to \bigwedge^p E$, using a local holomorphic frame field $s = (s_1, \cdots, s_r)$ of E, we can write

$$j'(\tau) = \sum_I \tau^I s_I, \qquad \text{where} \quad s_I = s_{i_1} \wedge \cdots \wedge s_{i_p} \quad \text{with} \quad i_1 < \cdots < i_p.$$

Let $\tilde{u}$ be an Hermitian structure for $\det \mathscr{F} \otimes [D]$ over the entire M, and set

$$f = u(\tau, \tau)/\tilde{u}(\tau, \tau) = \sum u_{I\bar{J}} \tau^I \bar{\tau}^J,$$

where

$$u_{I\bar{J}} = \bigwedge^p h(s_I, s_J)/\tilde{u}(\tau, \tau).$$

Clearly, f is defined independently of τ and is a smooth non-negative function on M vanishing exactly on V. (Since $(u_{I\bar{J}})$ is positive definite, f vanishes exactly where all τ^I vanish, i.e., where $j'(\tau) = 0$). Let $\mathscr{I}$ be the ideal sheaf generated by $\{\tau^I\}$. Then V is the zeros of $\mathscr{I}$. By Hironaka's resolution of singularities, there is a non-singular complex manifold M^* and a surjective holomorphic map $\pi: M^* \to M$ such that $\operatorname{codim}(\pi^{-1} V) = 1$ and that

$$\pi: M^* - \pi^{-1}V \longrightarrow M - V$$

is biholomorphic and

$$\pi^*\mathscr{I} = \mathscr{O}_{M^*}(-m(\pi^{-1}V)) \qquad \text{for some integer} \quad m>0\,,$$

where $\pi^*\mathscr{I}$ denotes the ideal sheaf generated by $\{\pi^*\tau^I\}$. If $\pi^{-1}V$ is defined locally by $\zeta=0$ in M^*, then $\pi^*\tau^I/\zeta^m$ are all holomorphic and, at each point, at least one of $\pi^*\tau^I/\zeta^m$ does not vanish. Since $(u_{I\bar{J}})$ is a C^∞ positive definite Hermitian matrix, we can write locally

$$\pi^*f = a|\zeta|^{2m}\,,$$

where a is a nowhere vanishing C^∞ function. By a theorem of Lelong [1]

$$\frac{i}{2\pi}d'd''\log\pi^*f = m\cdot\pi^{-1}V \qquad (\text{as a current on } M^*)\,.$$

In particular, since codim $V \geqq 2$, we have

$$\int_{M-V}\frac{i}{2\pi}d'd''\log f\wedge\Phi^{n-1} = \int_{M^*-\pi^{-1}V}\frac{i}{2\pi}d'd''\log\pi^*f\wedge\pi^*\Phi^{n-1}$$
$$= m\int_{\pi^{-1}V}\pi^*\Phi^{n-1} = m\int_V\Phi^{n-1} = 0\,.$$

On the other hand, from $u=f\tilde{u}$ we obtain

$$c_1(\det\mathscr{F}\otimes[D],\tilde{u}) = c_1(\det\mathscr{F}\otimes[D],u) + \frac{i}{2\pi}d'd''\log f \quad \text{on} \quad M-V.$$

Integrating this, we have

$$\int_M c_1(\det\mathscr{F}\otimes[D],\tilde{u})\wedge\Phi^{n-1} = \int_{M-V}c_1(\det\mathscr{F}\otimes[D],\tilde{u})\wedge\Phi^{n-1}$$
$$= \int_{M-V}c_1(\det\mathscr{F}\otimes[D],u)\wedge\Phi^{n-1}\,.$$

This proves (**).

Proof of (***) From the definition $u=j'^*(\wedge^p h)$, we have

$$c_1(\det\mathscr{F}\otimes[D],u) = c_1(\mathscr{F},h) \qquad \text{on} \quad M-W\,.$$

Since $(\mathscr{F},h)$ is an Hermitian subbundle of (E,h) over $M-W$, we have (see (8.2))

$$\frac{1}{p}c_1(\mathscr{F}, h) \wedge \Phi^{n-1} \leqq \frac{1}{r}c_1(E, h) \wedge \Phi^{n-1} \qquad \text{on} \quad M-W.$$

Hence,

$$\frac{1}{p}c_1(\det \mathscr{F} \otimes [D], u) \wedge \Phi^{n-1} \leqq \frac{1}{r}c_1(E, h) \wedge \Phi^{n-1} \qquad \text{on} \quad M-W.$$

Since both sides are of class C^∞ on $M-V$, the inequality holds on $M-V$. This proves (∗∗∗).

Finally, in order to prove that E is a direct sum of Φ-stable bundles, assume that

$$\frac{1}{p}\int_M c_1(\mathscr{F}) \wedge \Phi^{n-1} = \frac{1}{r}\int_M c_1(E) \wedge \Phi^{n-1}.$$

Then, $D=0$ by (∗). Moreover, the equality holds also in (∗∗∗). Then, over $M-W$, $\mathscr{F}$ is a subbundle of E with vanishing second fundamental form, and

$$E = \mathscr{F} + \mathscr{F}^\perp \qquad \text{(holomorphically)} \quad \text{on} \quad M-W,$$

where $\mathscr{F}^\perp$ is the orthogonal complement to $\mathscr{F}$, (see (I.6.4)). We consider the holonomy group of (E, h) over $M-W$. It is a subgroup of $U(p) \times U(r-p)$. Since W is a subvariety (of real codimension at least 2), the holonomy group of (E, h) over M is contained in the closure of the holonomy group of (E, h) over $M-W$. Hence, the above decomposition extends to M, i.e., we have

$$E = \mathscr{F} + \mathscr{F}^\perp \qquad \text{on} \quad M$$

with holomorphic subbundles $\mathscr{F}$ and $\mathscr{F}^\perp$.

It was pointed out by *T.* Mabuchi that the proof of (8.3) yields also the following

(8.6) **Theorem** *If a holomorphic vector bundle E over a compact Kähler manifold (M, g) admits an approximate Einstein-Hermitian structure, then E is Φ-semistable.*

Proof The only modification we have to make in the proof lies in the following two points. (i). Use (IV.5.3) in place of (IV.1.4) to see that $\bigwedge^p E \otimes (\det \mathscr{F})^*$ admits an approximate Einstein-Hermitian structure. (ii). Use the vanishing theorem (IV.5.6) instead of (III.1.9). Q.E.D.

(8.7) *Examples of stable bundles* In Sections 6 and 7 of Chapter IV, we gave several examples of irreducible Einstein-Hermitian vector bundles over compact Kähler manifolds. By (8.3), they are Φ-stable. In particular, the tangent and cotangent bundles of a compact irreducible Hermitian symmetric space, the symmetric tensor power $S^p(TP_n)$ and the exterior power $\bigwedge^p(TP_n)$ of the tangent bundle of the complex projective space P_n are all Φ-stable, (where Φ is the Kähler form of the canonical metric). More generally, the homogeneous vector bundles described in (IV.6.4) are Φ-stable; this fact has been directly established by Ramanan [1] and Umemura [5]. The null correlation bundle over P_{2n+1} described in (IV.6.5) is also Φ-stable; the case $n=1$ is well known, (see, for example, Okonek-Schneider-Spindler [1]). The projectively flat vector bundles over a torus described in (IV.7.54) are Φ-stable.

§ 9 *T*-stability of Bogomolov

In Section 1 we explained the concept of T-stability for vector bundles over compact Riemann surfaces. In order to extend this concept to the general case, we need to consider coherent sheaves as in the case of Φ-stability.

Let $\mathcal{S}$ be a torsion-free coherent analytic sheaf over a compact complex manifold M. A *weighted flag* of $\mathcal{S}$ is a sequence of pairs $\mathcal{F}=\{(\mathcal{S}_i, n_i);\ 1\leqq i\leqq k\}$ consisting of subsheaves

$$\mathcal{S}_1\subset\mathcal{S}_2\subset\cdots\subset\mathcal{S}_k\subset\mathcal{S}$$

with

$$0<\operatorname{rank}\mathcal{S}_1<\operatorname{rank}\mathcal{S}_2<\cdots<\operatorname{rank}\mathcal{S}_k<\operatorname{rank}\mathcal{S}$$

and positive integers $n_1, n_2, \cdots, n_k$. We set

$$r_i=\operatorname{rank}\mathcal{S}_i\,,\qquad r=\operatorname{rank}\mathcal{S}\,.$$

To such a flag $\mathcal{F}$ we associate a line bundle $T_{\mathcal{F}}$ by setting

$$(9.1)\qquad T_{\mathcal{F}}=\prod_{i=1}^{k}((\det\mathcal{S}_i)^r(\det\mathcal{S})^{-r_i})^{n_i}\,.$$

We say that $\mathcal{S}$ is *T-stable* if, for every weighted flag $\mathcal{F}$ of $\mathcal{S}$ and for every flat line bundle L over M, the line bundle $T_{\mathcal{F}}\otimes L$ admits no nonzero holomorphic sections. We say that $\mathcal{S}$ is *T-semistable* if, for every weighted flag $\mathcal{F}$ of $\mathcal{S}$ and for every flat line bundle L over M, every nonzero holomorphic section of the line bundle $T_{\mathcal{F}}\otimes L$ (if any) vanishes nowhere on M. (We note that if $T_{\mathcal{F}}\otimes L$ admits a nowhere vanishing holomorphic section, then it is a trivial line bundle and $T_{\mathcal{F}}$, being isomorphic to L^*, is flat.)

(9.2) **Proposition** *In the definition of T-stability and T-semistability, it suffices to consider only those flags* $\mathscr{F}=\{(\mathscr{S}_i, n_i)\}$ *for which the quotient sheaves* $\mathscr{S}/\mathscr{S}_i$ *are torsion-free.*

Proof As in the proof of (7.6), given an arbitrary flag $\{(\mathscr{S}_i, n_i)\}$ of $\mathscr{S}$, let $\mathscr{T}_i$ be the torsion subsheaf of the quotient $\mathscr{S}/\mathscr{S}_i$. We define a subsheaf $\tilde{\mathscr{S}}_i$ of $\mathscr{S}$ to be the kernel of the natural homomorphism $\mathscr{S}\to(\mathscr{S}/\mathscr{S}_i)/\mathscr{T}_i$. Then $\tilde{\mathscr{S}}_i/\mathscr{S}_i$ is isomorphic to $\mathscr{T}_i$ and the quotient sheaf $\mathscr{S}/\tilde{\mathscr{S}}_i$ is torsion-free. We shall show that it suffices to consider the new flag $\tilde{\mathscr{F}}=(\tilde{\mathscr{S}}_i, n_i)$ in place of $\mathscr{F}=(\mathscr{S}_i, n_i)$. Since $\mathscr{T}_i=\tilde{\mathscr{S}}_i/\mathscr{S}_i$, by (6.9) we have

$$\det\tilde{\mathscr{S}}_i=(\det\mathscr{S}_i)\otimes(\det\mathscr{T}_i)\,.$$

Hence,

$$T_{\tilde{\mathscr{F}}}\otimes L=T_{\mathscr{F}}\otimes L\otimes\prod_{i=1}^{k}(\det\mathscr{T}_i)^{n_i r}\,.$$

Our assertion follows from the fact that each $\det\mathscr{T}_i$ admits a nonzero holomorphic section, (see (6.14)). Q.E.D.

The following proposition is an analogue of (7.7).

(9.3) **Proposition** *Let* $\mathscr{S}$ *be a torsion-free sheaf over a compact complex manifold* M.

(a) *If* $\operatorname{rank}\mathscr{S}=1$, *then* $\mathscr{S}$ *is T-stable.*

(b) *Let* $\mathscr{L}$ *be* (*the sheaf of germs of holomorphic sections of*) *a line bundle over* M. *Then* $\mathscr{S}\otimes\mathscr{L}$ *is T-stable* (*resp. T-semistable*) *if and only if* $\mathscr{S}$ *is T-stable* (*resp. T-semistable*).

(c) $\mathscr{S}$ *is T-stable* (*resp. T-semistable*) *if and only if its dual* $\mathscr{S}^*$ *is T-stable* (*resp. T-semistable*).

Proof (a) is trivial. (b) follows from the fact that there is a natural correspondence between the flags $\mathscr{F}=\{(\mathscr{S}_i, n_i)\}$ of $\mathscr{S}$ and the flags $\mathscr{F}\otimes\mathscr{L}=\{(\mathscr{S}_i\otimes\mathscr{L}; n_i)\}$ of $\mathscr{S}\otimes\mathscr{L}$ and that $T_{\mathscr{F}}=T_{\mathscr{F}\otimes\mathscr{L}}$.

To prove (c), assume first that $\mathscr{S}^*$ is T-stable and let $\mathscr{F}=\{(\mathscr{S}_i, n_i)\}$ be a flag of $\mathscr{S}$ such that $\mathscr{S}/\mathscr{S}_i$ are all torsion-free. Then we obtain a flag $\mathscr{F}^*=\{((\mathscr{S}/\mathscr{S}_i)^*, n_i)\}$ of $\mathscr{S}^*$:

$$(\mathscr{S}/\mathscr{S}_k)^* \subset \cdots \subset (\mathscr{S}/\mathscr{S}_2)^* \subset \cdots \subset (\mathscr{S}/\mathscr{S}_1)^* \subset \mathscr{S}^* .$$

Then, using (6.9) and (6.12) we obtain

$$T_{\mathscr{F}^*} = \prod_{i=1}^{k} (((\det \mathscr{S})^{-1}(\det \mathscr{S}_i))^r (\det \mathscr{S})^{r-r_i})^{n_i} = T_{\mathscr{F}} .$$

This implies that $\mathscr{S}$ is T-stable. Similarly, if $\mathscr{S}^*$ is T-semistable, then $\mathscr{S}$ is T-semistable.

Conversely, assume that $\mathscr{S}$ is T-stable. Let $\mathscr{F} = \{(\mathscr{R}_i, n_i)\}$ be a flag of $\mathscr{S}^*$ such that the quotient sheaves $\mathscr{G}_i = \mathscr{S}^*/\mathscr{R}_i$ are all torsion-free. Dualizing the exact sequence

$$0 \longrightarrow \mathscr{R}_i \longrightarrow \mathscr{S}^* \longrightarrow \mathscr{G}_i \longrightarrow 0$$

we obtain an exact sequence

$$0 \longrightarrow \mathscr{G}_i^* \longrightarrow \mathscr{S}^{**} \longrightarrow \mathscr{R}_i^* .$$

Considering $\mathscr{S}$ as a subsheaf of $\mathscr{S}^{**}$ under the injection $\sigma: \mathscr{S} \to \mathscr{S}^{**}$, we set

$$\mathscr{S}_i = \mathscr{S} \cap \mathscr{G}_i^* .$$

We define torsion sheaves

$$\mathscr{T} = \mathscr{S}^{**}/\mathscr{S} , \qquad \mathscr{T}_i = \mathscr{G}_i^*/\mathscr{S}_i \subset \mathscr{T} .$$

As in the proof of (7.7), $\det(\mathscr{S}^{**}) = \det \mathscr{S}$ by (6.12). Hence $\det \mathscr{T} = \det(\mathscr{S}^{**}/\mathscr{S})$ is a trivial line bundle. In general, if

$$0 \longrightarrow \mathscr{T}' \longrightarrow \mathscr{T} \longrightarrow \mathscr{T}'' \longrightarrow 0$$

is an exact sequence of torsion sheaves and if $\det \mathscr{T}$ is a trivial line bundle, then both $\det \mathscr{T}'$ and $\det \mathscr{T}''$ are trivial line bundles by (6.9) and (6.14). Hence, $\det \mathscr{T}_i$ is a trivial line bundle so that

$$\det \mathscr{G}_i^* = \det \mathscr{S}_i .$$

Since $\mathscr{S}_k \subset \cdots \subset \mathscr{S}_2 \subset \mathscr{S}_1 \subset \mathscr{S}$ and $\operatorname{rank}(\mathscr{S}_i) = \operatorname{rank}(\mathscr{G}_i) = r - \operatorname{rank}(\mathscr{R}_i)$, it follows that $\mathscr{F}^* = \{(\mathscr{S}_i, n_i)\}$ is a flag of $\mathscr{S}$. Set $r_i = \operatorname{rank}(\mathscr{S}_i)$ so that $\operatorname{rank}(\mathscr{R}_i) = r - r_i$. Then

$$\begin{aligned} T_{\mathscr{F}} &= \prod ((\det \mathscr{R}_i)^r (\det \mathscr{S}^*)^{r_i - r})^{n_i} \\ &= \prod ((\det \mathscr{S}^*)^r (\det \mathscr{G}_i)^{-r} (\det \mathscr{S}^*)^{r_i - r})^{n_i} \\ &= \prod ((\det \mathscr{G}_i{}^*)^r (\det \mathscr{S}^*)^{r_i})^{n_i} \end{aligned}$$

$$=\prod((\det \mathscr{S}_i)^r(\det \mathscr{S})^{-r_i})^{n_i}=T_{\mathscr{F}^*}.$$

From $T_{\mathscr{F}}=T_{\mathscr{F}^*}$ it follows that $\mathscr{S}^*$ is T-stable if $\mathscr{S}$ is T-stable. Similarly, if $\mathscr{S}$ is T-semistable, then $\mathscr{S}^*$ is T-semistable. Q.E.D.

(9.4) **Theorem** *Let $\mathscr{S}$ be a torsion-free coherent sheaf over a compact Kähler manifold (M, g) with Kähler form Φ. If $\mathscr{S}$ is Φ-stable (resp. Φ-semistable), then it is T-stable (resp. T-semistable).*

Proof Let $\mathscr{F}=\{(\mathscr{S}_i, n_i)\}$ be a flag of $\mathscr{S}$. Let $r=\mathrm{rank}(\mathscr{S})$ and $r_i=\mathrm{rank}(\mathscr{S}_i)$. Then

$$\int_M c_1((\det \mathscr{S}_i)^r(\det \mathscr{S})^{-r_i})\wedge \Phi^{n-1}=\int_M (r\cdot c_1(\mathscr{S}_i)-r_i c_1(\mathscr{S}))\wedge \Phi^{n-1}\leqq 0$$

if $\mathscr{S}$ is Φ-semistable. The inequality is strict if $\mathscr{S}$ is Φ-stable. Let L be any flat line bundle. Then $c_1(L)=0$ in $H^2(M, \boldsymbol{R})$. It follows from the definition of $T_{\mathscr{F}}$ that if $\mathscr{S}$ is Φ-semistable,

$$\deg(T_{\mathscr{F}}\otimes L)=\int_M c_1(T_{\mathscr{F}}\otimes L)\wedge \Phi^{n-1}=\int_M c_1(T_{\mathscr{F}})\wedge \Phi^{n-1}\leqq 0\,. \tag{9.5}$$

The inequality is strict if $\mathscr{S}$ is Φ-stable. Now the theorem follows from the vanishing theorem (III.1.24). Q.E.D.

(9.6) **Corollary** *If (E, h) is an Einstein-Hermitian vector bundle over a compact Kähler manifold (M, g), then E is T-semistable and (E, h) is a direct sum of T-stable Einstein-Hermitian vector bundles $(E_1, h_1), \cdots, (E_k, h_k)$ with $\mu(E_1)=\cdots=\mu(E_k)=\mu(E)$.*

Proof This is immediate from (8.3) and (9.4). Q.E.D.

(9.7) **Corollary** *If a holomorphic vector bundle E over a compact Kähler manifold (M, g) admits an approximate Einstein-Hermitian structure, then it is T-semistable.*

Proof This is immediate from (8.6) and (9.4). Q.E.D.

The following is a partial converse to (9.4).

(9.8) **Theorem** *Let $\mathcal{S}$ be a torsion-free coherent sheaf over a compact Kähler manifold (M, g) with Kähler form Φ. Assume either*

(i) $\dim H^{1,1}(M, \boldsymbol{C}) = 1$,

or

(ii) *Φ represents an integral class (so that M is projective algebraic) and* $\mathrm{Pic}(M)/\mathrm{Pic}^0(M) = \boldsymbol{Z}$.

If $\mathcal{S}$ is T-stable (resp. T-semistable), then it is Φ-stable (resp. Φ-semistable).

In (ii), $\mathrm{Pic}(M)$ denotes the Picard group, i.e., the group $H^1(M, \mathcal{O}^*)$ of line bundles over M, and $\mathrm{Pic}^0(M)$ denotes the subgroup of $\mathrm{Pic}(M)$ consisting of line bundles with vanishing first Chern class.

Proof Let $\mathcal{S}'$ be a subsheaf of $\mathcal{S}$ with rank $r' < r = \mathrm{rank}(\mathcal{S})$. Consider the line bundle

$$F = (\det \mathcal{S}')^r (\det \mathcal{S})^{-r'} .$$

Then

$$c_1(F) = r \cdot c_1(\mathcal{S}') - r' c_1(\mathcal{S}) ,$$

$$\deg(F) = rr'(\mu(\mathcal{S}') - \mu(\mathcal{S})) .$$

Let $[\Phi]$ denote the cohomology class of Φ. Then under our assumption (i) or (ii), we have

$$c_1(F) = a[\Phi] \qquad \text{for some} \quad a \in \boldsymbol{R} .$$

(In case (ii), we use the fact that $[\Phi]$ is the Chern class of some line bundle.) Then $\deg(F) > 0$ (resp. $\deg(F) = 0$) if and only if $a > 0$ (resp. $a = 0$).

Assume that $\mu(\mathcal{S}') > \mu(\mathcal{S})$ so that $\deg(F) > 0$. Then $a > 0$ and F is ample. Hence, for some positive k, F^k admits a non-trivial section. This shows that if $\mathcal{S}$ is not Φ-semistable, then it is not T-semistable.

Assume that $\mu(\mathcal{S}') = \mu(\mathcal{S})$ so that $\deg(F) = 0$. Then $a = 0$ and F is a flat line bundle. Let $L = F^{-1}$ in the definition of T-stability. Being trivial, $F \otimes L$ admits a nonzero section. This shows that if $\mathcal{S}$ is not Φ-stable, then it is not T-stable. Q.E.D.

Both (9.4) and (9.8) are proved in Bogomolov [1] when M is algebraic. The differential geometric proofs given here is from Kobayashi [8].

§10 Stability in the sense of Gieseker

Since it is difficult to define higher dimensional Chern classes for sheaves over non-algebraic manifolds, we shall assume in this section that M is a projective algebraic manifold. We fix an ample line bundle H on M.

Let $\mathscr{S}$ be a torsion-free coherent analytic sheaf over M, and set

$$\mathscr{S}(k)=\mathscr{S}\otimes\mathcal{O}(H^k) \qquad \text{for} \quad k\in\boldsymbol{Z},$$

$$\chi(\mathscr{S}(k))=\sum(-1)^i\dim H^i(M,\mathscr{S}(k)),$$

$$p(\mathscr{S}(k))=\chi(\mathscr{S}(k))/\mathrm{rank}(\mathscr{S}).$$

We say that $\mathscr{S}$ is *Gieseker H-stable* (resp. *H-semistable*) if, for every coherent subsheaf $\mathscr{F}$ of $\mathscr{S}$ with $0<\mathrm{rank}(\mathscr{F})<\mathrm{rank}(\mathscr{S})$, the inequality

$$p(\mathscr{F}(k))<p(\mathscr{S}(k)) \qquad (\text{resp. } p(\mathscr{F}(k))\leqq p(\mathscr{S}(k))) \tag{10.1}$$

holds for sufficiently large integers k.

By the Riemann-Roch theorem, we can express $\chi(\mathscr{S}(k))$ in terms of Chern classes of M, $\mathscr{S}$ and H, (see (II.4.5)).

$$\chi(\mathscr{S}(k))=\int_M \mathrm{ch}(\mathscr{S})\cdot\mathrm{ch}(H^k)\cdot\mathrm{td}(M), \tag{10.2}$$

where

$$\mathrm{ch}(\mathscr{S})=r+c_1(\mathscr{S})+\frac{1}{2}(c_1(\mathscr{S})^2-2c_2(\mathscr{S}))+\cdots, \qquad (r=\mathrm{rank}(\mathscr{S})),$$

$$\mathrm{ch}(H^k)=1+kd+\frac{1}{2}k^2d^2+\cdots+\frac{1}{n!}k^nd^n, \qquad (d=c_1(H)),$$

$$\mathrm{td}(M)=1+\frac{1}{2}c_1(M)+\frac{1}{12}(c_1(M)^2+c_2(M))+\cdots.$$

Hence,

$$\chi(\mathscr{S}(k))=\int_M \frac{rk^n}{n!}d^n+\frac{k^{n-1}}{(n-1)!}d^{n-1}\left(c_1(\mathscr{S})+\frac{r}{2}c_1(M)\right)+\cdots.$$

Writing a similar formula for a subsheaf $\mathscr{F}$, we obtain

$$p(\mathscr{S}(k))-p(\mathscr{F}(k))=\frac{k^{n-1}}{(n-1)!}(\mu(\mathscr{S})-\mu(\mathscr{F}))+\cdots, \tag{10.3}$$

where the dots indicate terms of lower order (than k^{n-1}).

(10.4) **Proposition** *Let $\mathcal{S}$ be a torsion-free coherent sheaf over a projective algebraic manifold M with an ample line bundle H.*

(a) *If $\mathcal{S}$ is H-stable, then it is Gieseker H-stable;*

(b) *If $\mathcal{S}$ is Gieseker H-semistable, then it is H-semistable.*

This is evident from (10.3). From (8.3) and (10.4) we obtain

(10.5) **Corollary** *Let M be a projective algebraic manifold with an ample line bundle H. Choose a Kähler metric g such that its Kähler form Φ represents $c_1(H)$. Then every Einstein-Hermitian vector bundle (E, h) over (M, g) is a direct sum*

$$(E, h) = (E_1, h_1) \oplus \cdots \oplus (E_k, h_k)$$

of Gieseker H-stable Einstein-Hermitian vector bundles $(E_1, h_1), \cdots, (E_k, h_k)$ with $\mu(E) = \mu(E_1) = \cdots = \mu(E_k)$.

We shall pursue the analogy between H-stability and Gieseker H-stability further. The following lemma corresponds to (7.3).

(10.6) **Lemma** *If*

$$0 \longrightarrow \mathcal{S}' \longrightarrow \mathcal{S} \longrightarrow \mathcal{S}'' \longrightarrow 0$$

is an exact sequence of coherent sheaves over a projective algebraic manifold M with an ample line bundle H, then

$$r'(p(\mathcal{S}(k)) - p(\mathcal{S}'(k))) + r''(p(\mathcal{S}(k)) - p(\mathcal{S}''(k))) = 0$$

where $r' = \mathrm{rank}(\mathcal{S}')$ and $r'' = \mathrm{rank}(\mathcal{S}'')$.

Proof The proof is similar to that of (7.3). Replace deg by χ, and use the fact that

$$\chi(\mathcal{S}(k)) = \chi(\mathcal{S}'(k)) + \chi(\mathcal{S}''(k)) \,. \qquad \text{Q.E.D.}$$

The following proposition which corresponds to (7.4) follows from (10.6).

(10.7) **Proposition** *Let $\mathcal{S}$ be a torsion-free coherent sheaf over a projective algebraic manifold M with an ample line bundle H. Then $\mathcal{S}$ is Gieseker H-stable (resp. semistable) if and only if, for every quotient sheaf $\mathcal{S}''$ of $\mathcal{S}$ with $0 < \mathrm{rank}(\mathcal{S}'') < \mathrm{rank}(\mathcal{S})$, the inequality*

$$p(\mathscr{S}(k)) < p(\mathscr{S}''(k)) \qquad (resp.\ p(\mathscr{S}(k)) \leq p(\mathscr{S}''(k)))$$

holds for sufficiently large integers k.

The proof of the following proposition is similar to that of (7.11).

(10.8) **Proposition** *Let $\mathscr{S}_1$ and $\mathscr{S}_2$ be Gieseker H-semistable sheaves over a projective algebraic manifold M with an ample line bundle H. Let $f\colon \mathscr{S}_1 \to \mathscr{S}_2$ be a homomorphism.*

(a) *If $p(\mathscr{S}_1(k)) > p(\mathscr{S}_2(k))$ for $k \gg 0$, then $f = 0$;*

(b) *If $p(\mathscr{S}_1(k)) = p(\mathscr{S}_2(k))$ for $k \gg 0$ and if $\mathscr{S}_1$ is Gieseker H-stable, then* $\operatorname{rank}(\mathscr{S}_1) = \operatorname{rank}(f(\mathscr{S}_1))$ *and f is injective unless $f = 0$;*

(c) *If $p(\mathscr{S}_1(k)) = p(\mathscr{S}_2(k))$ for $k \gg 0$ and if $\mathscr{S}_2$ is Gieseker H-stable, then* $\operatorname{rank}(\mathscr{S}_2) = \operatorname{rank}(f(\mathscr{S}_1))$ *and f is generically surjective unless $f = 0$.*

(10.9) **Corollary** *Let E be a Gieseker H-stable vector bundle over a projective algebraic manifold M with an ample line bundle H. Then E is simple.*

Proof given an endomorphism $f\colon E \to E$, let a be an eigenvalue of $f\colon E_x \to E_x$ at an arbitrarily chosen point $x \in M$. Applying (10.8) to $f - aI_E$, we see that $f - aI_E$ is injective unless $f - aI_E$. If $f - aI_E$ is injective, it induces an injective endomorphism of the line bundle det E. But for a line bundle, such an endomorphism cannot have zeros. Hence, we must have $f - aI_E = 0$. Q.E.D.

Combining results in Sections 9 and 10, we have the following diagram of implications.

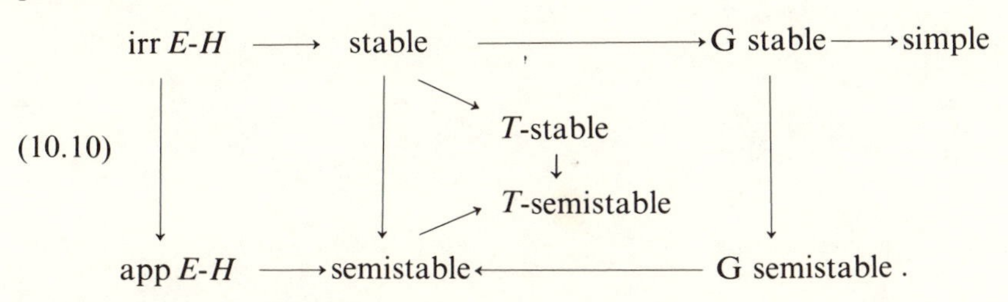

Here, we have used the following abbreviation.

"stable" (resp. semistable) means "Φ-stable" (resp. Φ-semistable);

"G stable" (resp. G semistable) means "Gieseker H-stable" (resp. Gieseker H-semistable), where Φ and H are related by the condition that Φ represents $c_1(H)$.

"irr E-H" stands for "irreducible Einstein-Hermitian structure", while "app E-H" means "approximate Einstein-Hermitian structure".

All three concepts of stability (resp. semistability) reduce to Mumford's stability (resp. semistability) when M is a compact Riemann surface.

Proposition (10.4) showing "stable"→"G stable"→"G semistable"→"semistable" is from Okonek-Schneider-Spindler [1].

Theorem (9.4) shows "stable"→"T-stable" and "semistable"→"T-semistable".

Theorem (8.3) says a little more than "irr E-H"→"stable" and was first announced in Kobayashi [5]. Theorem (2.6) is its converse when $\dim M = 1$ and is due to Narasimhan-Seshadri [1] with a direct proof by Donaldson [1]. The converse when M is an algebraic surface is due to Donaldson [2] and is stated in (10.19) of Chapter VI. The converse in general is probably true.

Mabuchi pointed out that the proof of (10.4) shows also "app E-H"→"semistable". Its converse for an algebraic manifold is in Donaldson [2] and will be proved in Chapter VI, (see (10.13)). It should be true for any compact Kähler manifold.

The implication "E-H"→"T-semistable" is in Kobayashi [3] and was the starting point of "the Einstein condition and stability".

Chapter VI

Existence of approximate Einstein-Hermitian structures

In this chapter we explain results of Donaldson [2] and prove the theorem to the effect that if M is an algebraic manifold with an ample line bundle H, then every H-semistable vector bundle E over M admits an approximate Einstein-Hermitian structure, (see (10.13)).

The partial differential equation expressing the Einstein condition is similar to that of harmonic maps. The best reference for analytic tools (in Sections 4 through 7) is therefore the lecture notes by Hamilton [1]. The reader who reads Japanese may find Nishikawa's notes (Nishikawa-Ochiai [1]) also useful.

§ 1 Space of Hermitian matrices

Let Herm(r) denote the space of $r \times r$ Hermitian matrices; it is a real vector space of dimension r^2. Let $\mathrm{Herm}^+(r)$ denote the set of positive definite Hermitian matrices of order r; it is a convex domain of Herm(r). The group $GL(r; \boldsymbol{C})$ which acts on Herm(r) by

$$(1.1) \qquad h \longmapsto {}^t\bar{a}ha\,, \qquad a \in GL(r; \boldsymbol{C})\,, \quad h \in \mathrm{Herm}(r)$$

is transitive on $\mathrm{Herm}^+(r)$. The isotropy subgroup at the identity matrix is the unitary group $U(r)$, which is a maximal compact subgroup of $GL(r; \boldsymbol{C})$. The domain $\mathrm{Herm}^+(r)$ can be identified with the quotient space $GL(r; \boldsymbol{C})/U(r)$:

$$(1.2) \qquad \mathrm{Herm}^+(r) \approx GL(r; \boldsymbol{C})/U(r)\,.$$

It is moreover a symmetric space. In fact, the symmetry at the identity matrix is given by

$$h \longmapsto h^{-1}\,.$$

The Cartan decomposition of the Lie algebra $\mathfrak{gl}(r; \boldsymbol{C})$ is given by

$$(1.3) \qquad \mathfrak{gl}(r; \boldsymbol{C}) = \mathfrak{u}(r) + \mathrm{Herm}(r)\,.$$

The exponential map

$$\exp\colon \mathrm{Herm}(r) \longrightarrow \mathrm{Herm}^+(r)$$

is a diffeomorphism. In particular, every element h of $\mathrm{Herm}^+(r)$ has a unique square root in $\mathrm{Herm}^+(r)$.

We shall now define an invariant Riemannian metric in $\mathrm{Herm}^+(r)$. Since $\mathrm{Herm}^+(r)$ is a domain in the vector space $\mathrm{Herm}(r)$, the tangent space at each point h of $\mathrm{Herm}^+(r)$ may be identified with $\mathrm{Herm}(r)$. We define an inner product in the tangent space $\mathrm{Herm}(r) \approx T_h(\mathrm{Herm}^+(r))$ by

$$(1.4) \qquad (v, w) = \sum h^{i\bar{k}} h^{l\bar{j}} v_{i\bar{j}} \bar{w}_{k\bar{l}} = \sum h^{i\bar{k}} h^{l\bar{j}} v_{i\bar{j}} w_{l\bar{k}}$$

$$\text{for} \quad v = (v_{i\bar{j}}), \quad w = (w_{i\bar{j}}) \in \mathrm{Herm}(r), \quad h = (h_{i\bar{j}}) \in \mathrm{Herm}^+(r),$$

where $(h^{i\bar{j}})$ denotes the inverse matrix h^{-1} of h. Thus, $\sum h^{i\bar{j}} h_{k\bar{j}} = \delta^i_k$. With the usual tensor notation, we rewrite (1.4) as follows:

$$(1.5) \qquad (v, w) = \sum v^j_i w^i_j,$$

where $v^j_i = \sum h^{j\bar{k}} v_{i\bar{k}}$ and $w^i_j = \sum h^{i\bar{k}} w_{j\bar{k}}$. In matrix notation, this can be expressed as

$$(1.6) \qquad (v, w) = \mathrm{tr}(h^{-1} v \cdot h^{-1} w).$$

It is obvious that this defines a $GL(r; \mathbf{C})$-invariant Riemannian metric on $\mathrm{Herm}^+(r)$ which coincides with the usual inner product at the identity matrix.

We consider now geodesics in the Riemannian manifold $\mathrm{Herm}^+(r)$. Fixing two points p and q in $\mathrm{Herm}^+(r)$, let $\Omega_{p,q}$ denote the space of piecewise differentiable curves $h = h(t)$, $a \leqq t \leqq b$, from p to q. Let $\partial_t h = dh/dt$ denote the velocity vector of h. Its Riemannian length or the speed of h is given by (see (1.6))

$$(1.7) \qquad |\partial_t h| = (\mathrm{tr}(h^{-1} \partial_t h \cdot h^{-1} \partial_t h))^{1/2}.$$

We define a functional E on $\Omega_{p,q}$, called the energy integral, by

$$(1.8) \qquad E(h) = \int_a^b |\partial_t h|^2 dt = \int_a^b \mathrm{tr}(h^{-1} \partial_t h \cdot h^{-1} \partial_t h) dt.$$

Then the geodesics (with affine parameter) from p to q are precisely the critical points of this functional E.

Let $h + sv$, $|s| < \delta$, be a variation of h with infinitesimal variation v. By definition, $v = v(t)$, $a \leqq t \leqq b$, is a curve in $\mathrm{Herm}(r)$ such that $v(a) = v(b) = 0$. Then

$$(1.9) \qquad E(h + sv) = \int_a^b \mathrm{tr}((h + sv)^{-1} \partial_t (h + sv) \cdot (h + sv)^{-1} \partial_t (h + sv)) dt.$$

We obtain easily

$$(1.10)\quad \left.\frac{dE(h+sv)}{ds}\right|_{s=0} = 2\int_a^b \mathrm{tr}(-h^{-1}v\cdot h^{-1}\partial_t h\cdot h^{-1}\partial_t h + h^{-1}\partial_t v\cdot h^{-1}\partial_t h)dt$$

$$= 2\int_a^b \left[\frac{d}{dt}\mathrm{tr}(h^{-1}v\cdot h^{-1}\partial_t h) - \mathrm{tr}(h^{-1}v(h^{-1}\partial_t^2 h - h^{-1}\partial_t h\cdot h^{-1}\partial_t h))\right]dt$$

$$= -2\int_a^b \mathrm{tr}(h^{-1}v(h^{-1}\partial_t^2 h - h^{-1}\partial_t h\cdot h^{-1}\partial_t h))dt\,.$$

Hence, h is a critical point of E (i.e., (1.10) vanishes for all v) if and only if $h^{-1}\partial_t^2 h - h^{-1}\partial_t h\cdot h^{-1}\partial_t h = 0$. This can be rewritten as

$$(1.11)\qquad \frac{d}{dt}(h^{-1}\partial_t h) = 0\,, \qquad \text{(equation of geodesics)}\,.$$

It follows that the geodesic from the identity matrix 1_r to e^a, $a\in \mathrm{Herm}(r)$, is given by

$$(1.12)\qquad h = e^{at}\,, \qquad 0 \leqq t \leqq 1\,.$$

Its arc-length, i.e., the distance from 1_r to e^a is given by

$$(1.13)\qquad \int_0^1 |\partial_t h|\, dt = \int_0^1 \mathrm{tr}(aa)^{1/2}dt = (\textstyle\sum \alpha_i^2)^{1/2}\,,$$

where $\alpha_1, \cdots, \alpha_r$ are the eigen-values of a. More generally, the distance $d(p, q)$ between $p, q \in \mathrm{Herm}^+(r)$ is given by

$$(1.14)\qquad d(p, q) = (\textstyle\sum (\log \lambda_i)^2)^{1/2}$$

where $\lambda_1, \cdots, \lambda_r$ are the eigen-values of $p^{-1}q$.

For applications to vector bundles, we have to formulate the above results in terms of Hermitian forms rather than Hermitian matrices.

Let V be a complex vector space of dimension r. Let $\mathrm{Herm}(V)$ denote the space of Hermitian bilinear forms $v\colon V\times V\to \boldsymbol{C}$ and $\mathrm{Herm}^+(V)$ the domain in $\mathrm{Herm}(V)$ consisting of positive definite Hermitian forms. We identify the tangent space at each point of $\mathrm{Herm}^+(V)$ with $\mathrm{Herm}(V)$.

Let $GL(V)$ be the group of complex automorphisms of V and $\mathfrak{gl}(V) = \mathrm{End}(V)$ its Lie algebra. Then $GL(V)$ acts on $\mathrm{Herm}(V)$ by

$$v \longmapsto a(v) = {}^t\bar{a}va\,, \quad \text{i.e.,} \qquad (a(v))(x, y) = v(ax, ay)\,,$$

where $v\in \mathrm{Herm}(V)$, $a\in GL(V)$, $x, y\in V$. Under this action, $GL(V)$ is tansitive on

$\mathrm{Herm}^+(V)$. We note that unless we choose a basis for V, there is no natural origin in $\mathrm{Herm}^+(V)$ like the identity matrix 1_r in $\mathrm{Herm}(r)$.

To define a $GL(V)$-invariant Riemannian metric on $\mathrm{Herm}^+(V)$, we define first a linear endomorphism $h^{-1}v \in \mathfrak{gl}(V)$ for $h \in \mathrm{Herm}^+(V)$ and $v \in \mathrm{Herm}(V)$ as follows:

$$(1.15) \qquad v(x, y) = h(x, (h^{-1}v)y) \qquad \text{for} \quad x, y \in V .$$

Given tangent vectors $v, v' \in \mathrm{Herm}(V)$ of $\mathrm{Herm}^+(V)$ at h, their inner product is defined by

$$(1.16) \qquad (v, v') = \mathrm{tr}(h^{-1}v \cdot h^{-1}v') .$$

A curve $h = h(t)$ in $\mathrm{Herm}^+(V)$ is a geodesic if and only if it satisfies (1.11), i.e., $h^{-1}\partial_t h$ is a fixed element (independent of t) of $\mathfrak{gl}(V)$.

§2 Space of Hermitian structures

Let E be a C^∞ complex vector bundle of rank r over an n-dimensional complex manifold M. Let $\mathrm{Herm}(E)$ denote the (infinite dimensional) vector space of all C^∞ Hermitian forms v on E; each element v defines an Hermitian bilinear form

$$v_x : E_x \times E_x \longrightarrow \boldsymbol{C}$$

at each point x of M. If $(s_1, \cdots, s_r)$ is a local frame field of E, then v is given by its components $v_{ij} = v(s_i, s_j)$, and the matrix (v_{ij}) is Hermitian.

Let $\mathrm{Herm}^+(E)$ denote the set of C^∞ Hermitian structures h in E; it consists of elements of $\mathrm{Herm}(E)$ which are positive definite everywhere on M. It is then clear that $\mathrm{Herm}^+(E)$ is a convex domain in $\mathrm{Herm}(E)$. We can regard $\mathrm{Herm}(E)$ as the tangent space of $\mathrm{Herm}^+(E)$ at each point h.

Let $GL(E)$ denote the group of (complex) automorphisms of E (inducing the identity transformation on the base space M); it is sometimes called the *complex gauge group* of E. Then $GL(E)$ acts on $\mathrm{Herm}(E)$ by

$$(2.1) \qquad v \longmapsto a(v) = {}^t\bar{a}va , \qquad v \in \mathrm{Herm}(E) , \quad a \in GL(E) ,$$

or

$$(a(v))(\xi, \eta) = v(a\xi, a\eta) \qquad \xi, \eta \in E_x .$$

The "Lie algebra" of $GL(E)$ consists of linear endomorphisms of E and will be denoted by $\mathfrak{gl}(E)$; it is the space of sections of the endomorphism bundle $\mathrm{End}(E)$.

We fix an Hermitian structure $k \in \mathrm{Herm}^+(E)$ as the "origin" of $\mathrm{Herm}^+(E)$. Let $U(E)$ denote the subgroup of $GL(E)$ consisting of transformations leaving k

invariant, i.e., unitary transformations with respect to k. It is sometimes called the *gauge group* of (E, k). Its Lie algebra $\mathfrak{u}(E)$ consists of endomorphisms of E which are skew-Hermitian with respect to k.

The action of $GL(E)$ on $\mathrm{Herm}^+(E)$ is transitive. In fact, given two Hermitian structures $k, h \in \mathrm{Herm}^+(E)$, there is a unique element $a \in GL(E)$ which is positive Hermitian with respect to k such that $h = a(k) = {}^t\bar{a}ka$. We can identify $\mathrm{Herm}^+(E)$ with the quotient space $GL(E)/U(E)$ and consider it as a symmetric space with the involution at k given by

$$h = {}^t\bar{a}ka \longmapsto a^{-1}k\,{}^t\bar{a}^{-1}. \tag{2.2}$$

Assume that M is a compact Kähler manifold with Kähler form Φ. We introduce an invariant Riemannian metric on $\mathrm{Herm}^+(E) = GL(E)/U(E)$ by defining an inner product in the tangent space $\mathrm{Herm}(E) = T_h(\mathrm{Herm}^+(E))$ at $h \in \mathrm{Herm}^+(E)$ as follows (see (1.16)):

$$(v, w) = \int_M \mathrm{tr}(h^{-1}v \cdot h^{-1}w)\Phi^n, \qquad v, w \in \mathrm{Herm}(E). \tag{2.3}$$

Since $\mathrm{tr}(h^{-1}v \cdot h^{-1}w)$ is invariant by $GL(E)$, the Riemannian metric thus introduced is also invariant by $GL(E)$.

Following the reasoning in Section 1 (see (1.9)–(1.11)), we see that a curve $h = h(t)$ in $\mathrm{Herm}^+(E)$ is a geodesic if and only if

$$\frac{d}{dt}(h^{-1}\partial_t h) = 0, \tag{2.4}$$

i.e., $h^{-1}\partial_t h$ is a fixed element (independent of t) of $\mathrm{Herm}(E)$.

§3 Donaldson's Lagrangian

Let E be a holomorphic vector bundle over a compact Kähler manifold M with Kähler form Φ. Let $\mathrm{Herm}^+(E)$ denote the space of C^∞ Hermitian structures in E. In the preceding section we defined a Riemannian structure in $\mathrm{Herm}^+(E)$. In this section, we construct a functional L on $\mathrm{Herm}^+(E)$ whose gradient flow is given by

$$\mathrm{grad}\, L = \hat{K} - ch. \tag{3.1}$$

Given two Hermitian structures $k, h \in \mathrm{Herm}^+(E)$, we connect them by a curve h_t, $0 \leqq t \leqq 1$, in $\mathrm{Herm}^+(E)$ so that $k = h_0$ and $h = h_1$. For each h_t, its curvature is denoted by $R_t = R(h_t) \in A^{1,1}(\mathrm{End}(E))$ as a $(1, 1)$-form with values in $\mathrm{End}(E)$. We set

$$(3.2) \qquad v_t = h_t^{-1}\partial_t h_t .$$

For each t, v_t is a field of endomorphisms of E, i.e., $v_t \in A^0(\mathrm{End}(E))$. We recall (see (2.4)) that h_t is a geodesic in $\mathrm{Herm}^+(E)$ if and only if v_t is independent of t, i.e., $\partial_t v_t = 0$.

We set

$$(3.3) \qquad Q_1(h, k) = \log(\det(k^{-1}h)) ,$$

$$(3.4) \qquad Q_2(h, k) = i\int_0^1 \mathrm{tr}(v_t \cdot R_t)dt ,$$

$$(3.5) \qquad L(h, k) = \int_M \left(Q_2(h, k) - \frac{c}{n} Q_1(h, k)\Phi\right) \wedge \frac{\Phi^{n-1}}{(n-1)!}$$

where c is the constant given by (see (IV.2.7))

$$c = 2n\pi \int c_1(E) \wedge \Phi^{n-1} \Big/ r \int \Phi^n .$$

We see from the next lemma that $L(h, k)$ does not depend on the choice of a curve h_t joining h to k.

(3.6) **Lemma** *Let h_t, $\alpha \leqq t \leqq \beta$, be a piecewise differentiable closed curve in* $\mathrm{Herm}^+(E)$, *(hence $h_\alpha = h_\beta$). Let $v_t = h_t^{-1}\partial_t h_t$. Then*

$$\int_\alpha^\beta \mathrm{tr}(v_t \cdot R_t)dt \in d'A^{0,1} + d''A^{1,0} .$$

Proof Let $\alpha = a_0 < a_1 < \cdots < a_k = \beta$ be the values of t where h_t is not differentiable. Fix an arbitrary Hermitian structure k as a reference point in $\mathrm{Herm}^+(E)$. It suffices then to prove the lemma for the closed curve consisting of a smooth curve (such as the line segment or the geodesic) from k to h_{a_j}, the curve h_t, $a_j \leqq t \leqq a_{j+1}$, and a smooth curve from $h_{a_{j+1}}$ back to k. We set

$$a = a_j , \qquad b = a_{j+1} ,$$

$$\Delta = \{(t, s);\ a \leqq t \leqq b,\ 0 \leqq s \leqq 1\} .$$

Let $h\colon \Delta \to \mathrm{Herm}^+(E)$ be a smooth mapping such that

$$h(t, 0) = k , \qquad h(t, 1) = h_t \qquad \text{for} \quad a \leqq t \leqq b ,$$

while $h(a, s)$ (resp. $h(b, s)$) describes the chosen curve from k to h_a (resp. h_b to k).

For example, if we use line segments to go from k to h_a and from h_b to k, then we can use $h(t,s)=sh_t+(1-s)k$.

We set

$$u=h^{-1}\partial_s h\,, \qquad v=h^{-1}\partial_t h\,,$$
$$R=d''(h^{-1}d'h)\,.$$

We define

$$\phi=i\cdot\operatorname{tr}(h^{-1}\tilde{d}hR)\,, \qquad \text{where} \quad \tilde{d}h=\partial_s h\cdot ds+\partial_t h\cdot dt\,. \tag{3.7}$$

We consider $\tilde{d}=(\partial/\partial s)ds+(\partial/\partial t)dt$ as exterior differentiation in the domain Δ as opposed to exterior differentiation d in the manifold M. Considering ϕ as a 1-form on the domain Δ, we apply the Stokes formula:

$$\int_{\partial\Delta}\phi=\int_{\Delta}\tilde{d}\phi\,. \tag{3.8}$$

We calculate first the left hand side:

$$\begin{aligned}\int_{\partial\Delta}\phi&=-\int_{t=a}^{t=b}\phi\,|_{s=0}+\int_{s=0}^{s=1}\phi\,|_{t=a}+\int_{t=a}^{t=b}\phi\,|_{s=1}-\int_{s=0}^{s=1}\phi\,|_{t=b}\\&=i\int_a^b\operatorname{tr}(v_t\cdot R_t)dt+Q_2(h_a,k)-Q_2(h_b,k)\,.\end{aligned} \tag{3.9}$$

It suffices therefore to show that the right hand side of (3.8) is in $d'A^{0,1}+d''A^{1,0}$.

We define a (0, 1)-form

$$\alpha=i\cdot\operatorname{tr}(v\cdot d''u)\,, \tag{3.10}$$

and we shall prove

$$\tilde{d}\phi=-(d'\alpha+d''\bar{\alpha})ds\wedge dt+id''d'\operatorname{tr}(v\cdot u)ds\wedge dt\,. \tag{3.11}$$

It is more convenient to use exterior covariant differentiation $D=D'+D''$ $(=D'+d'')$ of the Hermitian connection defined by $h(t,s)$ in place of $d=d'+d''$. Thus,

$$\alpha=i\cdot\operatorname{tr}(v\cdot D''u)\,, \tag{3.12}$$

$$d'\alpha=D'\alpha=i\cdot\operatorname{tr}(D'v\wedge D''u+v\cdot D'D''u)\,. \tag{3.13}$$

Since

$$\bar{u}=\bar{h}^{-1}\partial_s\bar{h}={}^t(\partial_s h\cdot h^{-1})={}^t(h\cdot u\cdot h^{-1})$$

$$\bar{v}={}^t(h\cdot v\cdot h^{-1})$$
$$\bar{\alpha}=-i\cdot\mathrm{tr}(\bar{v}\cdot D'\bar{u})=-i\,\mathrm{tr}({}^t(h\cdot v\cdot h^{-1})\cdot{}^t(h\cdot D'u\cdot h^{-1}))$$
$$=-i\cdot\mathrm{tr}(D'u\cdot v)\,,$$

we obtain

$$\bar{\alpha}=-i\cdot\mathrm{tr}(v\cdot D'u)\,, \tag{3.14}$$

$$d''\bar{\alpha}=D''\bar{\alpha}=-i\cdot\mathrm{tr}(D''v\wedge D'u+v\cdot D''D'u)\,. \tag{3.15}$$

In order to calculate $\tilde{d}\phi$, we need first $\partial_s R$, $\partial_t R$, $\partial_t u$ and $\partial_s v$. From the definition of R, we obtain

$$\partial_s R=d''(-h^{-1}\partial_s h\cdot h^{-1}d'h+h^{-1}d'\partial_s h)$$
$$=d''(-u\cdot\omega+d'(h^{-1}\partial_s h)+\omega\cdot u)=d''(-u\cdot\omega+d'u+\omega\cdot u)\,,$$

where $\omega=h^{-1}d'h$ is the connection form. Hence,

$$\partial_s R=d''D'u=D''D'u\,. \tag{3.16}$$

Similarly,

$$\partial_t R=d''D'v=D''D'v\,. \tag{3.17}$$

From the definition of u and v, we obtain

$$\partial_t u=-h^{-1}\partial_t h\cdot h^{-1}\partial_s h+h^{-1}\partial_t\partial_s h=-v\cdot u+h^{-1}\partial_t\partial_s h\,, \tag{3.18}$$

$$\partial_s v=-h^{-1}\partial_s h\cdot h^{-1}\partial_t h+h^{-1}\partial_s\partial_t h=-u\cdot v+h^{-1}\partial_s\partial_t h\,. \tag{3.19}$$

Hence,

$$\begin{aligned}\tilde{d}\phi&=i\cdot\mathrm{tr}(-\partial_t u\cdot R-u\cdot\partial_t R+\partial_s v\cdot R+v\cdot\partial_s R)ds\wedge dt\\&=i\cdot\mathrm{tr}(v\cdot u\cdot R-u\cdot v\cdot R+v\cdot D''D'u-u\cdot D''D'v)ds\wedge dt\\&=i\cdot\mathrm{tr}(-v(D'D''+D''D')u+v\cdot D''D'u-u\cdot D''D'v)ds\wedge dt\,.\end{aligned} \tag{3.20}$$

On the other hand,

$$\begin{aligned}d'\alpha+d''\bar{\alpha}&=i\cdot\mathrm{tr}(D'v\wedge D''u-D''v\wedge D'u+v\cdot D'D''u-v\cdot D''D'u)\\&=i\cdot\mathrm{tr}(-D''D'(vu)+D''D'v\cdot u+v\cdot D'D''u)\,.\end{aligned} \tag{3.21}$$

Comparing (3.20) with (3.21), we obtain (3.11). This completes the proof of the lemma. Q.E.D.

In the course of the proof (see (3.9)) we established also the following formula.

(3.22) **Lemma** *Let h_t, $a \leqq t \leqq b$, be any differentiable curve in* $\mathrm{Herm}^+(E)$ *and k any point of* $\mathrm{Herm}^+(E)$. *Then*

$$i\int_a^b \mathrm{tr}(v_t \cdot R_t)dt + Q_2(h_a, k) - Q_2(h_b, k) \in d'A^{0,1} + d''A^{1,0}.$$

The following lemma is also in Donaldson [2].

(3.23) **Lemma** *Let $h, h', h'' \in \mathrm{Herm}^+(E)$. Then*

(i) $L(h, h') + L(h', h'') + L(h'', h) = 0$,

(ii) $L(h, ah) = 0$ *for any positive constant* a.

(iii) $$id'd''\left(Q_2(h, h') - \frac{c}{n} Q_1(h, h')\Phi\right)$$
$$= -\frac{1}{2}\mathrm{tr}\left(\left(iR - \frac{c}{n}\Phi I\right)^2\right) + \frac{1}{2}\mathrm{tr}\left(\left(iR' - \frac{c}{n}\Phi I\right)^2\right),$$

where R and R' denote the curvature for h and h', respectively.

Proof (i) Clearly we have

$$Q_1(h, h') + Q_1(h', h'') + Q_1(h'', h) = 0.$$

Applying (3.6) to a triangle joining h, h', h'' in $\mathrm{Herm}^+(E)$, we obtain

$$Q_2(h, h') + Q_2(h', h'') + Q_2(h'', h) \equiv 0 \mod d'A^{0,1} + d''A^{1,0}.$$

Now, (i) follows from the definition (3.5) of L.

(ii) Clearly we have

$$Q_1(h, ah) = \log(1/a^r) = -r \cdot \log a.$$

To calculate $Q_2(h, ah)$, let $a = e^b$ and $h_t = e^{b(1-t)}h$. then $R_t = R$. Hence,

$$Q_2(h, ah) = -i\int_0^1 \mathrm{tr}(bR)dt = -ib \cdot \mathrm{tr}(R),$$

$$L(h, ah) = -\frac{ib}{n!}\int n \cdot \mathrm{tr}(R) \wedge \Phi^{n-1} - cr\Phi^n.$$

From the definition of the constant c (see (IV.2.7)), we obtain $L(h, ah) = 0$.

(iii) Join h' to h by a curve h_t, $0 \leqq t \leqq 1$, with $h' = h_0$ and $h = h_1$. Let $v_t = h_t^{-1}\partial_t h_t$. We shall prove

$$(3.24)\quad id'd''\left(Q_2(h_t, h') - \frac{c}{n}Q_1(h_t, h')\Phi\right) = \frac{1}{2}\operatorname{tr}\left(\left(R_t + \frac{ic}{n}\Phi I\right)^2 - \left(R_0 + \frac{ic}{n}\Phi I\right)^2\right).$$

Since this holds obviously for $t=0$, it suffices to show that both sides have the same derivative with respect to t. Using D' and D'' in place of d' and d'', we can write the derivative of the first term of (3.24) as follows:

$$\frac{d}{dt} id'd''(Q_2(h_t, h')) = d''d' \operatorname{tr}(v_t \cdot R_t) = \operatorname{tr}(D''D'v_t \wedge R_t).$$

Since $\det(h_t)$ defines an Hermitian structure in the line bundle $\det(E)$ whose curvature is $\operatorname{tr}(R_t)$, we obtain

$$d'd'' \log(\det(h_t)) = -\operatorname{tr}(R_t).$$

Hence, the derivative of the left hand side of (3.24) is given by

$$\operatorname{tr}(D''D'v_t \wedge R_t) + i\frac{c}{n}\operatorname{tr}(\partial_t R_t) \wedge \Phi.$$

On the other hand, the derivative of the right hand side of (3.24) is given by

$$\operatorname{tr}(\partial_t R_t \wedge R_t) + i\frac{c}{n}\operatorname{tr}(\partial_t R_t) \wedge \Phi.$$

Now, (iii) follows from $\partial_t R_t = D''D'v_t$ (see (3.17)). Q.E.D.

(3.25) **Lemma** *For any differentiable curve h_t in* $\operatorname{Herm}^+(E)$ *and any point k in* $\operatorname{Herm}^+(E)$, *we have*

(i) $\partial_t(Q_1(h_t, k)) = \operatorname{tr}(v_t)$, *where* $v_t = h_t^{-1}\partial_t h_t$;

(ii) $\partial_t(Q_2(h_t, k)) \equiv i \cdot \operatorname{tr}(v_t \cdot R_t) \mod d'A^{0,1} + d''A^{1,0}$.

Proof (i)

$$\begin{aligned}\partial_t Q_1(h_t, k) &= \partial_t \log(\det(k^{-1}h_t)) = \partial_t \log(\det h_t)\\ &= \partial_t(\det h_t)/\det h_t = \operatorname{tr}(h_t^{-1} \cdot \partial_t h_t).\end{aligned}$$

(ii) In (3.22), consider b as a variable. Differentiating (3.22) with respect to b, we obtain (ii). Q.E.D.

Fix an Hermitian structure $k \in \operatorname{Herm}^+(E)$ and let h_t be a differentiable curve in $\operatorname{Herm}^+(E)$. We shall prove the following formula:

(3.26)
$$\frac{dL(h_t, k)}{dt} = (\hat{K}_t - ch_t, \partial_t h_t)$$
$$= \int_M \sum h^{i\bar{l}} h^{k\bar{j}} (K_{i\bar{j}} - ch_{i\bar{j}}) \partial_t h_{k\bar{l}} \frac{\Phi^n}{n!},$$

where $(\hat{K}_t - ch_t, \partial_t h_t)$ denotes the Riemannian inner product defined by (2.3) and (1.4). The second line is nothing but an explicit expression of the inner product in components, (see (1.4)).

Since

$$L(h_t, k) = \int_M (Q_2(h_t, k) - \frac{c}{n} Q_1(h_t, k)\Phi) \wedge \frac{\Phi^{n-1}}{(n-1)!},$$

using (3.25) we obtain

$$dL(h_t, k)/dt = \int_M (i \cdot \mathrm{tr}(v_t \cdot R_t) - \frac{c}{n} \mathrm{tr}(v_t)\Phi) \wedge \frac{\Phi^{n-1}}{(n-1)!}.$$

Then making use of (see (IV.1.3))

$$iR_t \wedge \Phi^{n-1} = \frac{1}{n} K_t \Phi^n,$$

we obtain

(3.27)
$$dL(h_t, k)/dt = \int_M (\mathrm{tr}(v_t \cdot K_t) - c \cdot \mathrm{tr}(v_t)) \frac{\Phi^n}{n!}$$
$$= \int_M \mathrm{tr}((K_t - cI)v_t) \frac{\Phi^n}{n!}.$$

Since $v_t = h_t^{-1} \cdot \partial_t h_t$, (3.27) is equivalent to (3.26).

From (3.26) we obtain

(3.28) **Proposition** *Fix* $k \in \mathrm{Herm}^+(E)$ *and consider the functional* $L(h) = L(h, k)$ *on* $\mathrm{Herm}^+(E)$. *Then* h *is a critical point of this functional if and only if* $\hat{K} - ch = 0$, *i.e.,* h *is an Einstein-Hermitian structure in* (E, M, g).

For each fixed t, we consider $\partial_t h_t \in \mathrm{Herm}(E)$ as a tangent vector of $\mathrm{Herm}^+(E)$ at h_t. The differential dL of L evaluated at $\partial_t h_t$ is given by

$$dL(\partial_t h_t) = dL(h_t, k)/dt = (\hat{K}_t - ch_t, \partial_t h_t).$$

This means that the gradient of L, i.e., the vector field on $\mathrm{Herm}^+(E)$ dual to the 1-form dL with respect to the invariant Riemannian metric on $\mathrm{Herm}^+(E)$ is given by

$$\text{(3.29)} \qquad \operatorname{grad} L = \hat{K} - ch\,.$$

It is therefore natural to consider the evolution equation:

$$\text{(3.30)} \qquad dh_t/dt = -\operatorname{grad} L = -(\hat{K}_t - ch_t)\,,$$

or

$$\text{(3.31)} \qquad h_t^{-1}\partial_t h_t = -(K_t - cI)\,.$$

We calculate now the second variation $\partial_t^2 L(h_t, k)$. From (3.27) we obtain

$$\text{(3.32)} \qquad \partial_t^2 L(h_t, k) = \int_M \operatorname{tr}(\partial_t K \cdot v + (K - cI)\partial_t v)\frac{\Phi^n}{n!}\,.$$

But, according to (IV.2.16), we have

$$\partial_t K^i_j = -\sum g^{\alpha\bar{\beta}} v^i_{j\alpha\bar{\beta}}\,, \qquad (v^i_{j\alpha\bar{\beta}} = \nabla_{\bar{\beta}}\nabla_\alpha v^i_j)\,.$$

Hence, if h_0 is a critical point (i.e., $K - cI = 0$ at $t = 0$), then

$$\begin{aligned} \text{(3.33)} \qquad \partial_t^2 L(h_t, k)_{t=0} &= \int_M (-\sum g^{\alpha\bar{\beta}} v^i_{j\alpha\bar{\beta}} v^j_i)_{t=0}\frac{\Phi^n}{n!} \\ &= \int_M (\sum g^{\alpha\bar{\beta}} v^i_{j\alpha} v^j_{i\bar{\beta}})_{t=0}\frac{\Phi^n}{n!} \\ &= \|D'v\|^2_{t=0} \geqq 0\,. \end{aligned}$$

This shows that every critical point of $L(h, k)$, k fixed, is a local minimum.

Suppose h_0 is a critical point of $L(h, k)$. Let h_1 be any Hermitian structure in E. Join them by a geodesic h_t, $0 \leqq t \leqq 1$. The condition for h_t to be a geodesic is given by (see (2.4))

$$\partial_t v = 0\,, \qquad (\text{i.e., } \partial_t(h_t^{-1}\partial_t h_t) = 0)\,.$$

For such a geodesic h_t, we have

$$\text{(3.34)} \qquad \partial_t^2 L(h_t, k) = \int_M \operatorname{tr}(\partial_t K \cdot v) = \|D'v\|^2 \geqq 0 \qquad \text{for} \quad 0 \leqq t \leqq 1\,.$$

It follows that

$$\text{(3.35)} \qquad L(h_0, k) \leqq L(h_1, k)\,.$$

Assume that h_1 is also a critical point. Then, $L(h_0, k) = L(h_1, k)$ and

$$\partial_t^2 L(h_t, k) = \|D'v\|^2 = 0 \qquad \text{for all} \quad 0 \leqq t \leqq 1\,. \tag{3.36}$$

Since v is Hermitian, $D'v = 0$ implies $Dv = 0$, i.e., in particular, v is a holomorphic endomorphism of E. If the bundle E is simple, then $v = aI$ with $a \in \boldsymbol{R}$ and $h_1 = bh_0$ with $b \in \boldsymbol{R}$.

In summary, we have

(3.37) **Proposition** *For a fixed Hermitian structure k in E, the functional $L(h, k)$ on* Herm$^+(E)$ *possesses the following properties*:

(a) *h is a critical point of $L(h, k)$ if and only if h is Einstein-Hermitian i.e., $K - cI = 0$;*

(b) *If h_0 is a critical point of $L(h, k)$, then $L(h, k)$ attains an absolute minimum at h_0;*

(c) *If h_0 and h_1 are critical points of $L(h, k)$, then h_0 and h_1 define the same Hermitian connection in E;*

(d) *If E is a simple vector bundle and if h_0 and h_1 are critical points of $L(h, k)$, then $h_1 = ah_0$ for some $a \in \boldsymbol{R}$.*

§4 Maximum principle and uniqueness

In this section we shall prove the uniqueness of a solution for the evolution equation (3.30).

(4.1) **Lemma** (*Maximum Principle for Parabolic Equations*). *Let M be a compact Riemannian manifold and f: $M \times [0, a) \to \boldsymbol{R}$ a function of class C^1 with continuous Laplacian Δf satisfying the inequality*

$$\partial_t f + c\Delta f \leqq 0\,, \qquad (c > 0)\,.$$

Set $F(t) = \operatorname{Max}_{x \in M} f(x, t)$. Then $F(t)$ is monotone decreasing in t.

For later applications, we assume in the lemma above that the Laplacian Δf (defined in the distributional sense) is a continuous function instead of assuming that f is of class C^2.

Proof For small $e > 0$, let

$$f_e(x, t) = f(x, t) - et\,.$$

Then

$$\partial_t f_e + c\Delta f_e = \partial_t f - e + c\Delta f \leqq -e < 0 .$$

Let $0 \leqq t_1 < t_2$. Let $(\bar{x}, \bar{t})$ be a point of $M \times [t_1, t_2]$ where f_e achieves its maximum in $M \times [t_1, t_2]$. Then

$$\partial_t f_e(\bar{x}, \bar{t}) < -c\Delta f_e(\bar{x}, \bar{t}) \leqq 0 .$$

This shows that $\bar{t}$ cannot be an interior point of $[t_1, t_2]$. If $\bar{t} = t_2$, then $f_e(\bar{x}, t) \leqq f_e(\bar{x}, \bar{t})$ for $t_1 \leqq t \leqq t_2$, and hence

$$0 \leqq \partial_t f_e(\bar{x}, \bar{t}) < -c\Delta f_e(\bar{x}, \bar{t}) \leqq 0 .$$

This is a contradiction, and we must have $\bar{t} = t_1$. Hence,

$$\operatorname*{Max}_{x \in M} f_e(x, t_1) \geqq \operatorname*{Max}_{x \in M} f_e(x, t_2)$$

and

$$\operatorname*{Max}_{x \in M} f(x, t_1) - \operatorname*{Max}_{x \in M} f(x, t_2) \geqq e(t_1 - t_2) .$$

Let $e \to 0$. Then we obtain

$$F(t_1) \geqq F(t_2) . \qquad \text{Q.E.D.}$$

(4.2) **Corollary** *Let M be a compact Riemannian manifold and $f \colon M \times [0, a) \to \boldsymbol{R}$ a function of class C^1 with continuous Laplacian Δf satisfying the inequality*

$$\partial_t f + c\Delta f + c' f \leqq 0 ,$$

where $c > 0$ and c' are constants. If $f \leqq 0$ on $M \times \{0\}$, then $f \leqq 0$ on all of $M \times [0, a)$.

Proof Set $h = e^{c't} f$. Then

$$\partial_t h + c\Delta h = e^{c't}(\partial_t f + c\Delta f + c' f) .$$

Apply (4.1) to h. Q.E.D.

As an application of the maximum principle, we prove the uniqueness theorem for the evolution equation (3.30).

Given two curves h_t, k_t, $a \leqq t \leqq b$, in $\operatorname{Herm}^+(E)$, we consider a mapping f from the rectangle $\Delta = \{(t, s);\ a \leqq t \leqq b,\ 0 \leqq s \leqq 1\}$ into $\operatorname{Herm}^+(E)$ such that

$$h_t = f(t, 0)\,, \qquad k_t = f(t, 1)\,.$$

We set

$$u = f^{-1}\partial_s f\,, \qquad v = f^{-1}\partial_t f\,, \qquad R = d''(f^{-1}d'f)$$

as in Section 3. For each fixed (t, s), both u and v are fields of endomorphisms of the vector bundle E. Then we have (see (3.16–19))

$$(4.3)\qquad \begin{aligned} &\partial_s R = d''D'u = D''D'u\,, && \partial_t R = d''D'v = D''D'v\,.\\ &\partial_t u = -vu + f^{-1}\partial_t\partial_s f\,, && \partial_s v = -uv + f^{-1}\partial_s\partial_t f\,, \end{aligned}$$

where $D = D' + D''$ denotes the exterior covariant differentiation defined by f. Since $f = f(t, s)$ varies with (t, s), D' varies also with (t, s) while $D'' = d''$ always. We set

$$(4.4)\qquad e = e(t) = \frac{1}{2}\int_0^1 \operatorname{tr}(u^2)ds\,.$$

For each fixed t, $e(t)$ is a function on M, and the value of this function $e(t)$ at $x \in M$ is the energy of the path $f(t, s)$, $0 \leqq s \leqq 1$, from h_t to k_t in $\mathrm{Herm}^+(E_x)$ with respect to the invariant Riemannian metric. In particular, $e(t) \geqq 0$ with equality if and only if $h_t = k_t$.

(4.5) **Lemma** *If h_t and k_t are two solutions of the evolution equation* (3.30) *and if $f(t, s)$, $0 \leqq s \leqq 1$, is a geodesic in* $\mathrm{Herm}^+(E_x)$ *for each t and each x, then the energy function e satisfies*

$$\partial_t e + \Box e \leqq 0\,.$$

Proof Since $f(t, s)$, $0 \leqq s \leqq 1$, is a geodesic, u is independent of s (see (2.4)), i.e.,

$$\partial_s u = 0\,.$$

Making use of

$$\partial_s(f^{-1}\partial_t f) = -f^{-1}\partial_s f \cdot f^{-1}\partial_t f + f^{-1}\partial_s\partial_t f = -uv + f^{-1}\partial_s\partial_t f\,,$$

we obtain

$$\begin{aligned} \partial_t e &= \int_0^1 \operatorname{tr}(u\partial_t u)ds = \int_0^1 \operatorname{tr}(-uvu + uf^{-1}\partial_t\partial_s f)ds = \int_0^1 \operatorname{tr}(u\partial_s(f^{-1}\partial_t f))ds\\ &= \operatorname{tr}(uf^{-1}\partial_t f)\,\big|_{s=0}^{s=1} \qquad (\text{using } \partial_s u = 0)\\ &= \operatorname{tr}(u(k_t^{-1}\partial_t k_t - h_t^{-1}\partial_t h_t)) = \operatorname{tr}(u(K(h) - K(k))) \qquad (\text{using } (3.31))\,, \end{aligned}$$

where $K(h)$ and $K(k)$ denote the mean curvature transformations defined by $h=h_t$ and $k=k_t$, respectively.

On the other hand, we have

$$\begin{aligned} d''d'e=d''\int_0^1 \mathrm{tr}(uD'u)ds &= \int_0^1 \mathrm{tr}(D''u\wedge D'u+uD''D'u)ds \\ &= \int_0^1 \mathrm{tr}(D''u\wedge D'u)ds+\int_0^1 \mathrm{tr}(u\partial_s R)ds \\ &= -\int_0^1 \mathrm{tr}(D'u\wedge D''u)ds+\mathrm{tr}(uR)\Big|_{s=0}^{s=1} \\ &= -\int_0^1 \mathrm{tr}(D'u\wedge D''u)ds+\mathrm{tr}(u(R(k)-R(h)))\,. \end{aligned}$$

Taking the trace of both sides with respect to the Kähler metric g, we obtain

$$\Box e=-\int_0^1 |D'u|^2 ds+\mathrm{tr}(u(K(k)-K(h)))\,.$$

Hence,

$$\partial_t e+\Box e=-\int_0^1 |D'u|^2 ds\leqq 0\,. \qquad \text{Q.E.D.}$$

(4.6) **Corollary** *If h_t and k_t, $0\leqq t<T$ are two smooth solutions of the evolution equation* (3.30) *having the same initial condition $h_0=k_0$, then $h_t=k_t$ for $0\leqq t<T$.*

Proof This is immediate from (4.1) and (4.5). Q.E.D.

(4.7) **Corollary** *If a smooth solution h_t to the evolution equation* (3.30) *exists for $0\leqq t<T$, then h_t converge to a continuous Hermitian structure h_T uniformly as $t\to T$.*

Proof Let $x\in M$. The Riemannian distance $\rho=\rho(h_t(x), h_{t'}(x))$ between $h_t(x)$ and $h_{t'}(x)$ in $\mathrm{Herm}^+(E_x)$ and the energy $e=e(h_t(x), h_{t'}(x))$ of the (unique) geodesic path from $h_t(x)$ and $h_{t'}(x)$ in $\mathrm{Herm}^+(E_x)$ are related by

$$e=\frac{1}{2}\rho^2\,.$$

It suffices therefore to show that, given $\varepsilon>0$ there is δ such that

$$\operatorname*{Max}_{x \in M} e(h_t(x), h_{t'}(x)) < \varepsilon \qquad \text{for} \quad t, t' > T - \delta .$$

(This will imply that h_t is uniformly Cauchy as $t \to T$ with respect to the complete metric ρ and converges to a continuous Hermitian structure h_T). By continuity at $t = 0$, there is $\delta > 0$ such that

$$\operatorname*{Max}_{x \in M} e(h_0(x), h_a(x)) < \varepsilon \qquad \text{for} \quad 0 \leqq a < \delta .$$

Now, if $0 \leqq t < t' < T$ with $t' - t < \delta$, let $a = t' - t$ and consider the shifted curve $k_t = h_{t+a}$ $(= h_{t'})$. Then apply (4.5) to $e = e(h_t, k_t)$. By (4.1), $\operatorname*{Max}_M e(h_t, k_t)$ is monotone decreasing in t. Hence,

$$\operatorname*{Max}_M e(h_t, h_{t'}) = \operatorname*{Max}_M e(h_t, k_t) \leqq \operatorname*{Max}_M e(h_0, k_0) = \operatorname*{Max}_M e(h_0, h_a) . \qquad \text{Q.E.D.}$$

§ 5 Linear parabolic equations

The evolution equation (3.30) is non-linear. We have to study first the linearized equation. In this section we summarize results from the theory of linear parabolic differential equations which will be needed in the next sections 6 and 7.

Let M be an m-dimensional compact Riemannian manifold with metric g. Let E be a real vector bundle over M and $\tilde{E}$ its pull-back to $M \times [0, a]$. Let h be a fibre metric in $\tilde{E}$, i.e., a 1-parameter family of fibre metrics h_t, $0 \leqq t \leqq a$, in E. We consider a 1-parameter family of connections $D = \{D_t\}$ in E preserving h_t. We assume that D_t varies smoothly with t.

For integers $k \geqq 0$ and $1 < p < \infty$, we define the Sobolev space $L^p_k(M \times [0, a], \tilde{E})$ of sections of $\tilde{E}$ in the following manner. We write ∇ for covariant differentiation with respect to D. Given a smooth section f of $\tilde{E}$, we define its L^p_k-norm by

$$\|f\|_{p,k} = \left(\sum_{i+2j \leqq k} \int_{M \times [0,a]} |\nabla^i \partial_t^j f|^p dx dt \right)^{1/p}, \tag{5.1}$$

where dx stands for the volume element of (M, g) and $|\ |$ is the length measured by g and h. We note that the differentiation ∂_t in the time variable t is given twice the weight of the differentiation in the space variables. The completion of the space $C^\infty(M \times [0, a], \tilde{E})$ of C^∞ sections with respect to this norm is denoted by $L^p_k(M \times [0, a], \tilde{E})$. Let $C^\infty(M \times [0, a]/0, \tilde{E})$ denote the space of C^∞ sections f of $\tilde{E}$ such that $\nabla^i \partial_t^j f = 0$ on $M \times \{0\}$ for all $i + 2j \leqq k$. Its completion with respect to the norm $\|\ \|_{p,k}$ will be denoted by $L^p_k(M \times [0, a]/0, \tilde{E})$. When there is no ambiguity about which vector bundle E is being considered, we often write $L^p_k(M \times [0, a])$ and

$L^p_k(M\times[0,a]/0)$ omitting $\tilde{E}$.

For our later purpose, it is necessary to extend the definition of L^p_k to the situation where k is a real number. This can be done by considering fractional poweres of differential operators ∇ and ∂_t in (5.1). In our case it is more convenient to consider the pseudo-differential operator of weighted first order:

$$P=(1+\partial_t^2+\Delta^2)^{1/4}$$

and define

$$\|f\|_{p,k}=\left(\int_{M\times[0,a]}|P^k f|^p dxdt\right)^{1/p}, \tag{5.2}$$

When k is a positive integer, the norm (5.2) is equivalent to the norm (5.1). (We leave details to Hamilton [1] and Nishikawa-Ochiai [1].)

We define next the Hölder space $C^{k+\alpha}(M\times[0,a])=C^{k+\alpha}(M\times[0,a],\tilde{E})$ for an integer $k\geqq 0$ and a real number $0<\alpha\leqq 1$. It consists of sections f of class C^k of $\tilde{E}$ whose k-th derivatives $\nabla^i\partial_t^j f$, $(k=i+2j)$, are Hölder continuous with exponent α so that the norm

$$\|f\|_{k+\alpha}=\sum_{i+2j\leqq k}\sup_{M\times[0,a]}|\nabla^i\partial_t^j f|+\sum_{i+2j=k}\sup_{\substack{z,w\in M\times[0,a]\\ z\neq w}}\frac{|\nabla^i\partial_t^j f(z)-\nabla^i\partial_t^j f(w)|}{d(z,w)^\alpha} \tag{5.3}$$

is finite, where $d(z,w)$ denotes the distance between z and w. In the definition above, we cover M by a finite number of local coordinate neighborhoods and define the norm in terms of coordinates. But it would be too cumbersome to indicate that in (5.3).

All we need to know about Hölder spaces and Sobolev spaces in later sections can be summarized in the following five theorems.

These Sobolev and Hölder spaces are related by the following version of the Sobolev inequality and embedding theorem.

(5.4) **Theorem** *If $k-(m+2)/p\geqq 0$, write $k-(m+2)/p=l+\alpha$, where $l\geqq 0$ is an integer and $0<\alpha\leqq 1$. Then there is a constant c such that*

$$\|f\|_{l+\alpha}\leqq c\|f\|_{p,k} \qquad \textit{for} \quad f\in L^p_k(M\times[0,a]),$$

and we have a continuous inclusion

$$L^p_k(M\times[0,a])\longrightarrow C^{l+\alpha}(M\times[0,a])$$

and a compact inclusion

$$L_k^p(M\times[0,a]) \longrightarrow C^\gamma(M\times[0,a]) \qquad \textit{for} \quad \gamma<l+\alpha\,.$$

The following is due to Rellich and Kondrachov.

(5.5) **Theorem** *For integers $l<k$, the natural inclusion*

$$L_k^p(M\times[0,a]) \longrightarrow L_l^p(M\times[0,a])$$

is compact.

We need the following a priori estimate for the parabolic differential operator $A=\partial_t+\Delta-c$ operating on sections of E, where c is a constant and Δ is the Laplacian defined by $D=\{D_t\}$ and g.

(5.6) **Theorem** *Let $A=\partial_t+\Delta-c$, $1<p<\infty$, $-\infty<l<k$, and $0<t<a$. If $f\in L_l^p(M\times[0,a])$ and $Af\in L_{k-2}^p(M\times[0,a])$, then $f\in L_k^p(M\times[t,a])$ and*

$$\|f\|_{p,k,[t,a]}\leqq C(\|Af\|_{p,k-2}+\|f\|_{p,k})\,,$$

where C is a constant independent of f and $\|\ \|_{p,k,[t,a]}$ denotes the $\|\ \|_{p,k}$-norm on $M\times[t,a]$ while $\|\ \|_{p,k}$ on the right is the norm on $M\times[0,a]$.

We consider now an a priori estimate for a differential operator Q of the following type:

$$Qf=\sum a_{IJ}(x,t,f(x,t))(\nabla^{i_1}\partial_t^{j_1}f)\cdots(\nabla^{i_\nu}\partial_t^{j_\nu}f)\,, \tag{5.7}$$

where $I=(i_1,\cdots,i_\nu)$ and $J=(j_1,\cdots,j_\nu)$, and a_{IJ} are C^∞ functions. We expressed (5.7) symbolically omitting necessary indices for components of f. If $i_\alpha+2j_\alpha\leqq k$ and $\sum_\alpha(i_\alpha+2j_\alpha)\leqq N$, then we say that Q is a partial differential operator of polynomial type (N,k). Then we have

(5.8) **Theorem** *Let Q be a partial differential operator of polynomial type (N,k). Let $r\geqq 0$, $1<p,q<\infty$, $r+k<s$, and $p(r+N)<qs$. If $f\in C^0(M\times[0,a])\cap L_s^q(M\times[0,a])$, then $Qf\in L_r^p(M\times[0,a])$ and*

$$\|Qf\|_{p,r}\leqq C(1+\|f\|_{q,s})^{q/p}\,,$$

where C is a constant independent of f.

Finally we consider the heat equation operator $\partial_t+\Delta_0$, where the Laplacian Δ_0

is independent of t unlike the Laplacian Δ in (5.2) or (5.6).

(5.9) **Theorem** *Let* $1<p<\infty$ *and* $k>1/p$. *Then the heat operator*

$$\partial_t + \Delta_0 : L^p_k(M\times[0,a]/0) \longrightarrow L^p_{k-2}(M\times[0,a]/0)$$

is an isomorphism.

For all these we refer the reader to Hamilton [1] and Friedman [1]. Hamilton treats more generally the case where M has a boundary.

§6 Evolution equations—short time solutions

Let E be a holomorphic vector bundle of rank r over a compact Kähler manifold M with Kähler metric g. We consider the bundle H of Hermitian forms of E; its fibre H_x at $x\in M$ is the space $\mathrm{Herm}(E_x)$ of Hermitian forms on E_x. Thus, it is a real vector bundle of rank r^2 over M, and the space of C^∞ sections of H is nothing but $\mathrm{Herm}(E)$ considered in Section 2.

Let $a>0$. By pulling back H to $M\times[0,a]$ using the projection $M\times[0,a]\to M$, we consider H as a vector bundle over $M\times[0,a]$. Then a section of H over $M\times[0,a]$ may be considered as a curve v_t, $0\leqq t\leqq a$, in $\mathrm{Herm}(E)$.

Let $L^p_k(M\times[0,a],H)$ be the completion of the space $C^\infty(M\times[0,a],H)$ of C^∞ sections by the norm $\|\ \|_{p,k}$ as defined in Section 5. Since the bundle H is fixed throughout the section, we write often $L^p_k(M\times[0,a])$ for $L^p_k(M\times[0,a],H)$.

Let $C^\infty_k(M\times[0,a]/0,H)$ be the space of C^∞ sections of H over $M\times[0,a]$ whose derivatives up to order k vanish at $M\times 0$. Its closure in $L^p_k(M\times[0,a])$ will be denoted by $L^p_k(M\times[0,a]/0)$.

We fix a curve $h=h_t$, $0\leqq t\leqq a$, in $\mathrm{Herm}^+(E)$ and consider a nearby curve $h+v$, where $v=v_t$, $0\leqq t\leqq a$, is a curve in $\mathrm{Herm}(E)$. We may consider both h and v as sections of H over $M\times[0,a]$. Let $R(h+v)$ denote the curvature of the Hermitian structure $h+v$, and $\hat{K}(h+v)$ its mean curvature form. We consider the differential operator P:

$$h+v \longmapsto P(h+v)=\partial_t(h+v)+(\hat{K}(h+v)-c(h+v))\,. \tag{6.1}$$

We linearize P, i.e., find the differential of P at h by calculating

$$\lim_{s\to 0}\frac{P(h+sv)-P(h)}{s}=\partial_t v+\partial_s\hat{K}(h+sv)\,|_{s=0}-cv\,. \tag{6.2}$$

But, from (3.17) we see that

(6.3) $$\partial_s \hat{K}(h+sv)\big|_{s=0} = \square_h v\,,$$

where

$$(\square_h v)^i_j = -\sum g^{\alpha\bar{\beta}} v^i_{j\alpha\bar{\beta}}$$

in the notation of (III.2.43). We shall write simply $\square$ for $\square_h$. From (6.2) and (6.3) we see that the differential dP_h at h is given by

(6.4) $$dP_h(v) = \partial_t v + \square v - cv\,.$$

We consider dP_h as a linear mapping from $L^p_k(M \times [0, a]/0)$ into $L^p_{k-2}(M \times [0, a]/0)$.

(6.5) **Lemma** *Let $p > 2n+2$ and $k \geqq 2$. Then the linear mapping*

$$dP_h \colon L^p_k(M \times [0, a]/0) \longrightarrow L^p_{k-2}(M \times [0, a]/0)$$

is an isomorphism.

Proof Let v be in the kernel of dP_h. By (5.4), v is of class C^1. Let

$$f = \frac{1}{2}\operatorname{tr}(vv) = \frac{1}{2}|v|^2\,.$$

Then

$$\begin{aligned} 0 &= \operatorname{tr}(dP_h(v)v) = \operatorname{tr}((\partial_t v + \square v - cv)v) \\ &= \partial_t f + \square f - 2cf + |D'v|^2\,. \end{aligned}$$

Hence,

$$\partial_t f + \square f - 2cf \leqq 0\,.$$

Since $f \geqq 0$ and $f = 0$ at $t = 0$, the maximum principle (4.2) implies $f = 0$. Hence, $v = 0$, which shows that dP_h is injective.

To complete the proof, it suffices to show that the operator dP_h is a Fredholm operator of index 0. Since the Hermitian structure h varies with $t \in [0, a]$, so do covariant derivatives appearing in the definition of $\square v$. Let D_0 denote the Hermitian connection (and the corresponding covariant differentiation) defined by the Hermitian structure h at a fixed t, say $t = 0$, and let $\square_0$ be the Laplacian defined by D_0. Then the principal part of $\square_0$ coincides with that of $\square$ so that

(6.6) $$\square v = \square_0 v + Q(v)\,,$$

where Q is a linear differential operator of order 1 with smooth coefficients. We can write therefore

$$dP_h(v) = \partial_t v + \Box_0 v + Q_1(h)\,, \tag{6.7}$$

where $Q_1(h) = Q(h) - cv$. Since $Q_1: L^p_k(M \times [0, a]/0) \to L^p_{k-2}(M \times [0, a]/0)$ factors through the natural injection $L^p_{k-1}(M \times [0, a]/0) \to L^p_{k-2}(M \times [0, a]/0)$, Q_1 is a compact operator. On the other hand, the heat equation operator $\partial_t + \Box_0$ is known to be an isomorphism from $L^p_k(M \times [0, a]/0)$ onto $L^p_{k-2}(M \times [0, a]/0)$, (see (5.9)); it is then a Fredholm operator of index 0. Since the operator dP_h can be connected to $\partial_t + \Box_0$ by a 1-parameter family of Fredholm operators $\partial_t + \Box_0 + sQ_1$, $0 \leqq s \leqq 1$, it is also a Fredholm operator of index 0. Since we have already shown that dP_h is injective, it has to be also surjective. Q.E.D.

We shall now prove the existence of a short time solution for the evolution equation (3.30).

(6.8) **Theorem** *Let E be a holomorphic vector bundle of rank r over a compact Kähler manifold M of dimension n. Let $p > 2n + 2$. Given an Hermitian structure h_0 in E, there exist a positive number δ and $v \in L^p_2(M \times [0, \delta]/0)$ such that $h = h_0 + v$ satisfies the evolution equation*

$$P(h) = \partial_t h + K(h) - ch = 0\,.$$

Proof Extend h_0 to $M \times [0, a]$ as a constant curve $k = k_t$ in $\mathrm{Herm}^+(E)$ by setting $k_t = h_0$ for $0 \leqq t \leqq a$. Since $dP_k: L^p_2(M \times [0, a]/0) \to L^2_0(M \times [0, a])$ is an isomorphism by (6.5), the implicit function theorem implies that the mapping $v \longmapsto P(k + v)$ sends a neighborhood V of 0 in $L^p_2(M \times [0, a]/0)$ onto a neighborhood W of $P(k)$ in $L^2_0(M \times [0, a])$. Taking $\delta > 0$ small, we define an element w of $L^p_0(M \times [0, a])$ by setting

$$w = \begin{cases} 0 & \text{on} \quad M \times [0, \delta] \\ P(k) & \text{on} \quad M \times (\delta, a]\,. \end{cases}$$

If δ is sufficiently small, then w belongs to the neighborhood W. Then take $v \in V$ such that $P(k + v) = w$. Q.E.D.

§7 Regularity of solutions

We shall show that the solution obtained in (6.8) is smooth.

(7.1) **Theorem** *With the notation of* (6.8) *and under the assumption* $p > 2n+2$, *let* $v \in L_2^p(M \times [0, a]/0)$ *be a solution of*

$$P(h_0 + v) = 0 .$$

Then v *is of class* C^∞ *on* $M \times (0, a)$.

Proof We compare $P(h) = P(h_0 + v)$ with $dP_{h_0}(v)$. Since

$$P(h) = P(h_0 + v) = \partial_t v + \hat{K}(h_0 + v) - c(h_0 + v)$$

and

$$dP_{h_0}(v) = \partial_t v + \Box_0 v - cv ,$$

we have

$$P(h_0 + v) - dP_{h_0}(v) = \hat{K}(h_0 + v) - \Box_0 v - ch_0 .$$

Writing $\hat{K}(h)$ and $\Box_0 v$ explicitly in terms of coordinates, i.e.,

$$K_{i\bar{j}}(h) = -\sum g^{\alpha\bar{\beta}} \left(\frac{\partial^2 h_{i\bar{j}}}{\partial z^\alpha \partial \bar{z}^\beta} - \sum h^{p\bar{q}} \frac{\partial h_{i\bar{q}}}{\partial z^\alpha} \frac{\partial h_{p\bar{j}}}{\partial \bar{z}^\beta} \right)$$

$$\Box_0 v_{i\bar{j}} = -\sum g^{\alpha\bar{\beta}} \frac{\partial^2 v_{i\bar{j}}}{\partial z^\alpha \partial \bar{z}^\beta} + \cdots ,$$

we see that $P(h_0 + v) - dP_{h_0}(v)$ is a polynomial of degree $\leqq 2$ in the first partial derivatives $\partial v_{i\bar{j}} / \partial z^\alpha$, $\partial v_{i\bar{j}} / \partial \bar{z}^\beta$ of v whose coefficients are smooth functions of z^α, $\bar{z}^\beta$ and $v_{i\bar{j}}$. (After cancelation, it does not involve second partial derivatives of v.)

Set

$$F(v) = P(h_0 + v) - dP_{h_0}(v) . .$$

In the terminology of Section 5, $F(v)$ is a polynomial partial differential operator of type (2.1).

We write $v \in L_s^*$ if $v \in L_s^q(M \times [t_0, a])$ for an arbitrarily small $t_0 > 0$ and an arbitrary q, $1 < q < \infty$. We shall show that if v satisfies the assumption of the theorem, then the following three hold:

(A) $v \in L_2^*$,

(B) If $F(v) \in L_s^*$, then $v \in L_{s+2}^*$,

(C) If $v \in L_s^*$ and $s \geqq 2$, then $F(v) \in L_r^*$ for $r + 1 < s$.

Proof of (A) By assumption, $P(h_0 + v) = 0$ and $v \in L_2^p(M \times [0, a])$, where $p >$

$2n+2$. By (5.4), $v \in C^1(M \times [0, a])$. (We note that since our differentiability class is weighted (see Section 5), this does not mean that v is of class C^1 in the variable t in the usual sense.) Hence, from definition of $F(v)$ it follows that $F(v) \in C^0(M \times [0, a])$. Since $P(h_0 + v) = 0$ by assumption,

$$dP_{h_0}(v) = -F(v) \in C^0(M \times [0, a]) .$$

Hence,

$$dP_{h_0}(v) \in L_0^q(M \times [0, a]) \qquad \text{for all} \quad q , \quad 1 < q < \infty .$$

By (5.6),

$$v \in L_2^q(M \times [t_0, a]) \qquad \text{for all} \quad q , \quad 1 < q < \infty \quad \text{and} \quad t_0 > 0 .$$

Proof of (B) Assume that $F(v) \in L_s^q(M \times [t_0, a])$. Since $dP_{h_0}(v) = -F(v)$, we have $dP_{h_0}(v) \in L_s^q(M \times [t_0, a])$. By (5.6) again,

$$v \in L_{s+2}^q(M \times [t_1, a]) \qquad \text{for} \quad t_1 > t_0 .$$

Proof of (C) Since $F(v)$ is a polynomial of degree 2 in partial derivatives $\partial v_{i\bar{j}}/\partial z^\alpha$, $\partial v_{i\bar{j}}/\partial \bar{z}^\beta$, it follows from (5.8) that if $v \in L_s^q$ (and hence $v \in L_s^* \cap C^1(M \times [t_0, a])$, then $F(v) \in L_r^*$ for $r+1 < s$.

Now, from (A), (B) and (C) we conclude that $v \in L_s^*$ for all s. By (5.4), v is of class C^∞ on $M \times [t_0, a]$ for all $t_0 > 0$. Q.E.D.

§8 Solutions for all time

Suppose $h = h_t$, $0 \leqq t \leqq a$, is a family of Hermitian structures in E satisfying the evolution equation

$$\partial_t h = -(\hat{K}(h) - ch) . \tag{8.1}$$

Then the corresponding family of connections $\omega = h^{-1} d'h$ satisfies the following equation:

$$\begin{aligned}
\partial_t \omega &= -h^{-1} \partial_t h \cdot h^{-1} d'h + h^{-1} d' \partial_t h \\
&= h^{-1}(\hat{K} - ch) h^{-1} d'h - h^{-1} d'(\hat{K} - ch) \\
&= h^{-1}(\hat{K} - ch) h^{-1} d'h - d'(h^{-1}(\hat{K} - ch)) - h^{-1} d'h \cdot h^{-1}(\hat{K} - ch) \\
&= (K - cI)\omega - d'(K - cI) - \omega(K - cI) \\
&= -D'(K - cI) = -D'K .
\end{aligned}$$

Hence,

$$\partial_t \omega = -D'K. \tag{8.2}$$

Making use of the formula (2.39) of Chapter III (i.e., $\Lambda D' - D'\Lambda = \sqrt{-1}\delta''_h$), the Bianchi identity $DR=0$ and the formula $K=\sqrt{-1}\Lambda R$, we obtain

$$D'K = \delta''_h R. \tag{8.3}$$

Remark In terms of local coordinates, (8.3) may be obtained also as follows:

$$(D'K)^i_j = \sum K^i_{j,\gamma} dz^\gamma = \sum g^{\alpha\bar{\beta}} R^i_{j\alpha\bar{\beta},\gamma} dz^\gamma = \sum g^{\alpha\bar{\beta}} R^i_{j\gamma\bar{\beta},\alpha} dz^\gamma = (\delta''_h R)^i_j.$$

Using (8.3) we may rewrite (8.2) as follows:

$$\partial_t \omega = -\delta''_h R. \tag{8.4}$$

From (8.4) we obtain

$$\partial_t R = \partial_t d''\omega = d''\partial_t \omega = D''\partial_t \omega = -D''\delta''_h R = -\Box_h R. \tag{8.5}$$

We define functions $f=f_0, f_1, f_2, \cdots$ on $M \times [0, a]$ as follows.

$$f = |R|^2, \qquad f_k = |\nabla^k R|^2, \qquad k \geqq 0. \tag{8.6}$$

(8.7) **Lemma** *Suppose that* $h=h_t$, $0 \leqq t \leqq a$, *satisfies* (8.1). *Then*

(i) $(\partial_t + \Box)\,\mathrm{tr}(R) = 0$,

(ii) $(\partial_t + \Box) f \leqq c(f^{3/2} + f)$,

(iii) $(\partial_t + \Box)\|K\|^2 \leqq 0$,

(iv) $(\partial_t + \Box) f_k \leqq c_k f_k^{1/2} \left(\sum_{i+j=k} f_i^{1/2}(f_j^{1/2} + 1) \right)$,

where the constants c, c_k *depend only on the Kähler metric of* M.

Proof (i) Taking the trace of (8.5), we obtain (i).

(ii) In order to prove (ii), we use the following general formula for a (1, 1)-form F with values in the endomorphism bundle $\mathrm{End}(E)$. In terms of local coordinates, let

$$F^i_j = \sum F^i_{j\alpha\bar{\beta}} dz^\alpha \wedge d\bar{z}^\beta.$$

Then

$$(\delta_h'' F)^i_j = \sum g^{\gamma\bar\beta} F^i_{j\alpha\bar\beta,\gamma} dz^\alpha\,, \qquad (d''\delta_h'' F)^i_j = -\sum g^{\gamma\bar\beta} F^i_{j\alpha\bar\beta,\gamma,\bar\delta} dz^\alpha \wedge d\bar z^\delta$$

$$(d'' F)^i_j = -\sum (F^i_{j\alpha\bar\beta,\bar\delta} - F^i_{j\alpha\bar\delta,\bar\beta}) dz^\alpha \wedge d\bar z^\beta \wedge d\bar z^\delta\,,$$

$$(\delta_h'' d'' F)^i_j = -\sum g^{\gamma\bar\beta} (F^i_{j\alpha\bar\delta,\bar\beta,\gamma} - F^i_{j\alpha\bar\beta,\bar\delta,\gamma}) dz^\alpha \wedge d\bar z^\delta\,.$$

Hence, we obtain a formula for $\square_h F = (\delta_h'' d'' + d''\delta_h'')F$:

$$(8.8) \qquad (\square_h F)^i_j = -\sum g^{\gamma\bar\beta} F^i_{j\alpha\bar\delta,\bar\beta,\gamma} + \sum g^{\gamma\bar\beta} (F^i_{j\alpha\bar\beta,\bar\delta,\gamma} - F^i_{j\alpha\bar\beta,\gamma,\bar\delta})\,.$$

By the Ricci identity, the last term can be expressed in terms of F and the curvatures of (E,h) and (M, g). Explicitly,

$$F^i_{j\alpha\bar\beta,\bar\delta,\gamma} - F^i_{j\alpha\bar\beta,\gamma,\bar\delta} = \sum (R^i_{k\bar\delta\gamma} F^k_{j\alpha\bar\beta} - F^i_{k\alpha\bar\beta} R^k_{j\bar\delta\gamma}) - \sum (F^i_{j\epsilon\bar\beta} S^\epsilon_{\alpha\bar\delta\gamma} + F^i_{j\alpha\bar\epsilon} S^{\bar\epsilon}_{\bar\beta\bar\delta\gamma})\,,$$

where S denotes the curvature of (M, g) while R denotes the curvature of (E, h). But we do not need such a precise formula. It suffices to rewrite (8.8) as follows:

$$(8.9) \qquad \square_h F = \nabla''^* \nabla'' F + \{F, R\} + \{F, S\}\,,$$

where $\{\ ,\ \}$ denotes a certain bilinear expression.

We express $f = |R|^2$ as follows without using h.

$$(8.10) \qquad f = \sum g^{\alpha\bar\delta} g^{\gamma\bar\beta} R^i_{j\alpha\bar\beta} R^j_{i\gamma\bar\delta}\,.$$

Since g does not depend on t (although h does), we obtain from (8.10), (8.5) and (8.9) the following formula:

$$\begin{aligned}\partial_t f &= 2\langle \partial_t R, R\rangle = -2\langle \square_h R, R\rangle \\ &= -2\langle \nabla''^* \nabla'' R, R\rangle + \{R, R, R\} + \{R, R, S\}\,,\end{aligned}$$

where $\{\ ,\ ,\ \}$ denotes a certain trilinear expression. From (8.10) we obtain also

$$\square f = 2\langle \nabla''^* \nabla'' R, R\rangle - 2|\nabla'' R|^2\,.$$

Hence,

$$\begin{aligned}(\partial_t + \square) f &= \{R, R, R\} + \{R, R, S\} - 2|\nabla'' R|^2 \\ &\leqq \{R, R, R\} + \{R, R, S\} \leqq c(f^{3/2} + f)\,.\end{aligned}$$

(iii) From (8.5) we obtain

$$\partial_t K = -\square_h K\,.$$

Since K is a 0-form (with values in End(E)), we have

$$\square_h K = \delta_h'' d'' K = \nabla''^* \nabla'' K\,.$$

As in the proof of (ii), we express $|K|^2$ without using h:

$$|K|^2 = \sum K^i_j K^j_i .$$

Then

$$\partial_t |K|^2 = 2\langle \partial_t K, K \rangle = -2\langle \Box_h K, K \rangle ,$$
$$\Box |K|^2 = 2\langle \nabla''^* \nabla'' K, K \rangle - 2|\nabla'' K|^2 .$$

Hence,

$$(\partial_t + \Box)|K|^2 = -2|\nabla'' K|^2 \leqq 0 .$$

(iv) Since ∇ depends on t, it does not commute with ∂_t. Taking a local frame field $s = (s_1, \cdots, s_r)$ of E, we obtain

$$(\partial_t \circ \nabla - \nabla \circ \partial_t) s_i = \sum (\partial_t \omega^j_i) s_j .$$

This shows that $\partial_t \circ \nabla - \nabla \circ \partial_t$ is an algebraic derivation given by $\partial_t \omega$. By induction on k, we obtain

$$\text{(8.11)} \qquad \partial_t \circ \nabla^k = \nabla^k \circ \partial_t + \sum_{i+j=k-1} \nabla^i \circ \partial_t \omega \circ \nabla^j .$$

Applying (8.11) to R and making use of (8.4), (8.5) and (8.9), we obtain

$$\text{(8.12)} \qquad \partial_t \circ \nabla^k R = -\nabla^k(\nabla''^* \nabla'' R + \{R, R\} + \{R, S\}) + \sum_{i+j=k-1} \nabla^i(\{\nabla R, \nabla^j R\}) .$$

Similarly, by induction on k, for any section F of the bundle $\mathrm{End}(E) \otimes (T^*M)^{\otimes p}$, we can easily prove the following formula:

$$\text{(8.13)} \qquad \nabla^k \nabla''^* \nabla'' F = \nabla''^* \nabla'' \nabla^k F + \sum_{i+j=k} (\{\nabla^i F, \nabla^j R\} + \{\nabla^i F, \nabla^j S\}) .$$

Set $F = R$ in (8.13). This together with (8.12) implies

$$\text{(8.14)} \qquad (\partial_t + \nabla''^* \nabla'')(\nabla^k R) = \sum_{i+j=k} (\{\nabla^i R, \nabla^j R\} + \{\nabla^i R, \nabla^j S\}) .$$

Substituting (8.14) in

$$(\partial_t + \Box) f_k = 2\langle (\partial_t + \nabla''^* \nabla'')(\nabla^k R), \nabla^k R \rangle - 2|\nabla'' \nabla^k R|^2 ,$$

we obtain (iv). Q.E.D.

(8.15) **Lemma** *Suppose that* $h = h_t$, $0 \leqq t < a$, *is a smooth solution to the evolution equation* (8.1). *Then*

(i) *Both* $\sup_M |\mathrm{Tr}(R)|$ *and* $\sup_M |K|$ *are decreasing functions of* t *and hence are bounded for* $0 \leqq t < a$ *(even in the case* $a = \infty$*);*

(ii) *If the curvature* R *itself is bounded for* $0 \leqq t < a$*, i.e.,* $|R| \leqq B$ *on* $M \times [0, a)$*, then all its covariant derivatives are bounded, i.e.,* $|\nabla^k R| \leqq B_k$ *on* $M \times [0, a)$.

Proof (i) This follows from the maximum principle (4.1) and (i) and (iii) of (8.7).

(ii) The proof is by induction on k. For $k=1$, this is nothing but the hypothesis. Suppose $|\nabla^j R|$ are bounded for all $j<k$. By (iv) of (8.7), we obtain

$$(\partial_t + \Box) f_k \leqq A(1 + f_k) .$$

The equation

$$(\partial_t + \Box) u = (\partial_t + \Box)(1 + u) = A(1 + u) , \qquad u(0) = f_k(0) ,$$

which is linear in $1+u$, has a (unique) smooth solution u defined for all $0 < t < \infty$. By simple calculation we see that

$$(\partial_t + \Box)((f_k - u)e^{-At}) = e^{-At}((\partial_t + \Box) f_k - A(f_k + 1)) \leqq 0 .$$

Again, by the maximum principle, we conclude $f_k \leqq u$. Q.E.D.

(8.16) **Lemma** *Suppose that* $h = h_t$ *is a smooth solution to the evolution equation* (8.1) *for* $0 \leqq t < a$. *If the curvatures* R_t *of* h_t *are uniformly bounded in* $L^q(M)$ *for some* $q > 3n$, *then* R_t *are uniformly bounded (in* $L^\infty(M)$*).*

Proof We make use of the heat kernel $H(x, y, t)$ for the scalar equation $\partial_t + \Box$ on M and its asymptotic expansion

$$\text{(8.17)} \qquad H \underset{t \to 0^+}{\sim} (4\pi t)^{-n} e^{-r^2/4t}(u_0 + u_1 t + \cdots) ,$$

where u_j are smooth functions on $M \times M$ and $r = d(x, y)$ is the Riemannian distance between x and y. Using a change of variables

$$v^2 = pr^2/4t \qquad (\text{so that } r^{2n-1} dr = ct^n v^{2n-1} dv) ,$$

we obtain for $t < a$ and for each fixed $x \in M$ the following bound:

$$\| H(x, \cdot, t) \|_{L^p} \leqq \text{const.}\, t^{n(1-p)/p} .$$

Hence, for any $p < n/(n-1)$ we have

$$\int_0^a \|H(x,\cdot,t)\|_{L^p}dt \leqq c_p(a)\,. \tag{8.18}$$

Now we make use of the following formula:

$$u(x,t)=\int_{t_0}^t d\tau\int_M H(x,y,t-\tau)(\partial_\tau+\Box)u(y,\tau)dy + \int_M H(x,y,t-t_0)u(y,t_0)dy\,. \tag{8.19}$$

This is a direct consequence of the following "Green's formula" for a pair of functions $u(y,\tau)$ and $v(y,\tau)$:

$$\int_{t_0}^t d\tau\int_M \{v(\partial_\tau u+\Box u)+u(\partial_\tau v-\Box v)\}dy=\int_{t_0}^t d\tau\int_M \partial_\tau(uv)dy\,.$$

Since $H>0$, applying (8.19) to $u=f=|R|^2$ and making use of (ii) of (8.7), we obtain

$$f(x,t)\leqq c\int_0^t d\tau\int_M H(x,y,t-\tau)(f(y,\tau)^{3/2}+f(y,\tau))dy + \int_M H(x,y,t)f(y,0)dy\,. \tag{8.20}$$

The last term $\psi(x,t)=\int H(x,y,t)f(y,0)dy$ is a solution of the heat equation $(\partial_t+\Box)\psi=0$ satisfying the initial condition $\psi(x,0)=f(x,0)$. Hence,

$$\psi(x,t)\leqq \sup_M f(y,0)\,.$$

The first term on the right of (8.20) is bounded by

$$c\int_0^t \|H(x,\cdot,t-\tau\|_{L^p}\cdot\|f(\cdot,\tau)^{3/2}+f(\cdot,\tau)\|_{L^{p'}}d\tau \tag{8.21}$$

for conjugate indices p,p', $(1/p+1/p'=1)$. If $|R|$ is uniformly bounded in L^q for some $q>3n$ so that $f^{3/2}=|R|^3$ is uniformly bounded in $L^{p'}$ for some $p'>n$, then (8.18) implies that (8.21) is bounded for $t<a$. Q.E.D.

(8.22) **Lemma** *Let $h=h_t$, $0\leqq t<a$, be a one-parameter family of Hermitian structures in a holomorphic vector bundle E such that*

(i) *h_t converges in the C^0-topology to some continuous Hermitian structure h_a as*

$t \to a$,

(ii) $\sup_M |K(h_t)|$ *is uniformly bounded for* $t<a$.

Then h_t *is bounded in* C^1 *as well as in* L_2^p *and the curvature* $R(h_t)$ *is bounded in* L^p, $(p<\infty)$, *independent of* $t<a$.

Proof There is no natural C^1 norm on the space Herm(E). We can choose one Hermitian structure and use it to define a C^1 norm. Here we use local holomorphic trivializations of E and partial derivatives of the resulting matrix representations of h_t to define a C^1 norm.

Suppose that h_t is not bounded in C^1. Then there exists a sequence $t_i \to a$ such that, with $h_i = h_{t_i}$,

$$m_i = \sup_M |\partial h_i| \longrightarrow \infty .$$

Let $x_i \in M$ be a point where $|\partial h_i|$ attains its maximum value m_i. Taking a subsequence we may assume that x_i converges to a point x_0 in M. We shall now work within a coordinate neighborhood around x_0 with local coordinate system $z^1, \cdots, z^n$.

We fix a polydisc $|w^\alpha|<1$, $\alpha=1, \cdots, n$, in $\boldsymbol{C}^n$ and construct, for each i, a matrix valued function $\tilde{h}_i$ defined on this polydisc in the following manner. First, translating the coordinates z^α slightly we may assume that $|\partial h_i|$ attains its maximum value m_i at the origin $z=0$. Using a map

$$p_i : \{|w^\alpha|<1\} \longrightarrow \{|z^\alpha|<1/m_i\}$$

given by $w^\alpha = m_i z^\alpha$, we define $\tilde{h}_i$ to be the pull-back $p_i^*(h_i)$. Partial derivatives $\partial \tilde{h}_i$ are calculated in terms of the coordinates w^α. Hence,

$$\sup_w |\partial \tilde{h}_i| = 1 \qquad \text{is attained at} \quad w=0 .$$

We shall show that this will lead to a contradiction. We use the following formula expressing $\hat{K}$ in terms of h:

$$K_{r\bar{s}} = \Delta h_{r\bar{s}} - \sum g^{\alpha\bar{\beta}} h^{p\bar{q}} \partial_\alpha h_{r\bar{q}} \cdot \partial_{\bar{\beta}} h_{p\bar{s}} , \tag{8.23}$$

where $\partial_\alpha = \partial/\partial z^\alpha$, $\partial_{\bar{\beta}} = \partial/\partial \bar{z}^\beta$ and $\Delta = -\sum g^{\alpha\bar{\beta}} \partial_\alpha \partial_{\bar{\beta}}$. We apply (8.23) to each $\tilde{h}_i$. By condition (i), $\tilde{h}_i$ and $\tilde{h}_i^{-1}$ are bounded independent of i since $\tilde{h}_i = \tilde{h}_{t_i}$ approaches $h_a(0)$ as $i \to \infty$. Since $\tilde{h}_i$, $\tilde{h}_i^{-1}$ and $\partial \tilde{h}_i$ are bounded, condition (ii) and (8.23) imply that $\Delta \tilde{h}_i$ are also bounded independent of i in $|w^\alpha|<1$.

Now, we claim that $\tilde{h}_i$ are bounded, independent of i, in L_2^p, $1<p<\infty$, over a slightly smaller polydisc $|w^\alpha|<1-\delta$. To see this, we make use of the following L^p estimate for Δ:

$$\|f\|_{p,2} \leqq c_1 \|\Delta f\|_{p,0} + c_2 \|f\|_{1,0} \tag{8.24}$$

for all $f \in L^p_2$ vanishing near the boundary of the polydisc $|w^\alpha|<1$. We apply this estimate to $\rho \tilde{h}_i$, where ρ is a smooth function which is 1 in the smaller polydisc $|w^\alpha|<1-\delta$ and which decreases to 0 near the boundary of the polydisc $|w^\alpha|<1$.

Take $p>2n+2$. By (5.4), the mapping $L^p_2 \to C^1$ is compact. Since $\tilde{h}_i$ are bounded in L^p_2, a subsequence of the $\tilde{h}_i$ converges in C^1 to $\tilde{h}_\infty$ say. From condition (i) and the construction of $\tilde{h}_i$, it follows that $\tilde{h}_\infty$ has constant components $h_a(0)$ so that $\partial \tilde{h}_\infty = 0$. On the other hand, we have

$$|\partial \tilde{h}_\infty|_{w=0} = \lim |\partial \tilde{h}_i|_{w=0} = 1 \ .$$

This is a contradiction. Hence, h_t must be bounded in C^1.

Since h_t is bounded in C^1 and since $K(h_t)$ is uniformly bounded, the L^p estimate (8.24) implies that h_t is bounded in L^p_2. Hence the curvature $R(h_t)$ is bounded in L^p. Q.E.D.

(8.25) **Theorem** *Let E be a holomorphic vector bundle over a compact Kähler manifold M. Given any initial Hermitian structure h_0 in E, the evolution equation*

$$\partial_t h = -(\hat{K}(h) - ch)$$

has a unique smooth solution defined for all time $0 \leqq t < \infty$.

Proof We already know that a solution is unique ((4.6)) and smooth ((7.1)). We know also that a solution exists for a short time, (see (6.8)). Suppose that a solution exists only for a finite time $0 \leqq t < a$. By (4.7), h_t converges to a continuous Hermitian structure h_a uniformly as $t \to a$, thus satisfying condition (i) of (8.22). By (i) of (8.15), $\sup_M |K(h_t)|$ is bounded for $0 \leqq t < a$ so that condition (ii) of (8.22) is also satisfied. By (8.22) the curvature $R(h_t)$ is bounded in L^p for any $p<\infty$. Hence, by (8.16) it is uniformly bounded. By (ii) of (8.15) every covariant derivative of the curvature $R(h_t)$ is uniformly bounded for $0 \leqq t < a$.

Now we can show, using (8.23), that h_t is bounded in C^k for all k. We already know this for $k=1$, (see (8.22)). Assume that h_t is bounded in C^{k-1}. Then all first order partial derivatives h_t are bounded in C^{k-2}. We have shown above that $\hat{K}(h_t)$ is bounded in C^l for all l, in particular, for $l=k-2$. From (8.23) and the elliptic regularity, it follows that h_t is now bounded in C^k.

Thus, h_t, which converges to h_a in C^0 (by (8.22)), converges in C^∞ as $t \to a$. By the short time existence starting with h_a, we can extend the solution to $[0, a+\delta]$. Q.E.D.

§9 Properties of solutions of the evolution equation

We shall now study how the curvature of a solution of the evolution equation behaves as the time t goes to infinity.

(9.1) **Proposition** *Let $h = h_t$, $0 \leqq t < \infty$, be a 1-parameter family of Hermitian structures in E satisfying the evolution equation of* (8.25). *Then*

(i) *For any fixed Hermitian structure k of E, the Lagrangian $L(h_t, k)$ of* (3.5) *is a monotone decreasing function of t; in fact,*

$$\frac{d}{dt} L(h_t, k) = -\|K(h_t) - cI\|^2 \leqq 0\,;$$

(ii) $\operatorname*{Max}_M |K(h_t) - cI|^2$ *is a monotone decreasing function of t;*

(iii) *If $L(h_t, k)$ is bounded below, i.e., $L(h_t, k) \geqq A > -\infty$ for $0 \leqq t < \infty$, then* $\operatorname*{Max}_M |K(h_t) - cI|^2 \to 0$ *as $t \to \infty$.*

Proof (i) From (3.26) and the evolution equation (3.30), we obtain

$$\frac{d}{dt} L(h_t, k) = (\hat{K}(h_t) - ch_t, \partial_t h) = -\|\hat{K}(h_t) - ch_t\|^2\,.$$

(ii) Multiplying the evolution equation (3.30) by h_t^{-1} we rewrite it as follows:

$$v_t = -(K_t - cI)\,, \qquad \text{where} \quad v_t = h_t^{-1}\partial_t h_t \quad \text{and} \quad K_t = K(h_t)\,.$$

Then

$$D''D'v_t = -D''D'K_t\,.$$

On the other hand, by (3.17)

$$D''D'v_t = \partial_t R_t\,, \qquad \text{where} \quad R_t = R(h_t)\,.$$

Hence,

$$\partial_t R_t = D'D''K_t \qquad (\text{or } \partial_t R^i_{tj\alpha\bar{\beta}} = \nabla_\alpha \nabla_{\bar{\beta}} K^i_{tj})\,.$$

Contracting both sides with respect to the Kähler metric g, we obtain

$$\partial_t K_t = -\square_h K_t \qquad (\text{or } \partial_t K^i_{tj} = \sum g^{\alpha\bar{\beta}} \nabla_\alpha \nabla_{\bar{\beta}} K^i_{tj}) .$$

In order to calculate $\square |K_t - cI|^2$, we start with

$$D'D'' \operatorname{tr}((K_t - cI)(K_t - cI)) = 2 \operatorname{tr}((K_t - cI)(D'D''K_t)) + 2 \operatorname{tr}(D'K_t \cdot D''K_t) .$$

Taking the trace with respect to g (i.e., contracting with $g^{\alpha\bar{\beta}}$), we obtain

$$\begin{aligned} \square |K_t - cI|^2 &= 2 \operatorname{tr}((K_t - cI)(\square_h K_t)) - 2 |D'K_t|^2 \\ &= -2 \operatorname{tr}((K_t - cI)(\partial_t K_t)) - 2 |D'K_t|^2 \\ &= -2\partial_t(\operatorname{tr}((K_t - cI)(K_t - cI))) - 2 |D'K_t|^2 . \end{aligned}$$

Hence

$$(\partial_t + \square) |K_t - cI|^2 = -2 |D'K_t|^2 \leqq 0 . \tag{9.2}$$

Now (ii) follows from the maximum principle (4.1).

(iii) Integrating the equality in (i) from 0 to s, we obtain

$$L(h_s, k) - L(h_0, k) = -\int_0^s \|K_t - cI\|^2 dt .$$

Since $L(h_s, k)$ is bounded below by a constant independent of s, we have

$$\int_0^\infty \|K_t - cI\|^2 dt < \infty .$$

In particular,

$$\|K_t - cI\| \longrightarrow 0 \quad \text{as} \quad t \longrightarrow \infty . \tag{9.3}$$

Let $H(x, y, t)$ be the heat kernel for $\partial_t + \square$. Set

$$f(x, t) = (|K_t - cI|^2)(x) \qquad \text{for} \quad (x, t) \in M \times [0, \infty) .$$

Fix $t_0 \in [0, \infty)$ and set

$$u(x, t) = \int_M H(x, y, t - t_0) f(y, t_0) dy , \qquad \text{where} \quad dy = \frac{1}{n!} \Phi^n .$$

Then $u(x, t)$ is of class C^∞ on $M \times (t_0, \infty)$ and extends to a continuous function on $M \times [t_0, \infty)$. It satisfies

$$(\partial_t + \square) u(x, t) = 0 \qquad \text{for} \quad (x, t) \in M \times (t_0, \infty) ,$$

$$u(x, t_0) = f(x, t_0) \qquad \text{for} \quad x \in M .$$

Combined with (9.2), this yields

$$(\partial_t + \Box)(f(x, t) - u(x, t)) \leqq 0 \qquad \text{for} \quad (x, t) \in M \times (t_0, \infty).$$

By the maximum principle (4.1),

$$\operatorname*{Max}_{x \in M} (f(x, t) - u(x, t)) \leqq \operatorname*{Max}_{x \in M} (f(x, t_0) - u(x, t_0)) = 0 \qquad \text{for} \quad t \geqq t_0.$$

Hence,

$$\begin{aligned}\operatorname*{Max}_{x \in M} f(x, t_0 + a) &\leqq \operatorname*{Max}_{x \in M} u(x, t_0 + a) \\ &= \operatorname*{Max}_{x \in M} \int_M H(x, y, a) f(y, t_0) dy \\ &\leqq C_a \int_M f(y, t_0) dy \\ &= C_a \| K_{t_0} - cI \|^2,\end{aligned}$$

where $C_a = \operatorname*{Max}_{M \times M} H(x, y, a)$. Fix a, say $a = 1$, and let $t_0 \to \infty$. Using (9.3) we conclude

$$\operatorname*{Max}_{x \in M} f(x, t) \longrightarrow 0 \qquad \text{as} \quad t \longrightarrow \infty. \qquad \text{Q.E.D.}$$

§ 10 Semistable bundles and the Einstein condition

We begin with a proposition which will allow us to reduce problems in certain cases to those of vector bundles with lower ranks.

Let

$$(10.1) \qquad 0 \longrightarrow E' \longrightarrow E \longrightarrow E'' \longrightarrow 0$$

be an exact sequence of holomorphic vector bundles over a compact Kähler manifold M. As we have shown in Section 6 of Chapter I, an Hermitian structure h in E induces Hermitian structures h', h'' in E', E'' and the second fundamental form $A \in A^{1,0}(\mathrm{Hom}(E', E''))$ and its adjoint $B \in A^{0,1}(\mathrm{Hom}(E'', E'))$. To indicate the dependence of A, B on h, we write A_h, B_h. Since B is the adjoint of A, we can write $B_h^* = A_h$.

(10.2) **Proposition** *Given an exact sequence* (10.1) *and a pair of Hermitian structures h, k in E, the function Q_1 and the form Q_2 defined by* (3.3) *and* (3.4) *satisfy the following relations*:

(i) $Q_1(h, k) = Q_1(h', k') + Q_1(h'', k'')$,

(ii) $Q_2(h, k) \equiv Q_2(h', k') + Q_2(h'', k'') - i(\mathrm{tr}(B_h \wedge B_h^*) - \mathrm{tr}(B_k \wedge B_k^*))$

modulo $d'A^{0,1} + d''A^{1,0}$.

Proof (i) This is immediate from the definition of Q_1.

(ii) Given an Hermitian structure h, we have the following splitting of the exact sequence (10.1):

$$0 \longrightarrow E' \underset{\alpha}{\overset{\mu}{\rightleftarrows}} E \underset{\beta}{\overset{\lambda}{\leftrightarrows}} E'' \longrightarrow 0, \tag{10.3}$$

where λ and μ are C^∞ homomorphisms determined by the conditions that $\mathrm{Image}(\lambda) = \mathrm{Kernel}(\mu)$ is the orthogonal complement to $\alpha(E')$ and

$$\alpha \circ \mu + \lambda \circ \beta = I_E .$$

Then

$$B = \mu \circ d'' \circ \lambda .$$

Following the definition of Q_2 in Section 3, we consider a 1-parameter family of Hermitian structures $h = h_t$, $0 \leqq t \leqq 1$, such that $k = h_0$ and $h = h_1$. Corresponding to $h = h_t$, we have 1-parameter families of homomorphisms $\lambda = \lambda_t$ and $\mu = \mu_t$. We define a 1-parameter family of homomorphisms $S = S_t$: $E'' \to E'$ by

$$\lambda_t - \lambda_0 = \alpha \circ S_t , \qquad \text{or equivalently,} \quad \mu_t - \mu_0 = -S_t \circ \beta .$$

Since α and β are holomorphic and independent of t, we have

$$\begin{aligned}\partial_t B &= \partial_t \mu \circ d'' \circ \lambda + \mu \circ d'' \circ \partial_t \lambda = -\partial_t (S \circ \beta) \circ d'' \circ \lambda + \mu \circ d'' \partial_t (\alpha \circ S) \\ &= -\partial_t S \circ \beta \circ d'' \circ \lambda + \mu \circ \alpha \circ d'' \circ \partial_t S = -\partial_t S \circ d'' \circ (\beta \circ \lambda) + d'' \circ \partial_t S \\ &= -\partial_t S \circ d'' + d'' \circ \partial_t S = d''(\partial_t S) \in A^{0,1}(\mathrm{Hom}(E'', E')) .\end{aligned}$$

Let $e_1, \cdots, e_p$ (resp. $e_{p+1}, \cdots, e_r$) be a local field of orthonormal frames for E' with respect to h' (resp. for E'' with respect to h''). Then $\alpha(e_1), \cdots, \alpha(e_p)$, $\lambda(e_{p+1}), \cdots, \lambda(e_r)$ form a local field of orthonormal frames for E with respect to h. If $1 \leqq i \leqq p$ and $p+1 \leqq j \leqq r$, then

$$\begin{aligned}(\partial_t h)(\alpha(e_i), \lambda(e_j)) &= -h(\alpha(e_i), (\partial_t \lambda)(e_j)) \\ &= -h(\alpha(e_i), \alpha(\partial_t S)(e_j)) \\ &= -h'(e_i, (\partial_t S)(e_j)) .\end{aligned}$$

Hence, $v_t = h^{-1} \partial_t h$ is represented by the following matrix:

$$(10.4) \qquad v_t = \begin{bmatrix} v_t' & -\partial_t S \\ -(\partial_t S)^* & v_t'' \end{bmatrix}, \quad \text{where} \quad \begin{matrix} v_t' = h'^{-1}\partial_t h', \\ v_t'' = h''^{-1}\partial_t h''. \end{matrix}$$

On the other hand, from (I.6.12) it follows that the curvature form $R_t = R(h_t)$ is represented by the following matrix:

$$(10.5) \qquad R_t = \begin{bmatrix} R_t' - B \wedge B^* & D'B \\ -D''B^* & R_t'' - B^* \wedge B \end{bmatrix} \quad \text{where} \quad \begin{matrix} R_t' = R(h_t') \\ R_t'' = R(h_t''), \end{matrix}$$

and $B = B_{h_t}$.

From (10.4) and (10.5) we obtain

$$\begin{aligned} \mathrm{tr}(v_t R_t) - \mathrm{tr}(v_t' R_t') - \mathrm{tr}(v_t'' R_t'') = & -\mathrm{tr}(v_t' B \wedge B^*) - \mathrm{tr}(v_t'' B^* \wedge B) \\ & + \mathrm{tr}(\partial_t S \circ D'' B^*) - \mathrm{tr}((\partial_t S)^* \circ D' B). \end{aligned}$$

Since

$$\mathrm{tr}(\partial_t S \circ D'' B^*) = d''(\mathrm{tr}(\partial_t S \circ B^*)) - \mathrm{tr}(\partial_t B \wedge B^*),$$

$$\mathrm{tr}((\partial_t S)^* \circ D' B) = d'(\mathrm{tr}((\partial_t S)^* \circ B)) - \mathrm{tr}((\partial_t B)^* \wedge B),$$

we obtain

$$\begin{aligned} \mathrm{tr}(v_t R_t) - \mathrm{tr}(v_t' R_t') - \mathrm{tr}(v_t'' R_t'') \equiv & -\mathrm{tr}(v_t' B \wedge B^*) - \mathrm{tr}(v_t'' B^* \wedge B) \\ & - \mathrm{tr}(\partial_t B \wedge B^*) + \mathrm{tr}((\partial_t B)^* \wedge B) \end{aligned}$$

modulo $d'A^{0,1} + d''A^{1,0}$. The right hand side can be written as

$$-\mathrm{tr}(\partial_t B \wedge B^*) - \mathrm{tr}(B \wedge ((\partial_t B)^* + B^* v_t' - v_t'' B^*)).$$

We claim

$$(\partial_t B)^* + B^* v_t' - v_t'' B^* = \partial_t (B^*)$$

so that

$$\begin{aligned} \mathrm{tr}(v_t R_t) - \mathrm{tr}(v_t' R_t') - \mathrm{tr}(v_t'' R_t'') & \equiv -\mathrm{tr}(\partial_t B \wedge B^*) - \mathrm{tr}(B \wedge \partial_t (B^*)) \\ & = -\partial_t (\mathrm{tr}(B \wedge B^*)). \end{aligned}$$

In order to establish our claim, we apply ∂_t to

$$h'(\xi, B\eta) = h''(B^*\xi, \eta) \qquad \text{for} \quad \xi \in E', \quad \eta \in E''.$$

Then

$$(\partial_t h')(\xi, B\eta) + h'(\xi, \partial_t B\eta) = (\partial_t h'')(B^*\xi, \eta) + h''(\partial_t (B^*)\xi, \eta).$$

But

$$(\partial_t h')(\xi, B\eta) = h'(v'\xi, B\eta) = h''(B^* v'\xi, \eta),$$
$$h'(\xi, \partial_t B\eta) = h''((\partial_t B)^*\xi, \eta),$$
$$(\partial_t h'')(B^*\xi, \eta) = h''(v'' B^*\xi, \eta).$$

Hence,

$$h''(B^* v'\xi, \eta) + h''((\partial_t B)^*\xi, \eta) = h''(v'' B^*\xi, \eta) + h''(\partial_t(B^*)\xi, \eta).$$

This proves our claim. Thus,

$$\text{(10.6)} \qquad \operatorname{tr}(v_t R_t) \equiv \operatorname{tr}(v_t' R_t') + \operatorname{tr}(v_t'' R_t'') - \partial_t(\operatorname{tr}(B \wedge B^*))$$

modulo $d' A^{0,1} + d'' A^{1,0}$.

Multiplying (10.6) by i and integrating from $t=0$ to $t=1$, we obtain (ii). Q.E.D.

(10.7) **Corollary** *Given an exact sequence* (10.1) *such that*

$$\mu(E) = \mu(E') \qquad (\textit{where } \mu = \deg/\text{rank})$$

and a pair of Hermitian structures h, k in E, the Lagrangian L(h, k) defined in (3.5) *satisfies the following relation*:

$$L(h, k) = L(h', k') + L(h'', k'') + \|B_h\|^2 - \|B_k\|^2.$$

Proof From the assumption that E and E' have the same degree/rank ratio, i.e., $\mu(E) = \mu(E')$, we conclude also $\mu(E) = \mu(E') = \mu(E'')$. Then

$$L(h, k) = \int_M \left(Q_2(h, k) - \frac{c}{n} Q_1(h, k)\Phi \right) \wedge \frac{\Phi^{n-1}}{(n-1)!},$$
$$L(h', k') = \int_M \left(Q_2(h', k') - \frac{c}{n} Q_1(h', k')\Phi \right) \wedge \frac{\Phi^{n-1}}{(n-1)!},$$
$$L(h'', k'') = \int_M \left(Q_2(h'', k'') - \frac{c}{n} Q_1(h'', k'')\Phi \right) \wedge \frac{\Phi^{n-1}}{(n-1)!}.$$

We note that the constant factors c appearing above are all equal, i.e., $c(E) = c(E') = c(E'')$ since $c(E) = 2\pi\mu(E)/\text{vol}(M)$. Hence, from (10.2) we obtain

$$L(h, k) - L(h', k') - L(h'', k'') = -\int_M i(\operatorname{tr}(B_h \wedge B_h^*) - \operatorname{tr}(B_k \wedge B_k^*)) \wedge \frac{\Phi^{n-1}}{(n-1)!}.$$

By a direct calculation we see that

$$-i(\operatorname{tr}(B_h \wedge B_h^*)) \wedge \Phi^{n-1} = |B_h|^2 \frac{\Phi^n}{n}.$$

This implies the desired formula. Q.E.D.

We recall the following equation of Poincaré-Lelong. Let U be a domain in $\boldsymbol{C}^n$ and s a holomorphic function in U. Let V be the hypersurface in U defined by $s=0$. As current, V coincides with $(i/\pi)d'd'' \log|s|$, i.e.,

$$\int_V \eta = \frac{i}{\pi}\int_U (\log|s|)d'd''\eta \tag{10.8}$$

for every $(n-1, n-1)$-form η with compact support in U.

We extend this to an algebraic manifold M and a closed hypersurface V as follows. Let L be an ample line bundle over M with a global holomorphic section s such that V is defined by $s=0$. Let a be a C^∞ everywhere-positive section of $L \otimes \bar{L}$. We may consider a as an Hermitian structure in the line bundle L^{-1}. Since L is ample, we can always choose one such that its Chern form

$$\Phi = \frac{i}{2\pi} d'd'' \log a$$

is positive. Set

$$f = |s|^2/a.$$

Then f is a globally defined C^∞ function on M vanishing exactly at V. Then as current, we have the following equality:

$$\frac{i}{2\pi} d'd'' \log f = \frac{i}{\pi} d'd'' \log|s| - \Phi.$$

Using (10.8) we may write this as follows:

$$\frac{i}{2\pi}\int_M (\log f)d'd''\eta = \int_V \eta - \int_M \eta \wedge \Phi \tag{10.9}$$

for every $(n-1, n-1)$-form η on M.

We use the Chern form Φ of L as our Kähler form on M and its restriction $\Phi|_V$ as our Kähler form on V. Let E be a holomorphic vector bundle of rank r over M. We shall now denote the Lagrangian $L(h, k)$ by $L_M(h, k)$ to emphasize the base space M. For the restricted bundle $E|_V$ over V, we denote the corresponding

Lagrangian by $L_V(h, k)$. Using (10.9) we shall find a relationship between $L_M(h, k)$ and $L_V(h, k)$.

First, we observe that the constant c/n appearing in the definition (3.5) of $L_M(h, k)$ remains the same for $L_V(h, k)$ because

$$\int_M c_1(E) \wedge \Phi^{n-1} = \int_V c_1(E) \wedge \Phi^{n-2}$$

and

$$\int_M \Phi^n = \int_V \Phi^{n-1}$$

by (10.9). We set therefore

$$\lambda = \frac{c}{n}$$

Now, letting $\eta = (Q_2(h, k) - \lambda Q_1(h, k)\Phi) \wedge \Phi^{n-2}$ in (10.9), we obtain

$$\begin{aligned} (n-1)!\, L_M(h, k) &= \int_M (Q_2(h, k) - \lambda Q_1(h, k)) \wedge \Phi^{n-1} \\ &= \int_V (Q_2(h, k) - \lambda Q_1(h, k)) \wedge \Phi^{n-2} \\ &\quad - \frac{i}{2\pi} \int_M (\log f) d'd''(Q_2(h, k) - \lambda Q_1(h, k)) \wedge \Phi^{n-2}. \end{aligned}$$

Applying (iii) of (3.23) to the last term, we obtain

$$\begin{aligned} (n-1)!\, L_M(h, k) &= (n-2)!\, L_V(h, k) \\ &\quad + \frac{1}{4\pi} \int_M (\log f)(\mathrm{tr}(iR(h) - \lambda \Phi I)^2) \wedge \Phi^{n-2} \\ &\quad - \frac{1}{4\pi} \int_M (\log f)(\mathrm{tr}(iR(k) - \lambda \Phi I)^2) \wedge \Phi^{n-2}. \end{aligned} \tag{10.10}$$

Since we fix k, the last term on the right can be bounded by a constant independent of $h = h_t$. It is the middle term that we have to estimate.

We apply Hodge's primitive decomposition to $R(h)$. Then

$$iR(h) = \frac{1}{n} \Lambda(iR(h))\Phi + iS = \frac{1}{n} K(h)\Phi + iS,$$

where S is a primitive (1, 1)-form with values in End(E) so that

$$S \wedge \Phi^{n-1} = 0 .$$

Hence,

$$(iR(h) - \lambda \Phi I)^2 \wedge \Phi^{n-2} = \left(\frac{1}{n}(K(h) - cI)\Phi + iS\right)^2 \wedge \Phi^{n-2}$$

$$= \frac{1}{n^2}(K(h) - cI)^2 \Phi^n - S \wedge S \wedge \Phi^{n-2} .$$

It follows that

$$(\log f)(\mathrm{tr}(iR(h) - \lambda \Phi I)^2) \wedge \Phi^{n-2}$$

$$= \frac{1}{n^2}(\log f)(\mathrm{tr}(K(h) - cI)^2)\Phi^n - (\log f)\,\mathrm{tr}(S \wedge S) \wedge \Phi^{n-2} .$$

We shall show that $\mathrm{tr}(S \wedge S) \wedge \Phi^{n-2} \geqq 0$. Let $S^k_{j\alpha\bar{\beta}}$ be the components of S with respect to unitary bases for E and TM. Since $iR(h) = (1/n)K(h)\Phi + iS$, the components $S^k_{j\alpha\bar{\beta}}$ (and $S_{j\bar{k}\alpha\bar{\beta}}$) of S enjoy the same symmetry properties as the components of R, that is,

$$S_{j\bar{k}\alpha\bar{\beta}} = \bar{S}_{k\bar{j}\beta\bar{\alpha}} .$$

Hence, imitating the proof of (IV.4.3), we obtain

$$n(n-1)\,\mathrm{tr}(S \wedge S) \wedge \Phi^{n-2} = n(n-1) \sum S^k_j \wedge S^j_k \wedge \Phi^{n-2}$$

$$= -\sum (S_{k\bar{j}\alpha\bar{\alpha}} S_{j\bar{k}\beta\bar{\beta}} - S_{k\bar{j}\alpha\bar{\beta}} S_{j\bar{k}\beta\bar{\alpha}})\Phi^n$$

$$= \sum |S_{k\bar{j}\alpha\bar{\beta}}|^2 \Phi^n \geqq 0$$

since $\sum S_{k\bar{j}\alpha\bar{\alpha}} = 0$ by the primitivity of S.

Without loss of generality, we may assume that $f = |s|^2/a \leqq 1$ since we can multiply a by any positive constant without affecting $\Phi = (i/2\pi)d'd'' \log a$. Then $\log f \leqq 0$, so that

$$-(\log f)\,\mathrm{tr}(S \wedge S) \wedge \Phi^{n-2} \geqq 0 .$$

Hence,

$$(\log f)(\mathrm{tr}(iR(h) - \lambda \Phi I)^2) \wedge \Phi^{n-2} \geqq \frac{1}{n^2}(\log f)\,|K(h) - cI|^2 \Phi^n$$

and

$$\int_M (\log f)\,\mathrm{tr}(iR(h)-\lambda\Phi I)^2)\wedge\Phi^{n-2} \geqq -\frac{1}{n^2}\left(\int_M (-\log f)\Phi^n\right)\operatorname*{Max}_M |K(h)-cI|^2$$

We substitute this into the middle term of the right hand side of (10.10). The last term of (10.10) is independent of h. Hence, we have established the following:

(10.11) **Proposition** *Let E be a holomorphic vector bundle over a projective algebraic manifold M of dimension $n \geqq 2$. Let V be a non-singular hypersurface of M such that the line bundle L defined by the divisor V is ample. We use a positive closed $(1, 1)$-form Φ representing the Chern class of L as a Kähler form for M. Following (3.5) we define the Lagrangians $L_M(h, k)$ and $L_V(h, k)$ for E and $E|_V$, respectively. Then, for a fixed Hermitian structure k in E and for all Hermitian structures h in E, we have*

$$L_M(h, k) \geqq \frac{1}{n-1} L_V(h, k) - C\left(\operatorname*{Max}_M |K(h)-cI|^2\right) - C',$$

where C and C' are positive constants independent of h.

(10.12) **Theorem** *Let E be a semistable holomorphic vector bundle over a compact Riemann surface M. Then, for any fixed Hermitian structure $k \in \mathrm{Herm}^+(E)$, the set of numbers $\{L(h, k);\ h \in \mathrm{Herm}^+(E)\}$ is bounded below.*

Proof The proof is by induction on the rank of E. We know (see (V.2.6), Theorem of Narasimhan-Seshadri [1]) that if E is stable, then there exists an Einstein-Hermitian structure h_0. By (3.37), $L(h, k)$ attains its absolute minimum at h_0, i.e., $L(h, k) \geqq L(h_0, k)$ for all $h \in \mathrm{Herm}^+(E)$.

We assume therefore that E is semistable but not stable. Then there exists a proper subbundle E_1 of E with the same degree-rank ratio, i.e., $\mu(E_1)=\mu(E)$. Clearly, E_1 is semistable. If it is not stable, there is a proper subbundle E_2 of E_1 with $\mu(E_2)=\mu(E_1)$. By repeating this process we obtain a stable subbundle E' of E such that $\mu(E')=\mu(E)$. We set $E''=E/E'$. Then $\mu(E'')=\mu(E)$ and E'' is semistable. Applying (10.7) to the exact sequence

$$0 \longrightarrow E' \longrightarrow E \longrightarrow E'' \longrightarrow 0$$

and to a pair of Hermitian structures h, k in E, we obtain

$$L(h, k) = L(h', k') + L(h'', k'') + \|B_h\|^2 - \|B_k\|^2,$$

where h', k' (resp. h'', k'') are the Hermitian structures in E' (resp. E'') induced by h, k. By the inductive hypothesis, $L(h', k')$ (resp. $L(h'', k'')$) is bounded below by a constant depending only on k' (resp. k''). Hence, $L(h, k)$ is bounded below by a constant depending only on k. Q.E.D.

(10.13) **Theorem** *Let E be a holomorphic vector bundle over a compact Kähler manifold M with Kähler form Φ. Then we have implications* (1) $\Rightarrow$ (2) $\Rightarrow$ (3) *for the following statements*:

(1) *For any fixed Hermitian structure k in E, there exists a constant B such that $L(h,k) \geqq B$ for all Hermitian structures h in E;*

(2) *E admits an approximate Einstein-Hermitian structure, i.e., given any $\varepsilon > 0$, there exists an Hermitian structure h in E such that*

$$\operatorname*{Max}_M |\hat{K} - ch| < \varepsilon;$$

(3) *E is Φ-semistable.*

If there exists an ample line bundle H on M such that Φ represents the Chern class $c_1(H)$, then (1), (2) *and* (3) *are all equivalent.*

Proof (1) $\Rightarrow$ (2) This follows from (iii) of (9.1).

(2) $\Rightarrow$ (3) This was proved in (V.8.6).

Now, assume that there exists an ample line bundle H over M with $[\Phi] = c_1(H)$. We make use of the following result.

(10.14) **Theorem of Mumford and Mehta-Ramanathan** [1] *Let M be a projective algebraic manifold with an ample line bundle H. Let E be an H-semistable vector bundle over M. Then there exists a positive integer m such that, for a generic smooth $V \in |\mathcal{O}_M(m)|$, the bundle $E|_V$ is H-semistable.*

Using (10.11) and (10.14), we shall now prove the implication:

(3) $\Rightarrow$ (1) Let h be an arbitrary Hermitian structure in E and let h_t, $0 \leqq t < \infty$, be the solution of the evolution equation

$$\partial_t h_t = -(\hat{K}(h_t) - ch_t)$$

with the initial condition $h_0 = h$, (see (8.25)). First we prove

(10.15) **Lemma** *If E is H-semistable, then $\{L_M(h_t, k), 0 \leqq t < \infty\}$ is bounded below (by a constant depending on h, k).*

Proof Let V be a hypersurface of M as in (10.14). By (10.11), given an

Hermitian structure k, there exist positive constants C and C' such that

$$L_M(h_t, k) \geqq \frac{1}{n-1} L_V(h_t, k) - C\left(\operatorname*{Max}_M |K(h_t) - cI|^2\right) - C'$$

$$\geqq \frac{1}{n-1} L_V(h_t, k) - C\left(\operatorname*{Max}_M |K(h_0) - cI|^2\right) - C',$$

where the second inequality is a consequence of (ii) of (9.1). Since $L_V(h_t, k)$ is bounded below by the inductive hypothesis, so is $L_M(h_t, k)$. Q.E.D.

By (i) of (9.1), $L_M(h, k) \geqq L_M(h_t, k)$ for all $t \geqq 0$. By (iii) of (9.1), there exists t_1 such that

$$\operatorname*{Max}_M |K(h_t) - cI| < 1 \qquad \text{for} \quad t \geqq t_1 .$$

Hence,

$$L_M(h, k) \geqq L_M(h_t, k) \geqq \frac{1}{n-1} L_V(h_t, k) - C - C' \qquad \text{for} \quad t \geqq t_1 .$$

By the inductive hypothesis, $L_V(h_t, k)$ is bounded below by a constant (which depends on k but not on h). Hence, $\{L_M(h, k);\ h \in \mathrm{Herm}^+(E)\}$ is bounded below. This completes the proof of (3) $\Rightarrow$ (Γ). Q.E.D.

All conditions (1), (2) and (3) in (10.13) should be equivalent in general whether M is algebraic or not. One has to find the proof which does not make use of (10.14). The following theorem, proved by Maruyama [6] by an algebraic method, is a direct consequence of (IV.5.3) and (10.13). It should be also true whether M is algebraic or not.

(10.16) **Theorem** *Let M be a compact Kähler manifold with Kähler form Φ. Assume that Φ represents the first Chern class $c_1(H)$ of an ample line bundle H over M.*

(1) *If E is Φ-semistable, so are $E^{\otimes p} \otimes E^{*\otimes q}$, $\bigwedge^p E$ and $S^p E$;*

(2) *If E and E' are Φ-semistable, so is $E \otimes E'$.*

Note that if E is Φ-semistable, its dual E^* is Φ-semistable whether M is algebraic or not, (see (V.7.7)).

(10.17) *Remark* If E and E' are Φ-stable in (10.16), then $E \otimes E'$ is Φ-

semistable. But it should be a direct sum of Φ-stable bundles with $\mu=\mu(E)+\mu(E')$. This conjecture is consistent with (IV.1.4).

Using (10.13) and results of Uhlenbeck [1], [2], Donaldson [2] derived the following

(10.18) **Proposition** *Let M be an algebraic surface with an ample line bundle H and a Kähler form Φ representing the Chern class $c_1(H)$. Given a Φ-semistable bundle E over M, there exists a holomorphic vector bundle E' over M with the same rank and degree as E and such that*

(1) *E' admits an Einstein-Hermitian structure;*

(2) *there is a nonzero homomorphism $f\colon E \to E'$.*

As a consequence of (10.18), he obtained

(10.19) **Theorem** *Let M be as above and E a Φ-stable bundle. Then E admits an Einstein-Hermitian structure.*

In fact, E' is Φ-semistable by (V.8.3). The theorem follows from (V.7.12). It is the use of Uhlenbeck's results which limit (10.18) to the 2-dimensional case.

Chapter VII

Moduli spaces of vector bundles

Atiyah-Hitchin-Singer [1] constructed moduli spaces of self-dual Yang-Mills connections in principal bundles with compact Lie groups over 4-dimensional compact Riemannian manifolds and computed their dimensions. Itoh [1], [2] introduced Kähler structures in moduli spaces of anti-self-dual connections in $SU(n)$-bundles over compact Kähler surfaces. Kim [1] introduced complex structures in moduli spaces of Einstein-Hermitian vector bundles over compact Kähler manifolds.

Let E be a fixed C^∞ complex vector bundle over a compact Kähler manifold M, and h a fixed Hermitian structure in E. In Section 1 we relate the mod space of holomorphic structures in E with the moduli space of Einstein-Hermitian connections in (E, h).

Generalizing the elliptic complexes used by Atiyah-Hitchin-Singer and Itoh, Kim introduced an elliptic complex for $\mathrm{End}(E, h)$ using Einstein-Hermitian connections. This complex makes it possible to prove smoothness of the moduli space of Einstein-Hermitian connections under cohomological conditions on the tracefree part $\mathrm{End}^0(E, h)$ of $\mathrm{End}(E, h)$ rather than $\mathrm{End}(E, h)$ itself. This technical point is important in applications. In Section 2 we reproduce his results on this complex.

In Section 3 we construct the moduli space of simple holomorphic structures in E using the Dolbeault complex for $\mathrm{End}(E)$ and the Kuranishi map. The results in this section have been obtained also independently by Lübke-Okonek [1]. For a different analytic approach to moduli spaces of simple vector bundles, see Norton [1].

In Section 4, following Kim [1] we prove that the moduli space of Einstein-Hermitian connections in (E, h) has a natural complex structure and is open in the moduli space of holomorphic structures in E.

In Section 5 we prove the symplectic reduction theorem of Marsden-Weinstein [1] and its holomorphic analogue. The reduction theorem was used by Atiyah-Bott [1] to introduce Kähler structures in moduli spaces of stable vector bundles over compact Riemann surfaces. It will be used in Sections 6 and 7 for similar purposes.

In Section 6, generalizing Itoh [2] we introduce a Kähler structure in the moduli space of Einstein-Hermitian connections in (E, h).

Mukai [1] has shown that the moduli space of simple holomorphic structures in E over an abelian surface or $K3$ surface M carries a natural holomorphic symplectic structure. In Section 7 we generalize this to the case where M is an arbitrary compact Kähler manifold with a holomorphic symplectic structure.

In Section 8 we apply results of preceding sections to bundles over compact Kähler surfaces.

§1 Holomorphic and Hermitian structures

Let E be a C^∞ complex vector bundle of rank r over a complex manifold M. Let $GL(E)$ denote the group of C^∞ bundle automorphisms of E (which induce the identity transformations on the base M). The space of C^∞ sections of the endomorphism bundle $\mathrm{End}(E)=E\otimes E^*$ will be denoted by $\mathfrak{gl}(E)$ and is considered as the Lie algebra of $GL(E)$.

Let $A^{p,q}(E)$ be the space of (p, q)-forms on M with values in E, and set $A^r(E)=\sum_{p+q=r} A^{p,q}(E)$.

Let $\mathscr{D}''(E)$ denote the set of C-linear maps

$$D'': A^0(E) \longrightarrow A^{0,1}(E)$$

satisfying

$$(1.1) \qquad D''(fs)=(d''f)s+f\cdot D''s \qquad \text{for} \quad s\in A^0(E)\,, \quad f\in A^0\,.$$

Every D'' extends uniquely to a C-linear map

$$D'': A^{p,q}(E) \longrightarrow A^{p,q+1}(E)\,, \qquad p, q\geqq 0\,,$$

satisfying

$$(1.2) \qquad D''(\psi\sigma)=d''\psi\wedge\sigma+(-1)^{r+s}\psi\wedge D''\sigma \qquad \text{for} \quad \sigma\in A^{p,q}(E)\,, \quad \psi\in A^{r,s}\,.$$

If we fix $D''_0\in\mathscr{D}''(E)$, then for every $D''\in\mathscr{D}''(E)$ the difference $\alpha=D''-D''_0$ is a map

$$\alpha: A^0(E) \longrightarrow A^{0,1}(E)$$

which is linear over functions. Hence, α can be regarded as an element of $A^{0,1}(\mathrm{End}(E))$. Conversely, given any $D''_0\in\mathscr{D}''(E)$ and $\alpha\in A^{0,1}(\mathrm{End}(E))$, $D=D_0+\alpha$ is an element of $\mathscr{D}''(E)$. In other words, once $D_0\in\mathscr{D}''(E)$ is chosen and fixed, $\mathscr{D}''(E)$ can be identified with the (infinite dimensional) vector space $A^{0,1}(\mathrm{End}(E))$. Thus, $\mathscr{D}''(E)$ is an affine space.

Let $\mathscr{H}''(E) \subset \mathscr{D}''(E)$ be the subset consisting of D'' satisfying the integrability condition

$$(1.3) \qquad D'' \circ D'' = 0 .$$

We know that if E is a holomorphic vector bundle, then

$$d'' : A^0(E) \longrightarrow A^{0,1}(E)$$

can be well defined and is an element of $\mathscr{H}''(E)$. Conversely, every $D'' \in \mathscr{H}''(E)$ defines a unique holomorphic structure in E such that $D'' = d''$, (see (I.3.7)). Thus, $\mathscr{H}''(E)$ may be considered as the set of holomorphic bundle structures in E.

The group $GL(E)$ acts on $\mathscr{D}''(E)$ by

$$(1.4) \qquad D'' \longmapsto D''^{f} = f^{-1} \circ D'' \circ f = D'' + f^{-1} d'' f \qquad \text{for} \quad D'' \in \mathscr{D}''(E) , \quad f \in GL(E) .$$

Then $GL(E)$ sends $\mathscr{H}''(E)$ into itself. Two holomorphic structures D_1'' and D_2'' of E are considered *equivalent* if they are in the same $GL(E)$-orbit. The space $\mathscr{H}''(E)/GL(E)$ of $GL(E)$-orbits with the C^∞-topology is the *moduli space of holomorphic structures* in E. This space is, however, in general non-Hausdorff, (see Norton [1] on this point). This difficulty will be resolved by considering only (semi)stable holomorphic structures. For technical reasons we shall use later appropriate Sobolev spaces instead of the C^∞-topology.

Again, let E be a C^∞ complex vector bundle of rank r over a complex manifold M. We fix an Hermitian structure h in E. Let $U(E, h)$ denote the subgroup of $GL(E)$ consisting of unitary automorphisms of (E, h). Its Lie algebra, denoted by $\mathfrak{u}(E, h)$, consists of skew-Hermitian endomorphisms of (E, h).

Let $\mathscr{D}(E, h)$ be the set of connections D in E preserving h, i.e., $\boldsymbol{C}$-linear maps $D\colon A^0(E) \to A^1(E)$ satisfying

$$(1.5) \qquad \begin{aligned} & D(fs) = df \cdot s + f \cdot Ds && \text{for} \quad s \in A^0(E) \quad \text{and} \quad f \in A^0 , \\ & d(h(s, t)) = h(Ds, t) + h(s, Dt) && \text{for} \quad s, t \in A^0(E) . \end{aligned}$$

Every $D \in \mathscr{D}(E, h)$ extends to a unique $\boldsymbol{C}$-linear map

$$D\colon A^{p,q}(E) \longrightarrow A^{p+1,q}(E) + A^{p,q+1}(E)$$

satisfying

$$(1.6) \qquad D(\psi \wedge \sigma) = d\psi \wedge \sigma + (-1)^{r+s} \psi \wedge D\sigma \qquad \text{for} \quad \sigma \in A^{p,q}(E) , \quad \psi \in A^{r,s} .$$

Set

$$D = D' + D'' ,$$

where D': $A^0(E) \to A^{1,0}(E)$ and D'': $A^0(E) \to A^{0,1}(E)$. Then $D'' \in \mathscr{D}''(E)$. Thus, we have a natural map

$$(1.7) \qquad \mathscr{D}(E, h) \longrightarrow \mathscr{D}''(E) , \qquad D \longmapsto D'' .$$

The map (1.7) is bijective. In fact, given $D'' \in \mathscr{D}''(E)$, D' is determined by

$$(1.8) \qquad d''(h(s, t)) = h(D''s, t) + h(s, D't) \qquad \text{for} \quad s, t \in A^0(E) .$$

Then $D = D' + D''$ is in $\mathscr{D}(E, h)$.

Let End(E, h) be the vector bundle of skew-Hermitian endomorphisms of (E, h). If we fix a connection $D_0 \in \mathscr{D}(E, h)$, for every $D \in \mathscr{D}(E, h)$ the difference $\alpha = D - D_0$ satisfies

$$(1.9) \qquad h(\alpha s, t) + h(s, \alpha t) = 0$$

and can be regarded as an element of $A^1(\text{End}(E, h))$, i.e., a 1-form with values in the bundle End(E, h) of skew-Hermitian endomorphisms of (E, h). Conversely, given any such form α, $D = D_0 + \alpha$ is an element of $\mathscr{D}(E, h)$. Hence $\mathscr{D}(E, h)$ is an infinite dimensional affine space, which can be identified with the vector space $A^1(\text{End}(E, h))$ once an element $D_0 \in \mathscr{D}(E, h)$ is chosen as origin.

Let $\mathscr{H}(E, h)$ denote the subset of $\mathscr{D}(E, h)$ consisting of $D = D' + D''$ such that $D'' \circ D'' = 0$; in other words,

$$\mathscr{H}(E, h) = \{D \in \mathscr{D}(E, h);\ D'' \in \mathscr{H}''(E)\} .$$

If $D \in \mathscr{H}(E, h)$, then D'' defines a unique holomorphic structure in E such that $D'' = d''$. Hence, D is the Hermitian connection of (E, h) with respect to this holomorphic structure. Conversely, every holomorphic structure $D'' \in \mathscr{H}''(E)$ determines a unique connection $D = D' + D'' \in \mathscr{H}(E, h)$, which is nothing but the Hermitian connection of (E, h) with respect to the holomorphic structure D''. Thus we have a bijection $\mathscr{H}(E, h) \to \mathscr{H}''(E)$. Combining this with (1.7), we have the following diagram:

$$(1.10) \qquad \begin{array}{ccc} \mathscr{D}(E, h) & \longleftrightarrow & \mathscr{D}''(E) \\ \cup & & \cup \\ \mathscr{H}(E, h) & \longleftrightarrow & \mathscr{H}''(E) . \end{array}$$

The unitary automorphism group $U(E, h)$ acts on $\mathscr{D}(E, h)$ by

$$(1.11)\qquad D \longmapsto D^f = f^{-1}\circ D\circ f = D + f^{-1}df \qquad \text{for}\quad D\in\mathscr{D}(E,h),\quad f\in U(E,h)$$

and leaves the subset $\mathscr{H}(E,h)$ invariant.

Since $\mathscr{D}(E,h)$ is in one-to-one correspondence with $\mathscr{D}''(E)$, the group $GL(E)$ acting on $\mathscr{D}''(E)$ must act also on $\mathscr{D}(E,h)$. This corresponding action is given by

$$(1.12)\qquad D \longmapsto D^f = f^*\circ D'\circ f^{*-1} + f^{-1}\circ D''\circ f \qquad \text{for}\quad D\in\mathscr{D}(E,h),\quad f\in GL(E),$$

where f^* is the adjoint of f, i.e., $h(s, f^*t) = h(fs, t)$. To prove (1.12), set $D^f = D_1' + D_1''$. Since the correspondence $\mathscr{D}(E,h)\to\mathscr{D}''(E)$ is given by $D\mapsto D''$, we obtain $D_1'' = f^{-1}\circ D''\circ f$. Then using (1.8) we obtain

$$\begin{aligned} h(s, D_1' f^*t) &= d''(h(s, f^*t)) - h(D_1''s, f^*t)\\ &= d''(h(fs, t)) - h(D''(fs), t)\\ &= h(fs, D't) = h(s, f^*D't)\,. \end{aligned}$$

Hence, $D_1' = f^*\circ D'\circ f^{*-1}$

Since $GL(E)$ acting on $\mathscr{D}''(E)$ by (1.4) leaves $\mathscr{H}''(E)$ invariant, $GL(E)$ acting on $\mathscr{D}(E,h)$ by (1.12) must leave $\mathscr{H}(E,h)$ invariant.

The action of $U(E,h)$ on $\mathscr{D}(E,h)$ or $\mathscr{H}(E,h)$ is not effective. For $f\in U(E,h)$ and $D\in\mathscr{D}(E,h)$, the equality $D^f = D$ means $D\circ f = f\circ D$, i.e., the endomorphism f of E is parallel with respect to the connection D. Hence,

(1.13) **Proposition** *The subgroup of $U(E,h)$ fixing a given $D\in\mathscr{D}(E,h)$ consists of automorphisms f of E which are parallel with respect to D. It is therefore naturally isomorphic to the centralizer in the unitary group $U(r)$ of the holonomy group of the connection D. In particular, it is compact.*

If the holonomy group of D is irreducible, the subgroup in (1.13) consists of scalar multiplication by complex numbers λ of absolute value 1 and, hence, is isomorphic to $U(1) = \{\lambda I_E;\ \lambda\in\mathbf{C} \text{ and } |\lambda| = 1\}$. We remark that $U(1)$ is much smaller than the center of $U(E,h)$ which consists of automorphisms of the form λI_E where λ are *functions* with $|\lambda| = 1$.

(1.14) **Proposition** *The action of $U(E,h)$ on $\mathscr{D}(E,h)$ or $\mathscr{H}(E,h)$ is proper.*

Proof Set $U = U(E,h)$ and $\mathscr{D} = \mathscr{D}(E,h)$. Proposition means that the map

$$\mathscr{D} \times U \longrightarrow \mathscr{D} \times \mathscr{D}\,, \qquad (D, f) \longmapsto (D, D^f)$$

is proper. Let $D_j \in \mathscr{D}$ be a sequence of connections converging to $D_\infty \in \mathscr{D}$ and let $f_j \in U$ be a sequence of automorphisms of E such that $D_j^{f_j}$ converges to a connection $D^* \in \mathscr{D}$. The problem is to show that a subsequence of f_j converges to some element $f_\infty \in U$ and $D^* = D_\infty^{f_\infty}$.

We work in the principal $U(r)$-bundle P associated to (E, h). Fix $x_0 \in M$ and $v_0 \in P$ over x_0. Taking a subsequence, we have

$$f_j(v_0) \longrightarrow v_0^* \in P\,.$$

Since f_j commute with the right action of the structure group $U(r)$ of P, we have

$$f_j(v_0 a) = (f_j(v_0))a \longrightarrow v_0^* a \qquad \text{for all} \quad a \in U(r)\,.$$

We define f_∞ as follows. Let $x \in M$ and c a curve from x_0 to x. Let $\tilde{c}_j$ be the horizontal lift of c with respect to the connection D_j such that the starting point is v_0. Then $f_j(\tilde{c}_j)$ is the horizontal lift of c with respect to the connection $D_j^{f_j}$ such that its starting point is $f_j(v_0)$. Since $D_j \to D_\infty$, it follows that $\tilde{c}_j$ converges to $\tilde{c}_\infty$ (where $\tilde{c}_\infty$ is the horizontal lift of c with respect to D_∞ such that its starting point is v_0). Let $\tilde{c}^*$ be the horizontal lift of c with respect to D^* such that its starting point is v_0^*. Since $D_j^{f_j} \to D^*$ and $f_j(v_0) \to v_0^*$, it follows that $f_j(\tilde{c}_j) \to \tilde{c}^*$. We have to define f_∞ in such a way that $f_\infty(\tilde{c}_\infty) = \tilde{c}^*$. This condition together with the requirement that f_∞ commutes with the right action of the structure group $U(r)$ determines f_∞ uniquely. (Because if c is a closed curve in M and if a horizontal lift $\tilde{c}$ of c gives an element $a \in U(r)$ of the holonomy group, then for any $f \in U$, the horizontal lift $f(\tilde{c})$ gives the same element $a \in U(r)$). Q.E.D.

(1.15) **Corollary** *The quotient spaces $\mathscr{D}(E, h)/U(E, h)$ and $\mathscr{H}(E, h)/U(E, h)$ are Hausdorff.*

Assume that M is a compact Kähler manifold with Kähler metric g. Let $\mathscr{E}(E, h)$ denote the subset of $\mathscr{H}(E, h)$ consisting of Einstein-Hermitian connections D, i.e.,

$$\mathscr{E}(E, h) = \{D \in \mathscr{H}(E, h);\ K(D) = cI_E\}\,, \tag{1.16}$$

where $K(D)$ denotes the mean curvature of a connection D and c is a constant. We recall that the Einstein condition can be expressed in terms of the operator Λ as follows, (see (IV. 1.2)):

$$i\Lambda R(D) = cI_E\,.$$

In general, if $D \in \mathscr{D}(E, h)$ and $f \in U(E, h)$, then

$$R(D^f) = (f^{-1} \circ D \circ f) \circ (f^{-1} \circ D \circ f) = f^{-1} \circ R(D) \circ f \tag{1.17}$$

and

$$K(D^f) = f^{-1} \circ K(D) \circ f. \tag{1.18}$$

It follows that the action of $U(E, h)$ on $\mathscr{D}(E, h)$ leaves $\mathscr{E}(E, h)$ invariant.

Trivially, the natural map

$$\mathscr{E}(E, h)/U(E, h) \longrightarrow \mathscr{H}(E, h)/U(E, h)$$

is injective. By the uniqueness of Einstein-Hermitian connection (see (VI.3.37)), even the natural map

$$\mathscr{E}(E, h)/U(E, h) \longrightarrow \mathscr{H}''(E)/GL(E)$$

is injective. We call $\mathscr{E}(E, h)/U(E, h)$ the *moduli space of Einstein-Hermitian structures* in E.

(1.19) **Proposition** *The moduli space $\mathscr{E}(E, h)/U(E, h)$ of Einstein-Hermitian structures in E is Hausdorff and injects into the set $\mathscr{H}''(E)/GL(E)$ of holomorphic structure in E.*

The following chart organizes various spaces discussed in this section.

$$\begin{array}{ccccc}
 & & \mathscr{D}(E, h) & \longleftrightarrow & \mathscr{D}''(E) \\
 & & \cup & & \cup \\
\mathscr{E}(E, h) & \subset & \mathscr{H}(E, h) & \longleftrightarrow & \mathscr{H}''(E) \\
\downarrow & & \downarrow & & \downarrow \\
\mathscr{E}(E, h)/U(E, h) & \subset & \mathscr{H}(E, h)/U(E, h) & & \downarrow \\
 & \searrow^{\text{inj}} & \downarrow & & \downarrow \\
 & & \mathscr{H}(E, h)/GL(E) & \longleftrightarrow & \mathscr{H}''(E)/GL(E),
\end{array} \tag{1.20}$$

where inj indicates that the map is injective.

§2 Spaces of infinitesimal deformations

As in the preceding section, let E be a C^∞ complex vector bundle of rank r over a complex manifold M and $\mathscr{H}''(E)$ the set of holomorphic structures in E, (see (1.3)). We shall determine the space of infinitesimal deformations of a holomorphic structure D'' of E, i.e., the "tangent space" to $\mathscr{H}''(E)/GL(E)$ at D''. Let

$$D''_t = D'' + \alpha''_t, \qquad |t| < \delta,$$

be a curve in $\mathscr{H}''(E)$, where $\alpha''_t \in A^{0,1}(\mathrm{End}(E))$ and $\alpha''_0 = 0$. Then

$$(2.1) \qquad D''\alpha''_t + \alpha''_t \wedge \alpha''_t = 0.$$

For, if $s \in A^0(E)$, then

$$\begin{aligned} 0 = D''_t(D''_t(s)) &= D''(D''(s)) + D''(\alpha''_t(s)) + \alpha''_t(D''(s)) + (\alpha''_t \wedge \alpha''_t)(s) \\ &= (D''\alpha''_t)(s) + (\alpha''_t \wedge \alpha''_t)(s). \end{aligned}$$

Differentiating (2.1) with respect to t at $t=0$, we obtain

$$(2.2) \qquad D''\alpha'' = 0, \qquad \text{where} \quad \alpha'' = \partial_t \alpha''_t|_{t=0}.$$

If D''_t is obtained by a 1-parameter family of transformations $f_t \in GL(E)$, so that

$$D''_t = f_t^{-1} \circ D'' \circ f_t, \qquad \text{with} \quad f_0 = I_E,$$

then

$$(2.3) \qquad D''_t = D'' + f_t^{-1} \circ D'' f_t.$$

For, if $s \in A^0(E)$, then

$$D''_t(s) = f_t^{-1}(D''(f_t(s))) = D''s + (f_t^{-1} D'' f_t)(s).$$

Set

$$(2.4) \qquad \alpha''_t = f_t^{-1} D'' f_t, \qquad \alpha'' = \partial_t \alpha''_t|_{t=0}.$$

Then

$$(2.5) \qquad \alpha'' = D'' f, \qquad \text{where} \quad f = \partial_t f_t|_{t=0}.$$

From (2.2) and (2.5) we see that the tangent space to $\mathscr{H}''(E)/GL(E)$ is given by

$$(2.6) \qquad H^{0,1}(M, \mathrm{End}(E^{D''})) = \frac{\{\alpha'' \in A^{0,1}(\mathrm{End}(E));\ D''\alpha'' = 0\}}{\{D''f;\ f \in A^0(\mathrm{End}(E))\}}$$

provided that $\mathscr{H}''(E)/GL(E)$ is a manifold. (Here, $E^{D''}$ is the holomorphic vector bundle given by E and D''. $H^{0,1}(M, \mathrm{End}(E^{D''}))$ is the $(0,1)$-th D''-cohomology of the bundle $\mathrm{End}(E^{D''})$.

Assume that M is a compact Kähler manifold with Kähler metric g and Kähler form Φ. Given an Einstein-Hermitian connection $D \in \mathscr{E}(E, h)$ in E, we shall determine the tangent space to $\mathscr{E}(E, h)/U(E, h)$ at D. Let

$$D_t = D + \alpha_t, \qquad |t| < \delta$$

be a curve in $\mathscr{E}(E, h)$, where $\alpha_t \in A^1(\mathrm{End}(E, h))$ and $\alpha_0 = 0$. Then

$$D_t \circ D_t = D \circ D + D\alpha_t + \alpha_t \wedge \alpha_t . \tag{2.7}$$

Since the curvature $R(D_t) = D_t \circ D_t$ is of degree $(1, 1)$ for all t, differentiating (2.7) with respect to t at $t = 0$ we see that

$$D\alpha \in A^{1,1}(\mathrm{End}(E, h)), \qquad \text{where} \quad \alpha = \partial_t \alpha_t|_{t=0} . \tag{2.8}$$

Since each D_t is an Einstein-Hermitian connection, we have

$$i\Lambda R(D_t) = i\Lambda(D \circ D + D\alpha_t + \alpha_t \wedge \alpha_t) = cI_E , \tag{2.9}$$

where c is a constant (independent of t). Differentiating (2.9) with respect to t at $t = 0$, we obtain

$$\Lambda D\alpha = 0, \qquad \text{where} \quad \alpha = \partial_t \alpha_t|_{t=0} . \tag{2.10}$$

If D_t is obtained by a 1-parameter family of transformations $f_t \in U(E, h)$, i.e., $D_t = f_t^{-1} \circ D \circ f_t$, then

$$\alpha = Df, \qquad \text{where} \quad f = \partial_t f_t|_{t=0} . \tag{2.11}$$

The proof of (2.11) is the same as that of (2.5). It follows that the tangent space of $\mathscr{E}(E, h)/U(E, h)$ at D is given by the space H^1 defined by

$$H^1 = \frac{\{\alpha \in A^1(\mathrm{End}(E, h));\ D\alpha \in A^{1,1}(\mathrm{End}(E, h)) \text{ and } \Lambda D\alpha = 0\}}{\{Df;\ f \in A^0(\mathrm{End}(E, h))\}} \tag{2.12}$$

provided that $\mathscr{E}(E, h)/U(E, h)$ is a manifold.

Since $\{Df; f \in A^0(\mathrm{End}(E, h))\}$ represents the tangent space to the $U(E, h)$-orbit at D,

$$\{\alpha \in A^1(\mathrm{End}(E, h));\ D^*\alpha = 0\} \tag{2.13}$$

represents the normal space to the $U(E, h)$-orbit at D. Hence, H^1 is isomorphic to

$$\boldsymbol{H}^1 = \{\alpha \in A^1(\mathrm{End}(E, h));\ D\alpha \in A^{1,1}(\mathrm{End}(E, h)),\ \Lambda D\alpha = 0,\ D^*\alpha = 0\} . \tag{2.14}$$

Let

$$\alpha = \alpha' + \alpha'' \in A^1(\mathrm{End}(E, h)),$$

where α' (resp. α'') is of degree $(1,0)$ (resp. $(0,1)$). Since $\mathrm{End}(E, h)$ is the bundle of skew-Hermitian endomorphisms, we have

$$\alpha' = -{}^t\bar{\alpha}'' , \tag{2.15}$$

which shows that α'' determines α, i.e., $\alpha \longmapsto \alpha''$ gives an isomorphism $A^1(\mathrm{End}(E, h)) \approx A^{0,1}(\mathrm{End}(E))$. Corresponding to the conditions on α defining $\boldsymbol{H}^1$ in (2.14), we have the following conditions on α''.

$$\begin{aligned} & D\alpha \in A^{1,1}(\mathrm{End}(E, h)) \Longleftrightarrow D''\alpha'' = 0 . \\ & \left.\begin{aligned} \Lambda D\alpha = 0 &\Leftrightarrow \Lambda(D''\alpha' + D'\alpha'') = 0 \\ D^*\alpha = 0 &\Leftrightarrow \Lambda(D''\alpha' - D'\alpha'') = 0 \end{aligned}\right\} \Longleftrightarrow D''^*\alpha'' = 0 \end{aligned} \tag{2.16}$$

This follows from that formulas $\Lambda D''\alpha' = -iD'^*\alpha'$ and $\Lambda D'\alpha'' = iD''^*\alpha''$ which are consequences of (III.2.39).

Hence, the map $\alpha \longmapsto \alpha''$ gives an isomorphism of $\boldsymbol{H}^1$ onto the space of harmonic $(0,1)$-forms with values in $\mathrm{End}(E)$:

$$\boldsymbol{H}^{0,1}(\mathrm{End}(E^{D''})) = \{\alpha \in A^{0,1}(\mathrm{End}(E));\ D''\alpha'' = 0 \text{ and } D''^*\alpha'' = 0\} . \tag{2.17}$$

Since $H^{0,1}(M, \mathrm{End}(E^{D''}))$ is isomorphic to $\boldsymbol{H}^{0,1}(\mathrm{End}(E^{D''}))$, we have an isomorphism:

$$H^1 \approx H^{0,1}(M, \mathrm{End}(E^{D''})) . \tag{2.18}$$

In order to view the isomorphism (2.18) in a wider perspective, we consider, in addition to the Dolbeault complex (C^*), the following complex (B^*) introduced by Kim [1] which generalizes those considered by Atiyah-Hitchin-Singer [1] and Itoh [1]. Let

$$\begin{array}{ccccccccccccc} (B^*)\colon 0 & \longrightarrow & B^0 & \xrightarrow{D} & B^1 & \xrightarrow{D_+} & B^2_+ & \xrightarrow{D_2} & B^{0,3} & \xrightarrow{D''} \cdots \xrightarrow{D''} & B^{0,n} & \longrightarrow & 0 \\ & & \downarrow{\scriptstyle j_0} & & \downarrow{\scriptstyle j_1} & & \downarrow{\scriptstyle j_2} & & \downarrow{\scriptstyle j_3} & & \downarrow{\scriptstyle j_n} & & \\ (C^*)\colon 0 & \longrightarrow & C^{0,0} & \xrightarrow{D''} & C^{0,1} & \xrightarrow{D''} & C^{0,2} & \xrightarrow{D''} & C^{0,3} & \xrightarrow{D''} \cdots \xrightarrow{D''} & C^{0,n} & \longrightarrow & 0 , \end{array}$$

where

$$B^p = A^p_{\boldsymbol{R}}(\mathrm{End}(E, h)) = \{\text{real } p\text{-forms with values in } \mathrm{End}(E, h)\} ,$$

$$B^{p,q} = A^{p,q}(\mathrm{End}(E, h)) = A^{p,q} \otimes_{\boldsymbol{R}} B^0 = A^{p,q} \otimes_{\boldsymbol{R}} A^0_{\boldsymbol{R}}(\mathrm{End}(E, h)) ,$$

$$B^2_+ = B^2 \cap (B^{2,0} \oplus B^{0,2} \oplus B^0\Phi) = \{\omega + \bar{\omega} + \beta\Phi;\ \omega \in B^{2,0} \text{ and } \beta \in B^0\} ,$$

$$C^{0,q} = A^{0,q}(\mathrm{End}(E)) = A^{0,q} \otimes_{\boldsymbol{C}} A^{0,0}(\mathrm{End}(E)) .$$

Before we explain the mappings in the diagram above, we define the decomposition which is well known in Kähler geometry, (see Weil [1]):

$$B^2 = B^2_+ \oplus B^2_- ,$$

where B^2_- consists of effective (or primitive) real (1, 1)-forms with values in $\mathrm{End}(E, h)$, i.e.,

$$B^2_- = \{\omega \in A^{1,1}(\mathrm{End}(E, h);\ \omega = \bar{\omega} \text{ and } \Lambda\omega = 0\} .$$

Let

$$P_+ : B^2 \longrightarrow B^2_+ , \qquad P_- : B^2 \longrightarrow B^2_- ,$$

$$P^{2,0} : B^2 \longrightarrow B^{2,0} , \qquad P^{0,2} : B^2 \longrightarrow B^{0,2}$$

be all natural projections. Set

$$D_+ = P_+ \circ D , \qquad D_2 = D'' \circ P^{0,2} .$$

(2.19) **Proposition** *$D_+ \circ D = 0$ if D is Einstein-Hermitian.*

Proof For $f \in B^0 = A^0(\mathrm{End}(E, h))$, we have

$$D_+ \circ D(f) = P_+ \circ D \circ D(f) = P_+ \circ R(f) = P_+(R \circ f - f \circ R) = \frac{1}{n}(\Lambda R \circ f - f \circ \Lambda R)\Phi .$$

Since $\Lambda R = (ic/n)I_E$, we obtain $D_+ \circ D(f) = 0$. Q.E.D.

The vertical arrows $j_0, j_1, \cdots, j_n$ are all natural maps, and
j_0 is injective, $j_0^C : B^0 \times C \to C^{0,0}$ is bijective,
j_1 is bijective,
j_2 is surjective with kernel $\{\beta\Phi;\ \beta \in B^0\}$,
$j_3, \cdots, j_n$ are all bijective.

(2.20) **Lemma** *If $D \in \mathscr{H}(E, h)$, the complex (C^*) is elliptic. If $D \in \mathscr{E}(E, h)$, the complex (B^*) is elliptic.*

Proof This is well known for the Dolbeault complex (C^*); the proof is the same as and simpler than in the case of (B^*). To prove that (B^*) is an elliptic complex, we have to show that the sequence for (B^*) is exact at the symbol level. Let $w \neq 0$ be a real cotangent vector at a point x of M. Let

$$w = w' + w'' ,$$

where w' is a $(1,0)$-form at x and $w'' = \bar{w}'$. Then the symbol σ is given by

$$\sigma(D, w)a = w \wedge a ,$$

$$\sigma(D_+, w)a = P_+(w \wedge a) ,$$

$$\sigma(D_2, w)a = w'' \wedge P^{0,2}a ,$$

$$\sigma(D'', w)a = w'' \wedge a .$$

The symbol sequence for (B^*) is clearly exact at $B^0, B^4, \cdots, B^n$. To see that it is exact at B^1, suppose $\sigma(D_+, w)a = 0$ for $a = a' + a''$. Then $w'' \wedge a'' = 0$ and hence $a'' = w'' \wedge b$ for some $b \in \mathrm{End}(E, h)_x$. Then $a' = w' \wedge b$, so $a = w \wedge b = \sigma(D, w)b$.

Chasing the commutative diagram

$$\begin{array}{ccccc} B^2_+ & \longrightarrow & B^{0,3} & \longrightarrow & B^{0,4} \\ \downarrow & & \updownarrow & & \updownarrow \\ C^{0,2} & \longrightarrow & C^{0,3} & \longrightarrow & C^{0,4} \end{array}$$

we see that the symbol sequence is exact at $B^{0,3}$.

Although the exactness at B^2_+ can be shown explicitly, it can be established also in a indirect way from the vanishing of the following alternating sum:

$$\dim B^0_x - \dim B^1_x + \dim B^2_{+x} - \dim B^{0,3}_x + \cdots + (-1)^n \dim B^{0,n}_x$$

$$= r^2 \left\{ 1 - 2n + (n^2 - n + 1) + 2 \sum_{p=3}^{n} (-1)^p \binom{n}{p} \right\} = 0 .$$

(2.21) **Theorem** *Let $D \in \mathscr{E}(E, h)$. Let H^q be the q-th cohomology of the complex (B^*) and $H^{0,q}$ the q-th cohomology of the complex (C^*). Set $h^q = \dim_{\boldsymbol{R}} H^q$ and $h^{0,q} = \dim_{\boldsymbol{C}} H^{0,q}$. Then*

$$H^0 \otimes \boldsymbol{C} \approx H^{0,0} , \qquad h^0 = h^{0,0} ,$$

$$H^1 \approx H^{0,1} , \qquad h^1 = 2h^{0,1} ,$$

$$H^2 \approx H^{0,2} \oplus H^0 , \qquad h^2 = 2h^{0,2} + h^0 ,$$

$$H^q \approx H^{0,q} , \qquad h^q = 2h^{0,q} , \qquad \text{for} \quad q \geqq 3 .$$

In particular,

$$\sum(-1)^q h^q = 2\sum(-1)^q h^{0,q}.$$

Proof (0) Let $f_1, f_2 \in B^0$ and $f = f_1 + if_2 \in C^{0,0}$. Then

$$Df_1 = Df_2 = 0 \Longrightarrow D''f_1 = D''f_2 = 0 \Longrightarrow D''f = 0.$$

$$D''f = 0 \Longrightarrow D''f = 0 \quad \& \quad D'\bar{f} = 0 \Longrightarrow Df_1 = Df_2 = 0.$$

This proves

$$H^0 \otimes \boldsymbol{C} \approx H^{0,0}.$$

(1) Since the natural map $j_1 : B^1 \to C^{0,1}$ is bijective, the induced map $\tilde{j}_1 : H^1 \to H^{0,1}$ is surjective. The proof that $\tilde{j}_1$ is bijective is the same as that of (2.18). We define spaces $\boldsymbol{H}^1$ and $\boldsymbol{H}^{0,1}$ of harmonic forms by (2.14) and (2.17) and obtain isomorphisms:

$$H^1 \approx \boldsymbol{H}^1 \approx \boldsymbol{H}^{0,1} \approx H^{0,1}.$$

(2) In order to prove $H^2 = H^{0,2} \oplus H^0$, we define spaces $\boldsymbol{H}^0$, $\boldsymbol{H}^2$ and $\boldsymbol{H}^{0,2}$ of harmonic forms. We have

$$\boldsymbol{H}^0 = \{f \in B^0;\ Df = 0\}, \tag{2.22}$$

$$\boldsymbol{H}^2 = \{\omega \in B^2;\ D_2\omega = 0 \text{ and } D_+^*\omega = 0\}. \tag{2.23}$$

Given $\omega \in B_+^2$, write

$$\omega = \omega' + \omega'' + f\Phi, \qquad \text{where} \quad \omega' \in B^{2,0}, \quad \omega'' \in B^{0,2}, \quad f\Phi \in B^0.$$

Then

$$D_2\omega = 0 \Longleftrightarrow D''\omega'' = 0,$$

Since, for any $\theta = \theta' + \theta''$, ($\theta' \in B^{1,0}$ and $\theta'' \in B^{0,1}$) we have

$$\begin{aligned}(D_+^*\omega, \theta) &= (\omega', D_+\theta) + (\omega'', D_+\theta) + (f\Phi, D_+\theta)\\ &= (\omega', D'\theta') + (\omega'', D''\theta'') + \left(f\Phi, \frac{1}{n}(\Lambda D\theta)\Phi\right)\\ &= (D'^*\omega', \theta) + (D''^*\omega'', \theta) + (D^*(f\Phi), \theta),\end{aligned}$$

we obtain

$$D_+^*\omega = D'^*\omega' + D''^*\omega'' + D^*(f\Phi).$$

Hence, using (III.2.39), we obtain

$$D^*_+\omega=0 \Longleftrightarrow \begin{cases} D'^*\omega'+D''^*(f\Phi)=0 \\ D''^*\omega''+D'^*(f\Phi)=0 \end{cases} \Longleftrightarrow \begin{cases} D'^*\omega'+iD'f=0 \\ D''^*\omega''-iD''f=0 \end{cases}$$

$$\Longleftrightarrow \begin{cases} D'^*\omega'=D'f=0 \\ D''^*\omega''=D''f=0 \end{cases}$$

Therefore,

$$H^2=\{\omega=\omega'+\omega''+f\Phi\in B^2;\ D''\omega''=0,\ D'^*\omega'=D''^*\omega''=Df=0\}\,. \tag{2.24}$$

As is well known,

$$H^{0,2}=\{\omega''\in C^{0,2};\ D''\omega''=0 \text{ and } D''^*\omega''=0\}\,. \tag{2.25}$$

Comparing (2.22), (2.24) and (2.25), we obtain an exact sequence

$$0\longrightarrow H^0\longrightarrow H^2\longrightarrow H^{0,2}\longrightarrow 0\,, \tag{2.26}$$

where the map $H^0\to H^2$ is given by $f\longmapsto f\Phi$ while the map $H^2\to H^{0,2}$ is given by $\omega\longmapsto\omega''$. The sequence splits in a natural way; the splitting map $H^2\to H^0$ is given by $\omega\longmapsto f$.

(3) The isomorphism

$$H^q\longrightarrow H^{0,q} \qquad \text{for} \quad q\geqq 3$$

is easily seen. Q.E.D.

Let $\mathrm{End}^0(E)$ (resp. $\mathrm{End}^0(E,h)$) be the subbundle of $\mathrm{End}(E)$ (resp. $\mathrm{End}(E,h)$) consisting of endomorphisms (resp. skew-Hermitian endomorphisms) with trace 0. Then

$$\begin{aligned} &\mathrm{End}(E)=\mathrm{End}^0(E)+\boldsymbol{C}\,, \\ &\mathrm{End}(E,h)=\mathrm{End}^0(E,h)+\boldsymbol{R}\,. \end{aligned} \tag{2.27}$$

Let $(\tilde{C}^*)$ (resp. $(\tilde{B}^*)$) be the subcomplex of (C^*) (resp. (B^*)) consisting of elements with trace 0. Let $\tilde{H}^{0,q}$ (resp. $\tilde{H}^q$) be its q-th cohomology group. Then

$$\begin{aligned} &H^{0,q}=\tilde{H}^{0,q}+H^{0,q}(M,\boldsymbol{C}) && \text{for all} \quad q\,, \\ &H^0=\tilde{H}^0+H^0(M,\boldsymbol{R})\,, \\ &H^1=\tilde{H}^1+H^1(M,\boldsymbol{R})\,, \\ &H^2=\tilde{H}^2+H^2(M,\boldsymbol{C})\,, \\ &H^q=\tilde{H}^q+H^{0,q}(M,\boldsymbol{C}) && \text{for} \quad q\geqq 3\,, \end{aligned} \tag{2.28}$$

and

$$
(2.29)\qquad
\begin{aligned}
&\tilde{H}^0 \otimes \boldsymbol{C} \approx \tilde{H}^{0,0}\,, \\
&\tilde{H}^1 \approx \tilde{H}^{0,1}\,, \\
&\tilde{H}^2 \approx \tilde{H}^{0,2} + \tilde{H}^0\,, \\
&\tilde{H}^q \approx \tilde{H}^{0,q} && \text{for} \quad q \geqq 3\,.
\end{aligned}
$$

So far we have considered only C^∞ forms. In discussing harmonic forms, the Hodge decomposition and Green's operator, it is actually necessary to consider Sobolev forms. We shall summarize necessary results following Kuranishi [1]. Let V be a real or complex vector bundle over M, e.g., $V = E \otimes \bigwedge^p T^*M$ or $V = \mathrm{End}(E) \otimes \bigwedge^p T^*M$. By fixing an Hermitian structure and a compatible connection in V and a metric on M, we define the Sobolev space $L_k^p(V)$ by completing the space $C^\infty(V)$ of C^∞ sections with Sobolev norm $\|\ \|_{p,k}$. We are interested here in the case $p=2$ since $L_k^2(V)$ is a Hilbert space. We have natural inclusions

$$
(2.30)\qquad C^\infty(V) \subset \cdots \subset L_{k+1}^2(V) \subset L_k^2(V) \subset \cdots \subset L_0^2(V)\,,
$$

where each inclusion $L_{k+1}^2 \subset L_k^2$ is compact.

The Hölder space $C^{k+\alpha}(V)$ of sections can be defined also as in Section 5 of Chapter VI. These Sobolev and Hölder spaces are related by (VI.5.4).

Consider two vector bundles V and W over M. Let

$$
P\colon C^\infty(V) \longrightarrow C^\infty(W)
$$

be a linear differential operator of order b. Then it extends to a continuous linear operator

$$
P\colon L_k^2(V) \longrightarrow L_{k-b}^2(W)\,.
$$

Its adjoint

$$
P^*\colon C^\infty(W) \longrightarrow C^\infty(V)
$$

defined by

$$
(Pv, w) = (v, P^*w)
$$

is also a linear differential operator of order b.

Let $P\colon C^\infty(V) \to C^\infty(V)$ be a linear elliptic self-adjoint differential operator of order b. For example, the Laplacian $\square\colon A^{p,q}(E) \to A^{p,q}(E)$ is such an operator of order 2. Then the kernel of P is a finite dimensional subspace of $C^\infty(V)$ and agrees with the kernel of the extended operator $P: L_k^2(V) \to L_{k-b}^2(V)$. An element of Ker P is called P-harmonic. Let $(\text{Ker } P)^\perp$ be the orthogonal complement of Ker P in the Hilbert space $L_0^2(V)$. Then

$$(2.31) \qquad C^\infty(V) = \text{Ker } P \oplus ((\text{Ker } P)^\perp \cap C^\infty(V)) = \text{Ker } P \oplus P(C^\infty(V)) .$$

Let $H: C^\infty(V) \to \text{Ker } P$ be the projection given by the decomposition above. Then there is a unique continuous linear operator, called *Green's operator* for P,

$$G: (C^\infty(V), \| \ \|_{2,k}) \longrightarrow (C^\infty(V), \| \ \|_{2,k+b})$$

such that

$$(2.32) \qquad I = H + P \circ G , \qquad P \circ G = G \circ P , \qquad \text{Ker } G = \text{Ker } P .$$

In extends to a continuous linear operator

$$G: L_k^2(V) \longrightarrow L_{k+b}^2(V) .$$

Now we consider an elliptic complex of differential operators, i.e., a sequence of vector bundles V^j, $j = 0, 1, \cdots, m$, with linear differential operators of order b

$$D: C^\infty(V^j) \longrightarrow C^\infty(V^{j+1}) , \qquad j = 0, 1, \cdots, m ,$$

such that

(i) $D \circ D = 0$,

(ii) for each nonzero $\xi \in T_x^* M$, the symbol sequence

$$0 \longrightarrow V_x^0 \xrightarrow{\sigma(\xi)} V_x^1 \xrightarrow{\sigma(\xi)} \cdots \xrightarrow{\sigma(\xi)} V_x^m \longrightarrow 0$$

is exact.

In this section we have considered two such complexes, namely (B^*) and (C^*). Let D^* be the adjoint of D, and define

$$P = D \circ D^* + D^* \circ D: C^\infty(V^j) \longrightarrow C^\infty(V^j) , \qquad j = 0, 1, \cdots, m .$$

Then P is a self-adjoint elliptic linear differential operator of order $2b$ for each j. Let $\boldsymbol{H}^j$ denote the kernel of P on $C^\infty(V^j)$ and H the projection onto $\boldsymbol{H}^j$. Let G be Green's operator for P. Then

$$(2.33) \quad I = H + D \circ D^* \circ G + D^* \circ D \circ G , \qquad D \circ G = G \circ D , \quad D^* \circ G = G \circ D^* ,$$

$$(2.34) \quad \|Gv\|_{2,k+2b} \leqq c \|v\|_{2,k} , \qquad \|D^* \circ Gv\|_{2,k+b} \leqq c \|v\|_{2,k}$$

for all $v \in C^\infty(V^j)$.

If we denote the j-th cohomology of the elliptic complex $\{V^j, D\}$ by H^j, i.e.,

$$H^j = \frac{\{\operatorname{Ker} D\colon C^\infty(V^j) \to C^\infty(V^{j+1})\}}{\{\operatorname{Im} D\colon C^\infty(V^{j-1}) \to C^\infty(V^j)\}}, \tag{2.35}$$

then we have the usual isomorphisms

$$H^j \approx \boldsymbol{H}^j. \tag{2.36}$$

Finally, if we have elliptic complexes

$$D(s)\colon C^\infty(V^j) \longrightarrow C^\infty(V^{j+1})$$

depending infinitely differentiably on some parameter s, (where V^j's are independent of s), then $\dim \boldsymbol{H}^j(D(s))$ is upper semi-continuous in s, i.e.,

$$\overline{\lim_{s\to s_0}} \dim \boldsymbol{H}^j(D(s)) \leqq \dim \boldsymbol{H}^j(D(s_0)). \tag{2.37}$$

See also Kodaira [1] for some of the details.

§3 Local moduli for holomorphic structures

Let E be a C^∞ complex vector bundle of rank r over an n-dimensional compact Kähler manifold M with Kähler form Φ. We fix an Hermitian structure h in E.

If we fix a holomorphic structure $D'' \in \mathscr{H}''(E)$ of E, any other element of $\mathscr{D}''(E)$ is of the form

$$D'' + \alpha \qquad \text{with} \quad \alpha \in C^{0,1} = A^{0,1}(\operatorname{End}(E)), \tag{3.1}$$

and it is in $\mathscr{H}''(E)$ if and only if

$$0 = (D'' + \alpha) \circ (D'' + \alpha) = D''\alpha + \alpha \wedge \alpha. \tag{3.2}$$

Hence, the set $\mathscr{H}''(E)$ of holomorphic structures in E is given by

$$\mathscr{H}''(E) = \{D'' + \alpha;\ \alpha \in C^{0,1} \text{ and } D''\alpha + \alpha \wedge \alpha = 0\}. \tag{3.3}$$

For the elliptic complex $(C^*) = \{C^{0,q} = A^{0,q}(\operatorname{End}(E)), D''\}$ defined in the preceding section, we denote its Laplacian, Green's operator and harmonic projection by $\Box''_h$, G and H, respectively, so that (see (2.33))

$$\Box''_h = D'' \circ D''^* + D''^* \circ D'', \tag{3.4}$$

$$I = H + \Box''_h \circ G, \qquad G \circ D'' = D'' \circ G, \qquad D''^* \circ G = G \circ D''^*. \tag{3.5}$$

In the notation of (III.2.32), we may write d'' for D'' and δ''_h for D''^*.

We can extend these operators to appropriate Sobolev spaces $L^2_k(C^{0,q}) =$

$L^2_k(A^{0,q}(\mathrm{End}(E)))$ as explained in the preceding section. We set

$$(3.6)\qquad \begin{aligned} L^2_k(\mathscr{D}''(E)) &= \{D''+\alpha;\ \alpha\in L^2_k(C^{0,1})\}\,,\\ L^2_k(\mathscr{H}''(E)) &= \{D''+\alpha;\ \alpha\in L^2_k(C^{0,1}) \text{ and } D''\alpha+\alpha\wedge\alpha=0\}\,. \end{aligned}$$

Applying (3.5) to $D''\alpha+\alpha\wedge\alpha$, we obtain

$$\begin{aligned} D''\alpha+\alpha\wedge\alpha = H(D''\alpha+\alpha\wedge\alpha) &+ D''(D''^*\circ G\circ D''\alpha + D''^*\circ G(\alpha\wedge\alpha))\\ &+ D''^*\circ D''\circ G(\alpha\wedge\alpha)\,. \end{aligned}$$

But

$$\begin{aligned} D''^*\circ G\circ D''\alpha = D''^*\circ D''\circ G\alpha &= \Box''_h\circ G\alpha - D''\circ D''^*\circ G\alpha\\ &= \alpha - H\alpha - D''\circ D''^*\circ G\alpha\,. \end{aligned}$$

Substituting this into the equation above, we obtain

$$(3.7)\qquad D''\alpha+\alpha\wedge\alpha = H(\alpha\wedge\alpha) + D''(\alpha + D''^*\circ G(\alpha\wedge\alpha)) + D''^*\circ D''\circ G(\alpha\wedge\alpha)\,.$$

From this orthogonal decomposition we obtain

$$(3.8)\qquad D''\alpha+\alpha\wedge\alpha=0 \Longleftrightarrow \begin{cases} D''(\alpha+D''^*\circ G(\alpha\wedge\alpha))=0\\ D''^*\circ D''\circ G(\alpha\wedge\alpha)=0\\ H(\alpha\wedge\alpha)=0\,. \end{cases}$$

We consider a *slice* $D''+\mathscr{S}_{D''}$ in $\mathscr{H}''(E)$, where

$$(3.9)\qquad \mathscr{S}_{D''} = \{\alpha\in C^{0,1};\ D''\alpha+\alpha\wedge\alpha=0 \text{ and } D''^*\alpha=0\}\,.$$

While the first condition $D''\alpha+\alpha\wedge\alpha=0$ simply means that $D''+\alpha$ is in $\mathscr{H}''(E)$, the second condition $D''^*\alpha=0$ says that the slice $D''+\mathscr{S}_{D''}$ is perpendicular to the $GL(E)$-orbit at D''. In fact, for a 1-parameter group e^{tf}, $f\in A^0(\mathrm{End}(E))$, the tangent vector to its orbit at D'' is given by $D''f$, (see (2.5)), and

$$D''^*\alpha=0 \Longleftrightarrow (D''f,\alpha)=(f,D''^*\alpha)=0 \qquad \text{for all } f\,.$$

We define the *Kuranishi map* k by

$$(3.10)\qquad k\colon C^{0,1}\longrightarrow C^{0,1}\,,\qquad k(\alpha)=\alpha+D''^*\circ G(\alpha\wedge\alpha)\,.$$

If $D''+\alpha$ is in $\mathscr{H}''(E)$, then

$$(3.11)\qquad D''(k(\alpha))=0$$

by the first equation on the right of (3.8). On the other hand,

$$(3.12) \qquad D''^*(k(\alpha)) = D''^*\alpha .$$

If we write

$$(3.13) \qquad \boldsymbol{H}^{0,q} = \{\beta \in C^{0,q};\ \square''_h\beta = 0\} = \{\beta \in C^{0,q};\ D''\beta = 0 \ \&\ D''^*\beta = 0\} ,$$

then (3.10), (3.11) and (3.12) imply

$$(3.14) \qquad k(\mathscr{S}_{D''}) \subset \boldsymbol{H}^{0,1} .$$

We recall that $\mathrm{End}^0(E)$ denotes the bundle of trace-free endomorphisms of E, (see (2.27)) and that $(\tilde{C}^*) = \{\tilde{C}^{0,q} = A^{0,q}(\mathrm{End}^0(E)), D''\}$ is the subcomplex of (C^*) consisting of trace-free elements. The q-th cohomology of $(\tilde{C}^*)$ denoted by

$$(3.15) \qquad \tilde{H}^{0,q} = H^q(M, \mathrm{End}^0(E^{D''}))$$

is isomorphic (see (2.36)) to the space of harmonic forms

$$(3.16) \qquad \tilde{\boldsymbol{H}}^{0,q} = \{\beta \in \tilde{C}^{0,q};\ \square''_h\beta = 0\} .$$

We consider

$$\begin{aligned} \tilde{\boldsymbol{H}}^{0,0} &= \{\text{holomorphic sections of } \mathrm{End}^0(E^{D''})\} \\ &= \{\text{trace-free holomorphic endomorphisms of } E^{D''}\} . \end{aligned}$$

We see that the holomorphic vector bundle $E^{D''}$ is simple if and only if $\tilde{\boldsymbol{H}}^{0,0} = 0$. With this in mind, we state now

(3.17) **Theorem** *Let $D'' \in \mathscr{H}''(E)$ be a holomorphic structure in E such that the holomorphic vector bundle $E^{D''}$ is simple. Let*

$$p: \mathscr{S}_{D''} \longrightarrow \mathscr{H}''(E)/GL(E)$$

be the natural map which sends $\alpha \in \mathscr{S}_{D''}$ to the point $[D'' + \alpha]$ of $\mathscr{H}''(E)/GL(E)$ represented by $D'' + \alpha \in \mathscr{H}''(E)$. Then p gives a homeomorphism of a neighborhood of 0 in $\mathscr{S}_{D''}$ onto a neighborhood of $[D''] = p(0)$ in $\mathscr{H}''(E)/GL(E)$.

Proof We prove first that p is locally surjective at 0. Let $D'' + \beta \in \mathscr{H}''(E)$, where $\beta \in C^{0,1}$ is sufficiently close to 0. We must show that there is a transformation $f \in GL(E)$ such that

$$(D'' + \beta)^f = f^{-1} \circ (D'' + \beta) \circ f \in D'' + \mathscr{S}_{D''} .$$

Set

(3.18) $$\alpha = f^{-1} \circ (D'' + \beta) \circ f - D''$$

with f yet to be chosen. Then

$$D'' + \alpha = f^{-1} \circ (D'' + \beta) \circ f \in \mathscr{H}''(E) \qquad \text{for every} \quad f \in GL(E),$$

so that (see (3.3))

$$D''\alpha + \alpha \wedge \alpha = 0,$$

which is one of the two defining conditions for $\mathscr{S}_{D''}$. The problem is then to show that, for a suitable $f \in GL(E)$, α satisfies the other condition for $\mathscr{S}_{D''}$, (see (3.9)):

$$D''^{*}\alpha = 0.$$

We consider $C^{0,0} = A^0(\mathrm{End}(E))$ as the Lie algebra of the group $GL(E)$. Since $E^{D''}$ is simple, $H^{0,0} = H^0(M, \mathrm{End}(E^{D''}))$ is 1-dimensional and consists of transformations cI_E, $c \in \boldsymbol{C}$. Let S be the orthogonal complement of $\boldsymbol{H}^{0,0}$ in $C^{0,0}$, i.e.,

(3.19) $$S = \left\{ u \in C^{0,0};\ \int_M (\mathrm{tr}\, u)\Phi^n = 0 \right\},$$

so that

$$C^{0,0} = \boldsymbol{H}^{0,0} \oplus S.$$

Then S is an ideal of the Lie algebra $C^{0,0}$ since $\mathrm{tr}([u, v]) = 0$ for all $u, v \in C^{0,0}$.

We define a map

$$F\colon S \times C^{0,1} \longrightarrow S$$

by

(3.20) $$F(u, \beta) = D''^{*}(e^{-u} \circ (D'' + \beta) \circ e^{u} - D''), \qquad (u, \beta) \in S \times C^{0,1}.$$

We extend F to a map

$$F\colon L^2_{k+1}(S) \times L^2_k(C^{0,1}) \longrightarrow L^2_{k-1}(S) \qquad \text{for} \quad k > n.$$

For $\beta = 0$, we consider the derivative of $F(\cdot, 0)\colon S \to S$ at $u = 0$. Then

$$\frac{\partial F}{\partial u}(0, 0) = D''^{*} \circ D'' = \square''_h\colon L^2_{k+1}(S) \longrightarrow L^2_{k-1}(S)$$

is an isomorphism. By the implicit function theorem, there is a smooth map

$$(v, \beta) \longmapsto u = u(v, \beta) \in S$$

defined in a neighborhood of $(0,0)$ in $S \times C^{0,1}$ such that

$$F(u(v, \beta), \beta) = v .$$

Thus the mapping $(u, \beta) \longmapsto F(u, \beta)$ fibers a neighborhood of $(0,0)$ in $S \times C^{0,1}$ over a neighborhood of 0 in S; the fibre over $v \in S$ consists of points $(u(v, \beta), \beta)$ where β runs through a neighborhood of 0 in $C^{0,1}$.

In particular, for $v=0$ and β small, we obtain $u=u(0, \beta)$ such that $F(u, \beta)=0$. We set $f=e^u$. Then

$$\begin{aligned} 0 = F(u, \beta) &= D''^*(f^{-1}D''f + f^{-1} \circ \beta \circ f) \\ &= f^{-1}(\square''_h f + \text{lower order terms}) . \end{aligned}$$

This shows that f is a solution to an elliptic differential equation and hence is of class C^∞. Thus we have found $f \in GL(E)$ such that

$$\alpha = f^{-1} \circ (D'' + \beta) \circ f - D'' \in \mathscr{S}_{D''} .$$

Next, we shall show that p is locally injective at 0. We have to show that if two holomorphic structures $D'' + \alpha_1$ and $D'' + \alpha_2$ with $\alpha_1, \alpha_2 \in \mathscr{S}_{D''}$ are sufficiently close to D'' (i.e., α_1 and α_2 are sufficiently close to 0) and if there is a transformation $f \in GL(E)$ such that

$$f^{-1} \circ (D'' + \alpha_1) \circ f = D'' + \alpha_2 , \tag{3.21}$$

then $\alpha_1 = \alpha_2$. When f is sufficiently close to the identity, i.e., $f=e^u$, where $u \in C^{0,0}$ is near 0, this follows from the implicit function theorem stated above. In fact, since $\alpha_1, \alpha_2 \in \mathscr{S}_{D''}$, we have

$$F(u, \alpha_1) = D''^*\alpha_2 = 0 , \qquad F(0, \alpha_1) = D''^*\alpha_1 = 0 .$$

Hence, $u=0$ and $\alpha_1 = \alpha_2$.

If f is an arbitrary element of $GL(E)$, we rewrite (3.21) as

$$D''f = f \circ \alpha_2 - \alpha_1 \circ f . \tag{3.22}$$

Applying the decomposition $C^{0,0} = H^{0,0} + S$ to f, we write

$$f = cI_E + f_0 , \qquad \text{where} \quad c \in \boldsymbol{C} , \quad f_0 \in S .$$

Since

$$\left(\frac{1}{c} f\right)^{-1} \circ (D'' + \alpha_1) \circ \left(\frac{1}{c} f\right) = f^{-1} \circ (D'' + \alpha_1) \circ f ,$$

replacing f by $(1/c)f$ we may assume that

$$f = I_E + f_0, \qquad f_0 \in S.$$

We apply the following estimate (see (2.34))

$$\|D''^* \circ G\psi\|_{2,k+1} \leqq c\|\psi\|_{2,k} \qquad \psi \in C^{0,1}$$

to $\psi = D''f_0$. Then we have

$$\|f_0\|_{2,k+1} = \|\Box''_h \circ G f_0\|_{2,k+1} = \|D''^* \circ G \circ D''f_0\|_{2,k+1} \leqq c\|D''f_0\|_{2,k}.$$

On the other hand, from (3.22) we obtain

$$\begin{aligned} \|D''f_0\|_{2,k} = \|D''f\|_{2,k} &\leqq c'\|f\|_{2,k+1}(\|\alpha_1\|_{2,k} + \|\alpha_2\|_{2,k}) \\ &\leqq c'(\|I_E\|_{2,k+1} + \|f_0\|_{2,k+1})(\|\alpha_1\|_{2,k} + \|\alpha_2\|_{2,k}). \end{aligned}$$

Hence,

$$\|f_0\|_{2,k+1} \leqq c''(\|I_E\|_{2,k+1} + \|f_0\|_{2,k+1})(\|\alpha_1\|_{2,k} + \|\alpha_2\|_{2,k}),$$

which gives

$$\|f_0\|_{2,k+1} \leqq \frac{c''\|I_E\|_{2,k+1}(\|\alpha_1\|_{2,k} + \|\alpha_2\|_{2,k})}{1 - c''(\|\alpha_1\|_{2,k} + \|\alpha_2\|_{2,k})}.$$

This shows that if $\|\alpha_1\|_{2,k}$ and $\|\alpha_2\|_{2,k}$ are small, then f_0 is near 0 and f is near the identity. Then the special case considered above applies. Q.E.D.

(3.23) **Theorem** *Let $D'' \in \mathscr{H}''(E)$ be a holomorphic structure such that*

$$\tilde{H}^{0,2} = H^2(M, \operatorname{End}^0(E^{D''})) = 0.$$

Then the Kuranishi map k gives a homeomorphism of a neighborhood of 0 *in the slice $\mathscr{S}_{D''}$ onto a neighborhood of* 0 *in $\boldsymbol{H}^{0,1} \approx H^1(M, \operatorname{End}(E^{D''}))$.*

We note that by (2.28) the condition $\tilde{H}^{0,2} = 0$ is equivalent to

$$H^2(M, \operatorname{End}(E^{D''})) = H^{0,2}(M, \boldsymbol{C}). \tag{3.24}$$

Proof Since the differential of the Kuranishi map $k\colon L^2_k(C^{0,1}) \to L^2_k(C^{0,1})$ at 0 is the identity map, by the inverse function theorem there is an inverse k^{-1} defined in a neighborhood of 0 in $L^2_k(C^{0,1})$. Let $\beta \in \boldsymbol{H}^{0,1} \subset C^{0,1}$ be near 0 so that $k^{-1}(\beta)$ is defined, and set

$$\alpha = k^{-1}(\beta).$$

Then

$$\beta = k(\alpha) = \alpha + D''^* \circ G(\alpha \wedge \alpha)\,. \tag{3.25}$$

Applying $\Box''_h$ to (3.25) we obtain

$$\begin{aligned} 0 = \Box''_h \beta = \Box''_h \alpha + D''^* \circ \Box''_h \circ G(\alpha \wedge \alpha) = \Box''_h \alpha + D''^*(\alpha \wedge \alpha - H(\alpha \wedge \alpha)) \\ = \Box''_h \alpha + D''^*(\alpha \wedge \alpha)\,, \end{aligned}$$

showing that α is a solution to an elliptic differential equation and hence is of class C^∞. Also from (3.25) we obtain

$$0 = D''\beta = D''\alpha + D'' \circ D''^* \circ G(\alpha \wedge \alpha) \tag{3.26}$$

$$0 = D''^*\beta = D''^*\alpha\,. \tag{3.27}$$

Applying (3.5) to $\alpha \wedge \alpha$, we obtain

$$\begin{aligned} D''\alpha + \alpha \wedge \alpha &= D''\alpha + D'' \circ D''^* \circ G(\alpha \wedge \alpha) + D''^* \circ D'' \circ G(\alpha \wedge \alpha) + H(\alpha \wedge \alpha) \\ &= D''^* \circ D'' \circ G(\alpha \wedge \alpha) + H(\alpha \wedge \alpha) \end{aligned} \tag{3.28}$$

by (3.26). Since $\alpha \wedge \alpha$ is trace-free, $H(\alpha \wedge \alpha)$ is in the trace-free part $\tilde{\boldsymbol{H}}^{0,2}$ of $\boldsymbol{H}^{0,2}$. By our assumption, $H(\alpha \wedge \alpha) = 0$. Denoting the left hand side of (3.28) by γ, we have

$$\begin{aligned} \gamma = D''\alpha + \alpha \wedge \alpha &= D''^* \circ D'' \circ G(\alpha \wedge \alpha) \\ &= D''^* \circ G(D''\alpha \wedge \alpha - \alpha \wedge D''\alpha) \\ &= D''^* \circ G(\gamma \wedge \alpha - \alpha \wedge \gamma)\,. \end{aligned} \tag{3.29}$$

Apply the following estimate (see (2.34))

$$\|D''^* \circ Gv\|_{2,k+1} \leqq c\|v\|_{2,k}$$

to $v = \gamma \wedge \alpha - \alpha \wedge \gamma$. Then

$$\begin{aligned} \|\gamma\|_{2,k} \leqq \|\gamma\|_{2,k+1} &= \|D''^* \circ G(\gamma \wedge \alpha - \alpha \wedge \gamma)\|_{2,k+1} \\ &\leqq c\|\gamma\|_{2,k} \cdot \|\alpha\|_{2,k}\,. \end{aligned} \tag{3.30}$$

Taking α sufficiently close to 0 so that $\|\alpha\|_{2,k} < 1/c$, we conclude $\gamma = 0$ from (3.30). This, together with (3.27), shows that α is in $\mathscr{S}_{D''}$. Q.E.D.

We know (see (2.1)–(2.5)) that, given a 1-parameter family $D''_t \in \mathscr{H}''(E)$ of holomorphic structure with $D'' = D''_0$, the induced infinitesimal variation $\partial_t D''_t|_{t=0}$

defines an element of $\boldsymbol{H}^{0,1} \approx H^{0,1}$. Conversely, we have

(3.31) **Corollary** *Let $D'' \in \mathscr{H}''(E)$. Under the same assumption as in (3.23), every element of $\boldsymbol{H}^{0,1} \approx H^1(M, \mathrm{End}(E^{D''}))$ comes from a 1-parameter family of variations $D''_t \in \mathscr{H}''(E)$ of D''.*

Proof Let $\beta \in \boldsymbol{H}^{0,1}$. For each small t, there is a unique $\alpha_t \in \mathscr{S}_{D''}$ such that

$$t\beta = k(\alpha_t) = \alpha_t + D''^* \circ G(\alpha_t \wedge \alpha_t) .$$

Then

$$t\beta = H(t\beta) = H(\alpha_t) .$$

Differentiating this equation with respect to t at $t=0$, we obtain

$$\beta = H(\partial_t \alpha_t |_{t=0}) .$$

This shows that β comes from the 1-parameter family of holomorphic structures $D''_t = D'' + \alpha_t$. Q.E.D.

From (3.17) and (3.23) we obtain

(3.32) **Corollary** *The moduli space $\mathscr{H}''(E)/GL(E)$ of holomorphic structures in E is a nonsingular complex manifold in a neighborhood of $[D''] \in \mathscr{H}''(E)/GL(E)$ if the holomorphic vector bundle $E^{D''}$ is simple and if $\tilde{H}^{0,2} = H^2(M, \mathrm{End}^0(E^{D''})) = 0$. Then its tangent space at $[D'']$ is naturally isomorphic to $\boldsymbol{H}^{0,1} \approx H^1(M, \mathrm{End}(E^{D''}))$.*

Proof Since the Kuranishi map $k: C^{0,1} \to C^{0,1}$ is a (quadratic) polynomial map and is non-degenerate at 0, by (3.23) the slice $\mathscr{S}_{D''}$ is a complex submanifold of $C^{0,1}$. By (3.17), a neighborhood of 0 in $\mathscr{S}_{D''}$ can be considered as a coordinate neighborhood of $[D'']$ in $\mathscr{H}''(E)/GL(E)$. Q.E.D.

Now we consider a holomorphic structure $D'' \in \mathscr{H}''(E)$ in E such that $E^{D''}$ is a simple vector bundle with $\tilde{H}^{0,2} = H^2(M, \mathrm{End}^0(E^{D''})) \neq 0$. In general, at such a point D'', the Kuranishi map

$$k: \mathscr{S}_{D''} \longrightarrow \boldsymbol{H}^{0,1} \approx H^1(M, \mathrm{End}(E^{D''}))$$

is only injective but not necessarily surjective at the origin. Since $k: L^2_k(C^{0,1}) \to L^2_k(C^{0,1})$ is a biholomorphic map in a neighborhood of 0 in $L^2_k(C^{0,1})$, $k^{-1}(\boldsymbol{H}^{0,1})$ is a nonsingular complex submanifold in a neighborhood of 0. Since the condition

$D''^*\alpha=0$ in (3.9) is automatically satisfied by α in $k^{-1}(\boldsymbol{H}^{0,1})$, (see (3.12)), the slice $\mathscr{S}_{D''}$ is given by

$$\mathscr{S}_{D''}=\{\alpha\in k^{-1}(\boldsymbol{H}^{0,1});\ D''\alpha+\alpha\wedge\alpha=0\}\ . \tag{3.33}$$

It follows that $\mathscr{S}_{D''}$ is a (possibly non-reduced) analytic subset of a complex manifold $k^{-1}(\boldsymbol{H}^{0,1})$ in a neighborhood of 0; it is, in fact, defined by quadratic polynomials.

Let $\hat{\mathscr{H}}''(E)$ denote the set of simple holomorphic structures in E. The group $GL(E)$ acting on $\mathscr{H}''(E)$ leaves $\hat{\mathscr{H}}''(E)$ invariant, and we can speak of the *moduli space* $\hat{\mathscr{H}}''(E)/GL(E)$ *of simple holomorphic structures* in E. We have shown

(3.34) **Theorem** *The moduli space $\hat{\mathscr{H}}''(E)/GL(E)$ of simple holomorphic structures in E is a (possibly non-Hausdorff and non-reduced) complex analytic space. It is nonsingular at $[D'']$ if $\tilde{H}^{0,2}=H^2(M, \mathrm{End}^0(E^{D''}))=0$, and its tangent space at such a nonsingular point is naturally isomorphic to $\boldsymbol{H}^{0,1}\approx H^1(M, \mathrm{End}(E^{D''}))$.*

We shall now consider the question of Hausdorff property for moduli spaces. The following argument is due to Okonek.

(3.35) **Lemma** *Let D'', $\tilde{D}''\in\mathscr{H}''(E)$ be two holomorphic structures in E representing two distinct points $[D'']$ and $[\tilde{D}'']$ of $\mathscr{H}''(E)/GL(E)$. If every neighborhood of $[D'']$ in $\mathscr{H}''(E)/GL(E)$ intersects every neighborhood of $[\tilde{D}'']$, then there are nonzero (sheaf) homomorphisms*

$$\varphi: E^{\tilde{D}''}\longrightarrow E^{D''} \quad \text{and} \quad \psi: E^{D''}\longrightarrow E^{\tilde{D}''}\ .$$

Proof There are sequences $D''_i\in\mathscr{H}''(E)$ and $f_i\in GL(E)$ such that

$$D''_i\longrightarrow D'' \quad \text{and} \quad D''^{f_i}_i=f_i^{-1}\circ D''_i\circ f_i\longrightarrow\tilde{D}''$$

as $i\to\infty$. Write

$$\tilde{D}''_i=D''^{f_i}_i\ .$$

Since $f_i\circ\tilde{D}''_i=D''_i\circ f_i$, the map

$$f_i: E^{\tilde{D}''_i}\longrightarrow E^{D''_i}$$

is holomorphic, i.e., $f_i\in H^0(M, \mathrm{Hom}(E^{\tilde{D}''_i}, E^{D''_i}))$. By the upper semi-continuity of cohomology (see (2.37)), we have

$$\dim H^0(M, \mathrm{Hom}(E^{\tilde{D}''}, E^{D''}))\geqq\limsup\dim H^0(M, \mathrm{Hom}(E^{\tilde{D}''_i}, E^{D''_i}))\geqq 1\ .$$

Hence, there is a nonzero homomorphism $\varphi: E^{\tilde{D}''} \to E^{D''}$. By interchanging the roles of D'' and $\tilde{D}''$ in the argument above, we obtain a nonzero homomorphism $\psi: E^{D''} \to E^{\tilde{D}''}$. Q.E.D.

(3.36) **Lemma** *If $E^{D''}$ is simple in* (3.35), *then*

$$\varphi \circ \psi = 0 .$$

Proof Since $\varphi \circ \psi$ is an endomorphism of a simple bundle, we have $\varphi \circ \psi = cI$ with $c \in \boldsymbol{C}$. Since $E^{D''}$ and $E^{\tilde{D}''}$ have the same rank, it follows that φ and ψ are isomorphisms if $c \neq 0$. Since $[D''] \neq [\tilde{D}'']$ by assumption, we must have $c = 0$. Q.E.D.

(3.37) **Proposition** *Let D'', $\tilde{D}'' \in \mathscr{H}''(E)$ represent two distinct points $[D'']$ and $[\tilde{D}'']$ of $\mathscr{H}''(E)/GL(E)$. If $E^{D''}$ is Φ-stable and $E^{\tilde{D}''}$ is Φ-semistable, then $[D'']$ and $[\tilde{D}'']$ have disjoint neighborhoods in $\mathscr{H}''(E)/GL(E)$.*

Proof Assume the contrary. Let φ and ψ be the nonzero homomorphisms obtained in (3.35). By (2) of (V.7.11), ψ is an isomorphism. Hence, $\varphi \circ \psi \neq 0$, contradicting (3.36). Q.E.D.

(3.38) *Remark* Let $\mathscr{S}''(E)$ denote the set of Φ-stable holomorphic structures D'' in E. Then $\mathscr{S}''(E) \subset \hat{\mathscr{H}}''(E)$, and by (3.37) $\mathscr{S}''(E)/GL(E)$ is Hausdorff. But there does not seem to exist an analytic proof showing that $\mathscr{S}''(E)/GL(E)$ is open in $\hat{\mathscr{H}}''(E)/GL(E)$. In the algebraic case, this is known for Gieseker stable bundles, (see Maruyama [2]).

§4 Moduli of Einstein-Hermitian structures

In this section we present results of H-J. Kim [1]. Let E be a C^∞ complex vector bundle of rank r over an n-dimensional compact Kähler manifold M with Kähler form Φ. We fix an Hermitian structure h in E. We continue to use the notation in Sections 2 and 3.

We fix an Einstein-Hermitian connection $D \in \mathscr{E}(E, h)$. Using the projection $P_+ : B^2 \to B^2_+$ introduced in Section 2, the Einstein condition can be expressed by

$$P_+(R) = \frac{1}{n}(\Lambda R)\Phi = -\frac{ic}{n} I_E \Phi . \tag{4.1}$$

Any h-connection in E is of the form $D+\alpha$, where $\alpha \in B^1 = A^1(\mathrm{End}(E, h))$. Its curvature $R(D+\alpha)$ is given by

$$R(D+\alpha)=(D+\alpha)\circ(D+\alpha)=R+D\alpha+\alpha\wedge\alpha\,. \tag{4.2}$$

The connection $D+\alpha$ is Einstein-Hermitian if and only if

$$P_+(R(D+\alpha))=-\frac{ic}{n}I_E=P_+(R)\,,$$

or equivalently, if and only if $P_+(D\alpha+\alpha\wedge\alpha)=0$. Hence the set $\mathscr{E}(E, h)$ of Einstein-Hermitian connections is given by

$$\mathscr{E}(E,h)=\{D+\alpha;\ \alpha\in B^1 \text{ and } D_+\alpha+P_+(\alpha\wedge\alpha)=0\}\,. \tag{4.3}$$

For the elliptic complex (B^*), we denote its Laplacian, Green's operator and harmonic projection by Δ, G and H, respectively, so that (see (2.33))

$$I=H+\Delta\circ G\,. \tag{4.4}$$

The commutation relations $G\circ D''=D''\circ G$ and $D''^*\circ G=G\circ D''^*$ of (3.5) must be modified by replacing D'' with D, D_+ and D_2 at appropriate dimensions, (see the definition of (B^*) in Section 2). In particular, on B^2_+, the Laplacian Δ is given by

$$\Delta=D_+\circ D_+^*+D_2^*\circ D_2\,. \tag{4.5}$$

As in Section 3, we can extend these operators to appropriate Sobolev spaces.

Applying (4.4) and (4.5) to $P_+(\alpha\wedge\alpha)$, we obtain the following orthogonal decomposition of $D_+\alpha+P_+(\alpha\wedge\alpha)$:

$$\begin{aligned} D_+\alpha+P_+(\alpha\wedge\alpha) &= D_+(\alpha+D_+^*\circ G\circ P_+(\alpha\wedge\alpha)) \\ &\quad +D_2^*\circ D_2\circ G\circ P_+(\alpha\wedge\alpha)+H\circ P_+(\alpha\wedge\alpha)\,. \end{aligned}$$

Hence,

$$D_+\alpha+P_+(\alpha\wedge\alpha)=0 \Longleftrightarrow \begin{cases} D_+(\alpha+D_+^*\circ G\circ P_+(\alpha\wedge\alpha))=0 \\ D_2^*\circ D_2\circ G\circ P_+(\alpha\wedge\alpha)=0 \\ H\circ P_+(\alpha\wedge\alpha)=0 \end{cases} \tag{4.6}$$

We consider a *slice* $D+\mathscr{S}_D$ in $\mathscr{E}(E, h)$, where

$$\mathscr{S}_D=\{\alpha\in B^1;\ D_+\alpha+P_+(\alpha\wedge\alpha)=0,\ D^*\alpha=0\}\,. \tag{4.7}$$

We note that the second condition $D^*\alpha=0$ above says that the slice is perpendicular to the $U(E, h)$-orbit of D while the first condition is nothing but the condition for $D+\alpha$ to be in $\mathscr{E}(E, h)$, see (4.3). In fact, for a 1-parameter group $e^{tf}, f \in A^0(\mathrm{End}(E, h))$, the tangent vector to its orbit at D is given by Df, (see (2.11)), and

$$D^*\alpha = 0 \Longleftrightarrow (Df, \alpha) = (f, D^*\alpha) = 0 \qquad \text{for all} \quad f.$$

We define the *Kuranishi map* k by

$$k : B^1 \longrightarrow B^1, \qquad k(\alpha) = \alpha + D_+^* \circ G \circ P_+(\alpha \wedge \alpha). \tag{4.8}$$

If $D+\alpha$ is in $\mathscr{E}(E, h)$, then

$$D_+(k(\alpha)) = 0 \tag{4.9}$$

by the first equation on the right hand side of (4.6). On the other hand,

$$D^*(k(\alpha)) = D^*\alpha. \tag{4.10}$$

Hence, if we write

$$\boldsymbol{H}^q = \{\beta \in B^q;\ \Delta\beta = 0\}, \tag{4.11}$$

then

$$k(\mathscr{S}_D) \subset \boldsymbol{H}^1 = \{\beta \in B^1;\ D_+\beta = 0 \text{ and } D^*\beta = 0\}. \tag{4.12}$$

We recall that $\mathrm{End}^0(E, h)$ is the bundle of trace-free skew-Hermitian endomorphisms of (E, h), (see (2.27)) and that $(\tilde{B}^*) = \{\tilde{B}^q = A^q(\mathrm{End}^0(E, h)), D''\}$ is the subcomplex of (B^*) consisting of trace-free elements. (Here, D'' should be replaced by D, D_+ and D_2 in dimension $q = 0$, 1 and 2, (see Section 2).) The q-th cohomology of $(\tilde{B}^*)$ denoted by $\tilde{H}^q$ is isomorphic to the space of harmonic forms

$$\tilde{\boldsymbol{H}}^q = \{\beta \in \tilde{B}^q;\ \Delta\beta = 0\}. \tag{4.13}$$

We consider

$$\begin{aligned} \tilde{\boldsymbol{H}}^0 &= \{\beta \in B^0;\ D\beta = 0\} \\ &= \{\text{trace-free parallel sections of } \mathrm{End}(E, h)\}. \end{aligned}$$

Since $\tilde{H}^{0,0} = \tilde{H}^0 \otimes \boldsymbol{C}$, (see (2.29)), we have

(4.14) **Proposition** *For $D \in \mathscr{E}(E, h)$, the following conditions are equivalent.*

(a) $\tilde{\boldsymbol{H}}^0 = \{\beta \in A^0(\mathrm{End}^0(E, h));\ D\beta = 0\} = 0$;

(b) $\tilde{\boldsymbol{H}}^{0,0} \approx H^0(M, \mathrm{End}^0(E^{D''})) = 0$;

(c) *the holomorphic vector bundle $E^{D''}$ is simple*;

(d) *the connection D is irreducible in the sense that its holonomy group is* *irreducible.*

With this in mind, we state

(4.16) **Theorem** *Let $D \in \mathscr{E}(E, h)$ be irreducible. Let*

$$p: \mathscr{S}_D \longrightarrow \mathscr{E}(E, h)/U(E, h)$$

be the natural map which sends $\alpha \in \mathscr{S}_D$ to the point $[D+\alpha]$ of $\mathscr{E}(E, h)/U(E, h)$ represented by the connection $D+\alpha \in \mathscr{E}(E, h)$. Then p gives a homeomorphism of a neighborhood of 0 in $\mathscr{S}_D$ onto a neighborhood of $[D]=p(0)$ in $\mathscr{E}(E, h)/U(E, h)$.

The proof is analogous to that of (3.17). Leaving details to Kim [1], we shall only indicate how the proof of (3.17) should be adapted. Generally, we have to replace $C^{0,0}$, $C^{0,1}$, D'' and $\square''_h$ by B^0, B^1, D or D_+ and Δ. We have to replace $\alpha \wedge \alpha$ by $P_+(\alpha \wedge \alpha)$. The most notable change is in the decomposition of an element $f \in U(E, h)$. Since f is not necessarily in B^0, we complexify the decomposition $B^0 = \boldsymbol{H}^0 + S$ to obtain the decomposition $C^{0,0} = \boldsymbol{H}^{0,0} + S_C$, where $S_C = S \otimes \boldsymbol{C}$, and decompose $f \in C^{0,0}$ accordingly. Then the remainder of the proof will be the same.

(4.17) **Theorem** *Let $D \in \mathscr{E}(E, h)$. If*

$$H^0(M, \mathrm{End}^0(E^{D''}))=0 \quad \textit{and} \quad H^2(M, \mathrm{End}^0(E^{D''}))=0\,,$$

then the Kuranishi map k gives a homeomorphism of a neighborhood of 0 in the slice $\mathscr{S}_D$ onto a neighborhood of 0 in $\boldsymbol{H}^1 \approx H^1(M, \mathrm{End}(E^{D''}))$.

Since $\tilde{\boldsymbol{H}}^2 \approx \tilde{\boldsymbol{H}}^{0,0} + \tilde{\boldsymbol{H}}^{0,2}$, (see (2.29)), the assumption in (4.17) is equivalent to $\tilde{\boldsymbol{H}}^2 = 0$. In the proof, the assumption is used in the form of $\tilde{\boldsymbol{H}}^2 = 0$. We note also that the isomorphism $\boldsymbol{H}^1 \approx \boldsymbol{H}^{0,1}$ is in (2.21).

The proof is analogous to that of (3.23), see Kim [1] for details. We have to replace D'' by either D, D_+ or D_2. As in the case of (4.16), we must replace $\alpha \wedge \alpha$ by $P_+(\alpha \wedge \alpha)$.

We know (see (2.7)–(2.11)) that, given a 1-parameter family $D_t \in \mathscr{E}(E, h)$ of connections with $D = D_0$, the induced infinitesimal variation $\partial_t D_t|_{t=0}$ defines an element of $\boldsymbol{H}^1 \approx H^1$. The proof of the following converse is similar to that of (3.31).

(4.18) **Corollary** *Let $D \in \mathscr{E}(E, h)$. Under the same assumption as in (4.17), every element of $\boldsymbol{H}^1 \approx H^1(M, \mathrm{End}(E^{D''}))$ comes from a 1-parameter family of* *variations* $D_t \in \mathscr{E}(E, h)$ *of* D.

From (4.16) and (4.17) we obtain

(4.19) **Corollary** *The moduli space* $\mathscr{E}(E, h)/U(E, h)$ *of Einstein-Hermitian connections in* (E, h) *is a nonsingular complex manifold in a neighborhood of* $[D] \in \mathscr{E}(E, h)/U(E, h)$ *if the holomorphic vector bundle* $E^{D''}$ *is simple and if* $H^2(M, \mathrm{End}^0(E^{D''})) = 0$.

Let $\hat{\mathscr{E}}(E, h)$ denote the set of irreducible Einstein-Hermitian connections in (E, h). In analogy with (3.34) we have

(4.20) **Theorem** *The moduli space* $\hat{\mathscr{E}}(E, h)/U(E, h)$ *of irreducible Einstein-Hermitian connections in* (E, h) *is a (possibly non-reduced) complex analytic space. It is nonsingular at* $[D]$ *if* $\tilde{H}^{0,2} = H^2(M, \mathrm{End}^0(E^{D''})) = 0$, *and its tangent space at such a nonsingular point is naturally isomorphic to* $H^{0,1} = H^1(M, \mathrm{End}(E^{D''}))$.

Comparing (4.20) with (3.34) we can see that $\hat{\mathscr{E}}(E, h)/U(E, h)$ is open in $\hat{\mathscr{H}}''(E)/GL(E)$ at its nonsingular points. However, we can prove the following stronger result (due to Kim [1]) more directly.

(4.21) **Theorem** *The moduli space* $\mathscr{E}(E, h)/U(E, h)$ *of Einstein-Hermitian connections in* (E, h) *is open in the moduli space* $\mathscr{H}''(E)/GL(E)$ *of holomorphic structures in* E.

Proof Let $D \in \mathscr{E}(E, h)$ and $D'' \in \mathscr{H}''(E)$ the corresponding point. We shall show that given $\tilde{D}'' \in \mathscr{H}''(E)$ sufficiently close to D'', there exists $D_1 \in \mathscr{E}(E, h)$ and $f_1 \in GL(E)$ such that $D_1'' = f_1^{-1} \circ \tilde{D}'' \circ f_1$.

Let

$$P = \{f \in GL(E);\ h(f\xi, \eta) = h(\xi, f\eta),\ h(f\xi, \xi) \geqq 0,\ \det f = 1\}, \tag{4.22}$$

in other words, P consists of positive definite Hermitian automorphisms of E with determinant 1. Its tangent space $T_I P$ at the identity transformation I is given by

$$Q = T_I P = \{a \in A^0(\mathrm{End}(E));\ h(a\xi, \eta) = h(\xi, a\eta),\ \mathrm{tr}\, a = 0\}, \tag{4.23}$$

i.e., Q consists of Hermitian endomorphisms of E with trace 0.

We consider a general h-connection $D \in \mathscr{D}(E, h)$. Then, for $f \in P$, we have (see (1.12))

$$(4.24)\qquad D^f = f^* \circ D' \circ f^{*-1} + f^{-1} \circ D'' \circ f = f \circ D' \circ f^{-1} + f^{-1} \circ D'' \circ f .$$

Hence, the curvature $R(D^f)$ of D^f is given by

$$(4.25)\qquad R(D^f) = f \circ D' \circ D' \circ f^{-1} + f^{-1} \circ D'' \circ D'' \circ f + (f \circ D' \circ f^{-1} \circ f^{-1} \circ D'' \circ f + f^{-1} \circ D'' \circ f \circ f \circ D' \circ f^{-1}) .$$

Its mean curvature is denoted by $K(D^f)$. We set

$$(4.26)\qquad K^0(D^f) = K(D^f) - \frac{1}{r}(\operatorname{tr} K(D^f))I .$$

Since $K(D^f)$ is an Hermitian endomorphism of E, its trace-free part $K^0(D^f)$ belongs to Q. We define a mapping

$$(4.27)\qquad F\colon \mathscr{D}(E, h) \times P \longrightarrow Q , \qquad F(D, f) = K^0(D^f) .$$

We recall that the tangent space $T_D(\mathscr{D}(E, h))$ to $\mathscr{D}(E, h)$ at D is naturally isomorphic to $A^1(\operatorname{End}(E, h))$. We wish to calculate the differential of F at (D, I):

$$(4.28)\qquad dF_{(D,I)}\colon A^1(\operatorname{End}(E, h)) \times Q \longrightarrow Q .$$

Let $a \in Q$ and $f(t) = e^{at}$. From (4.25) it follows that the $(1, 1)$-component $R^{(1,1)}(D^{f(t)})$ of the curvature $R(D^{f(t)})$ is given by

$$(4.29)\qquad R^{(1,1)}(D^{f(t)}) = e^{at} \circ D' \circ e^{-2at} \circ D'' \circ e^{at} + e^{-at} \circ D'' \circ e^{2at} \circ D' \circ e^{-at} .$$

Differentiating (4.29) with respect to t at $t = 0$, we obtain

$$(4.30)\qquad \partial_t R^{(1,1)}(D^{f(t)})\big|_{t=0} = D'D''a - D''D'a .$$

Taking the trace of (4.30) with respect to the Kähler metric g, we obtain

$$(4.31)\qquad \partial_t K(D^{f(t)})\big|_{t=0} = \Delta a .$$

From (4.27) it follows that

$$(4.32)\qquad dF_{(D,I)}(0, a) = \Delta a \qquad \text{for} \quad a \in Q .$$

We extend F to a smooth map

$$F\colon L^2_k(\mathscr{D}(E, h)) \times L^2_{k+1}(P) \longrightarrow L^2_{k-1}(Q) , \qquad (k > n) ,$$

so that

$$dF_{(D,I)}(0, a) = \Delta a \qquad \text{for} \quad a \in L^2_{k+1}(Q) .$$

Now assume that D is irreducible in the sense that its holonomy group is an irreducible subgroup of $U(r)$. We shall show that $\Delta: L^2_{k+1}(Q) \to L^2_{k-1}(Q)$ is an isomorphism. Since $I = H + G \circ \Delta$, it suffices to show that Δ is injective. If $\Delta a = 0$, then a is of class C^∞ and

$$0 = (\Delta a, a) = (D^* Da, a) = (Da, Da),$$

i.e., a is parallel. Since D is irreducible, a must be of the form cI_E with c constant. On the other hand, $\mathrm{tr}(a) = 0$. Hence, $a = 0$.

Assume in addition that D is in $\mathscr{E}(E, h)$, (see (1.16)), so that, in particular

$$F(D, I) = K^0(D) = 0.$$

Since

$$a \in L^2_{k+1}(Q) \longmapsto dF_{(0,I)}(0, a) = \Delta a \in L^2_{k-1}(Q)$$

is an isomorphism, the implicit function theorem implies that if $\tilde{D}'' \in \mathscr{H}''(E)$ is sufficiently close to $D'' \in \mathscr{H}''(E)$ so that the corresponding $\tilde{D} \in \mathscr{D}(E, h)$ is sufficiently close to D, then there exists a unique $f \in L^2_{k+1}(P)$ near the identity I_E such that

$$F(\tilde{D}, f) = K^0(\tilde{D}^f) = 0. \tag{4.33}$$

Linearizing the differential equation (4.33) for f, we obtained the elliptic equation ($\tilde{\Delta} a = 0$, (see (4.32)), where $\tilde{\Delta}$ is the Laplacian defined by $\tilde{D}$. Hence, (4.33) is an elliptic equation. By the elliptic regularity, f is of class C^∞.

We shall now prove that there exists $D_1 \in \mathscr{E}(E, h)$ and $f_1 \in GL(E)$ such that $D_1'' = f_1^{-1} \circ \tilde{D}'' \circ f_1$. Since $GL(E)$ acting on $\mathscr{D}(E, h)$ by (1.12) leaves $\mathscr{H}(E, h)$ invariant, $\tilde{D} \in \mathscr{H}(E, h)$ implies $\tilde{D}^f \in \mathscr{H}(E, h)$. On the other hand, the equation $K^0(\tilde{D}^f) = 0$ of (4.33) means

$$K(\tilde{D}^f) = \varphi I_E,$$

where φ is a function on M, (see (4.26)). By (IV.2.4) there is a positive function b on M such that the Hermitian structure $h' = b^2 h$, together with the holomorphic structure $\tilde{D}''^f \in \mathscr{H}''(E)$ (corresponding to $\tilde{D}^f \in \mathscr{H}(E, h)$), satisfies the Einstein condition. Let $\hat{D}$ be the Hermitian connection defined by $h' = b^2 h$ and the holomorphic structure $\tilde{D}''^f$, i.e., $\hat{D} h' = 0$ and $\hat{D}'' = \tilde{D}''^f$. Since it satisfies the Einstein condition, $\hat{D}$ is by definition an element of $\mathscr{E}(E, h')$. Since

$$\begin{aligned}d(h(\xi,\eta)) &= d(h'(b^{-1}\xi, b^{-1}\eta))\\ &= h'(\hat{D}(b^{-1}\xi), b^{-1}\eta) + h'(b^{-1}\xi, \hat{D}(b^{-1}\eta))\\ &= h(b\hat{D}(b^{-1}\xi), \eta) + h(\xi, b\hat{D}(b^{-1}\eta)),\end{aligned}$$

we set

$$D_1 = b \circ \hat{D} \circ b^{-1} \qquad (\text{where } b \text{ denotes } bI_E).$$

Then $D_1 h = 0$ and $D_1 \circ D_1 = \hat{D} \circ \hat{D}$. Hence, $D_1 \in \mathscr{E}(E, h)$. Moreover,

$$D_1'' = b \circ \hat{D}'' \circ b^{-1} = b \circ \tilde{D}''^{f} \circ b^{-1} = (fb^{-1})^{-1} \circ \tilde{D}'' \circ (fb^{-1}) = f_1^{-1} \circ \tilde{D}''_{\cdot} \circ f_1,$$

where $f_1 = fb^{-1}$. Q.E.D.

§5 Symplectic structures

In this section, let V be a Banach manifold although, in later applications, we use only Hilbert manifolds. By a *symplectic form* ω on V we mean a 2-form ω satisfying the following conditions:

(5.1)
- (a) For each $x \in V$, $\omega_x\colon T_xV \times T_xV \to \boldsymbol{R}$ is continuous;
- (b) For each $x \in V$, ω_x is non-degenerate, i.e., if $\omega_x(u, v) = 0$ for all $v \in T_xV$, then $u = 0$;
- (c) ω_x is C^∞ in x;
- (d) ω is closed.

For short, we say that ω is a non-degenerate closed 2-form on V. The form ω_x defines a continuous linear map $T_xV \to T_x^*V$. The non-degeneracy condition (b) means that this linear map is injective. We do not assume that it is bijective.

A transformation f of V is called *symplectic* if $f^*\omega = \omega$. A vector field a on V is called an *infinitesimal symplectic transformation* if the Lie derivation $L_a = d \circ i_a + i_a \circ d$ annihilates ω, i.e., $L_a\omega = 0$. Since ω is closed, the condition $L_a\omega = 0$ reduces to $d \circ i_a\omega = 0$, i.e., the 1-form $i_a\omega$ is closed. We shall be soon interested in the situation where this 1-form is actually exact.

Let G be a Banach Lie group acting on V as a group of symplectic transformations. Let $\mathfrak{g}$ be the Banach Lie algebra of G, and $\mathfrak{g}^*$ its dual Banach space.

A *momentum map* for the action of G on V is a map $\mu\colon V \to \mathfrak{g}^*$ such that

$$(5.2) \qquad \langle a, d\mu_x(v)\rangle = \omega(a_x, v) \qquad \text{for} \quad a \in \mathfrak{g},\ v \in T_xV,\ x \in V,$$

where $d\mu_x\colon T_xV \to \mathfrak{g}^*$ is the differential of μ at x, $a_x \in T_xV$ is the vector defined by $a \in \mathfrak{g}$ through the action of G, and $\langle\ ,\ \rangle$ denotes the dual pairing between $\mathfrak{g}$ and $\mathfrak{g}^*$.

A momentum map μ may or may not exist. Its existence means that the closed 1-form $i_a\omega$ is exact and is equal to $d(\langle a, \mu\rangle)$. It is not hard to see that a momentum map is unique up to an additive factor; if μ' is another momentum map, then $\mu'-\mu$ is a (constant) element of $\mathfrak{g}^*$.

We shall now impose the following three conditions (a), (b) and (c) on our momentum map μ.

(a) Assume that μ is equivariant with respect to the coadjoint action of G in the sense that

$$\mu(g(x)) = (ad\, g)^*(\mu(x)) \quad \text{for} \quad g\in G, \quad x\in V. \tag{5.3}$$

Then G leaves $\mu^{-1}(0)\subset V$ invariant. The quotient space

$$W=\mu^{-1}(0)/G \tag{5.4}$$

is called the *reduced phase space* in symplectic geometry. More generally, for any $\alpha\in\mathfrak{g}^*$ we can define

$$W_\alpha=\mu^{-1}(\alpha)/G_\alpha,$$

where $G_\alpha=\{g\in G;\ (ad\, g)^*\alpha=\alpha\}$. However, we shall be concerned only with $W=W_0$. The following diagram organizes V, $\mu^{-1}(0)$ and W:

$$\begin{array}{ccc} \mu^{-1}(0) & \xrightarrow{j} & V \\ \downarrow & & \\ W=\mu^{-1}(0)/G & & \end{array} \tag{5.5}$$

where j is the natural injection and π is the projection.

(b) Assume that $0\in\mathfrak{g}^*$ is a weakly regular value of μ in the sense that

(5.6) $\mu^{-1}(0)$ is submanifold of V;

(5.7) for every $x\in\mu^{-1}(0)$, the inclusion $T_x(\mu^{-1}(0))\subset \mathrm{Ker}(d\mu_x)$ is an equality.

If $d\mu_x\colon T_xV\to\mathfrak{g}^*$ is surjective for every $x\in\mu^{-1}(0)$, then the implicit function theorem guarantees (b).

(c) Assume that the action of G on $\mu^{-1}(0)$ is free and that at each point $x\in\mu^{-1}(0)$ there is a *slice* $S_x\subset\mu^{-1}(0)$ for the action, i.e., a submanifold S_x of $\mu^{-1}(0)$ through x which is transversal to the orbit $G(x)$ in the sense that

$$T_x(\mu^{-1}(0)) = T_x(S_x) + T_x(G(x)).$$

If we take S_x sufficiently small, then the projection $\pi\colon\mu^{-1}(0)\to W$ defines a homeomorphism of S_x onto an open set $\pi(S_x)$ of W. This introduces a local coordinate system in W and makes W into a manifold, which may or may not be

Hausdorff. In order to have a Hausdorff manifold, we have to further assume that the action of G on $\mu^{-1}(0)$ is proper. For our later applications we have to consider the case where the action of G may not be proper.

We are now in a position to state the *reduction theorem* (Marsden-Weinstein [1]).

(5.8) **Theorem** *Let V be a Banach manifold with a symplectic form ω_V. Let G be a Banach Lie group acting on V. If there is a momentum map $\mu: V \to \mathfrak{g}^*$ satisfying* (a), (b) *and* (c), *there is a unique symplectic form ω_W on the reduced phased space W such that*

$$\pi^* \omega_W = j^* \omega_V \qquad \text{on} \quad \mu^{-1}(0).$$

Proof In order to define ω_W, we take $u, v \in T_x(\mu^{-1}(0))$ and set

$$\omega_W(\pi u, \pi v) = \omega_V(u, v).$$

The fact that ω_W is well defined is a consequence of the following three facts:

(i) If $u' \in T_{x'}(\mu^{-1}(0))$ and $\pi u' = u$, then $u' = g(u + a_x)$ for some $g \in G$ and $a \in \mathfrak{g}$;

(ii) ω_V is G-invariant, i.e., $\omega_V(g(u), g(v)) = \omega_V(u, v)$ for $v \in G$;

(iii) $i_a \omega_V = 0$, i.e., $\omega_V(a_x, \cdot) = 0$ for every $a \in \mathfrak{g}$.

This establishes the existence of a 2-form ω_W such that $\pi^* \omega_W = j^* \omega_V$. The uniqueness follows from the fact that $\pi: T_x(\mu^{-1}(0)) \to T_{\pi(x)}(W)$ is surjective.

Since $\pi: S_x \to \pi(S_x) \subset W$ is a diffeomorphism and since

$$\pi^* \omega_W|_{S_x} = \omega_V|_{S_x},$$

it follows that ω_W is also smooth and closed.

To see that ω_W is non-degenerate, let $u \in T_x(\mu^{-1}(0))$ be such that $\omega_V(u, v) = 0$ for all $v \in T_x(\mu^{-1}(0))$. We have to show that $\pi u = 0$, i.e., $u = a_x$ for some $a \in \mathfrak{g}$. This will follow from the lemma below (Marsden-Ratiu [1]):

(5.9) **Lemma** *Let X be a Banach space and $\omega: X \times X \to \boldsymbol{R}$ a continuous skew-symmetric form which is non-degenerate in the sense that $\omega(u, v) = 0$ holds for all $v \in X$ if and only if $u = 0$. For any closed subspace Y of X, set*

$$Y^\omega = \{v \in X;\ \omega(u, v) = 0 \text{ for all } u \in Y\}.$$

Then $(Y^\omega)^\omega = Y$.

We show first that this lemma implies the desired result. Let $X = T_x V$, $Y = \{a_x;\ a \in \mathfrak{g}\} = \{u \in T_x(\mu^{-1}(0));\ \pi u = 0\}$. Using (5.2) and (5.7) we obtain

$$\begin{aligned} Y^\omega &= \{v \in T_xV;\ \omega_V(a_x, v)=0 \text{ for all } a \in \mathfrak{g}\} \\ &= \{v \in T_xV;\ d\mu(v)=0\} \\ &= T_x(\mu^{-1}(0)) . \end{aligned}$$

By Lemma,

$$Y=(Y^\omega)^\omega=\{u \in T_xV;\ \omega_V(u, v)=0 \text{ for all } v \in T_x(\mu^{-1}(0))\} ,$$

and this is exactly what we wanted to prove.

Proof of (5.9) Given X and ω, we consider the following family of seminorms $\{p_v\}$ on X:

$$p_v(u)=|\omega(u, v)| .$$

With these seminorms X becomes a locally convex topological vector space; it is Hausdorff since ω is non-degenerate.

Let i: $X \to X^*$ be the injection defined by

$$i(v)=i_v\omega=\omega(\cdot, v) \qquad \text{for} \quad v \in X .$$

Then *the dual of* X *as a locally convex topological vector space is* $i(X) \subset X^*$. *That is, a linear map* α: $X \to \boldsymbol{R}$ *is continuous in the locally convex topology given by* $\{p_v\}$ *if and only if there exists an element* $v \in X$ *such that* $\alpha(u)=\omega(u, v)$ *for all* $u \in X$.

To prove this statement, let α: $X \to \boldsymbol{R}$ be continuous with respect to the family of seminorms $\{p_v\}$. Then there exist $v_1, \cdots, v_n \in X$ such that

$$|\alpha(u)| \leqq C \cdot \max p_{v_j}(u) \qquad \text{for all} \quad u \in X$$

for some positive constant C. Then α vanishes on

$$E=\bigcap \ker i(v_j)=(\operatorname{span}(v_1, \cdots, v_n))^\omega .$$

Since E is a closed subspace of finite codimension $\leqq n$, there is a finite dimensional complement F so that $X=E \oplus F$, (a Banach space direct sum). Since $i(v_1)|_F, \cdots, i(v_n)|_F$ span F^*, there exist $a_1, \cdots, a_n \in \boldsymbol{R}$ such that

$$\alpha=a_1 \cdot i(v_1)+ \cdots +a_n \cdot i(v_n) \qquad \text{on} \quad F .$$

Since both sides of this equality vanishes on E, we have

$$\alpha=a_1 \cdot i(v_1)+ \cdots +a_n \cdot i(v_n) .$$

This proves the statement above.

We shall now complete the proof of (5.9). Trivially we have $Y \subset (Y^\omega)^\omega$. To prove the opposite inclusion, let $v \notin Y$. By the Hahn-Banach theorem for locally convex topological vector spaces, there exists a linear functional $\alpha: X \to \boldsymbol{R}$ (continuous in the locally convex topology given by $\{p_v\}$) such that $\alpha = 0$ on Y and $\alpha(v) = 1$. By the statement above, there is an element $w \in X$ such that $\alpha(u) = \omega(u, w)$ for all $u \in X$. Then $\omega(v, w) \neq 0$ and $\omega(u, w) = 0$ for all $u \in Y$. Then $w \in Y^\omega$. Since $\omega(v, w) \neq 0$, $v \notin (Y^\omega)^\omega$. Thus, $(Y^\omega)^\omega \subset Y$. This completes the proof of (5.9). Q.E.D.

We need also a holomorphic analogue of (5.8). Let V be a complex Banach manifold of infinite dimension with a holomorphic symplectic form ω_V, where ω_v is a non-degenerate closed holomorphic 2-form on V satisfying the conditions analogous to those of (5.1). Let G be a complex Banach Lie group acting holomorphically on V as a group of symplectic transformations, i.e., leaving ω_V invariant. Let $\mathfrak{g}$ be the Banach Lie algebra of G and $\mathfrak{g}^*$ its dual Banach space. A *momentum map* for the action of G on V is a holomorphic map $\mu: V \to \mathfrak{g}^*$ such that

$$(5.10) \qquad \langle a, d\mu_x(v)\rangle = \omega(a_x, v) \qquad \text{for} \quad a \in \mathfrak{g}, \quad v \in T_x V, \quad x \in V,$$

which should be interpreted in the same way as (5.2). We note that if V is compact, then a momentum map cannot exist since there is no holomorphic map $V \to \mathfrak{g}^*$ other than the constant maps.

We shall now impose the following conditions (a′), (b′) and (c′) which are analogous to the conditions (a), (b) and (c) imposed on the real momentum map.

(a′) Assume that μ is equivariant with respect to the coadjoint action of G, i.e., μ satisfies (5.3).

The reduced phased space $W = \mu^{-1}(0)/G$ is defined in the same way as in the real case.

(b′) Assume that $0 \in \mathfrak{g}^*$ is a weakly regular value of μ in the sense that (5.6) and (5.7) are satisfied. We note that $\mu^{-1}(0)$ is now a complex submanifold of V.

(c′) Assume that the action of G on $\mu^{-1}(0)$ is free and that at each point $x \in \mu^{-1}(0)$ there is a *holomorphic slice* S_x for the action, i.e., a complex submanifold S_x of $\mu^{-1}(0)$ through x which is transversal to the orbit $G(x)$. The requirement that $\pi: S_x \to \pi(S_x) \subset W$ be biholomorphic makes W into a complex manifold, which may be non-Hausdorff in general. Now the holomorphic analogue of (5.8) states as follows:

(5.11) **Theorem** *Let V be a complex Banach manifold with a holomorphic symplectic form ω_V. Let G be a complex Banach Lie group acting on V leaving ω_V invariant. If there is a holomorphic momentum map $\mu: V \to \mathfrak{g}^*$ satisfying* (a′), (b′) *and* (c′), *then there is a unique holomorphic symplectic form ω_W on the reduced phase* *space* $W=\mu^{-1}(0)/G$ *such that*

$$\pi^*\omega_W = j^*\omega_V \qquad on \quad \mu^{-1}(0).$$

The proof is identical to that of (5.8).

§6 Kähler structures on moduli spaces

Let E be a C^∞ complex vector bundle of rank r over a compact Kähler manifold M of dimension n. We fix an Hermitian structure h in E. Let $\hat{\mathscr{E}}(E, h)/U(E, h)$ be the moduli space of irreducible Einstein-Hermitian connections in (E, h). We know that it is an open subset of the moduli space $\hat{\mathscr{H}}''(E)/GL(E)$ of simple holomorphic vector bundle structures in E and is a Hausdorff complex space, see (4.21). We shall now construct a Kähler metric on the non-singular part of $\hat{\mathscr{E}}(E, h)/U(E, h)$.

First we clarify linear algebra of Hermitian forms involved. Let X be a real vector space (possibly an infinite dimensional real Banach space) with a complex structure J, $J^2=-I$. Extend J to the complexification $X\otimes \boldsymbol{C}$ as a $\boldsymbol{C}$-linear endomorphism and write

$$X\otimes \boldsymbol{C} = Z + \bar{Z}, \tag{6.1}$$

where

$$Z=\{z\in X\otimes \boldsymbol{C};\ Jz=iz\}, \qquad \bar{Z}=\{z\in X\otimes \boldsymbol{C};\ Jz=-iz\}. \tag{6.2}$$

Then we have a linear isomorphism

$$X \longrightarrow Z, \qquad x \longmapsto z=\frac{1}{2}(x-iJx). \tag{6.3}$$

Under this isomorphism, J in X corresponds to the multiplication by i in Z.

Let h be an Hermitian inner product in Z, i.e.,

$$h: Z\times Z \longrightarrow \boldsymbol{C} \tag{6.4}$$

such that

(i) $h(z, w)$ is $\boldsymbol{C}$-linear in z,

(ii) $h(z, w)=\overline{h(w, z)}$,

(iii) $h(z, z)>0$ for every nonzero $z\in Z$.

Then h induces a real inner product g in X, i.e.,

$$g(x, u)=h(z, w)+h(w, z)=2\,\mathrm{Re}(h(z, w)) \tag{6.5}$$

if $z, w\in Z$ correspond to $x, u\in X$ under the isomorphism (6.3). Then

(6.6) $$g(Jx, Ju) = g(x, u)\,.$$

Conversely, every real inner product g in X satisfying (6.6) arises from a unique Hermitian inner product h in Z. The Hermitian form h induces also a non-degenerate skew-symmetric 2-form ω on X. Namely, we set

(6.7) $$\omega(x, u) = \frac{1}{i}(h(z, w) - h(w, z)) = 2\,\mathrm{Im}(h(z, w))\,.$$

Then

(6.8) $$\omega(x, u) = g(x, Ju)\,.$$

All these are familiar relations between a Kähler metric and the corresponding Riemannian metric and Kähler form. We shall use these relations in the infinite dimensional case as well as in the finite dimensional case.

Let

(6.9) $$X = L^2_k(A^1(\mathrm{End}(E, h)))\,, \qquad k > n\,,$$

i.e., the L^2_k-space of 1-forms over M with values in the skew-Hermitian endomorphisms of (E, h). Given $\xi \in X$, we decompose ξ:

(6.10) $$\xi = \xi' + \xi''\,,$$

where ξ' (resp. ξ'') is a $(1,0)$-form (resp. $(0,1)$-form). Then the condition that ξ is skew-Hermitian, i.e., ${}^t\bar{\xi} = -\xi$, is equivalent to

(6.11) $$\xi' = -{}^t\bar{\xi}''\,.$$

Define a complex structure J on X by

(6.12) $$J\xi = -i\xi' + i\xi''$$

so that

(6.13) $$Z = L^2_k(A^{0,1}(\mathrm{End}\,E))\,, \qquad \bar{Z} = L^2_k(A^{1,0}(\mathrm{End}\,E))\,.$$

The isomorphism $X \to Z$ is given by

(6.14) $$\xi \longmapsto \xi''\,.$$

In order to define an Hermitian inner product $\hat{h}$ in Z, we consider first the local inner product (α, β), where $\alpha, \beta \in Z$. In terms of an orthonormal basis $s_1, \cdots, s_r$ of the fibre of E and an orthonormal coframe $\theta^1, \cdots, \theta^n$ of M, we write

$$\alpha(s_j) = \sum a^i_{j\bar\lambda}\bar\theta^\lambda s_i\,, \qquad \beta(s_j) = \sum b^i_{j\bar\lambda}\bar\theta^\lambda s_i\,, \tag{6.15}$$

i.e., $a^i_{j\bar\lambda}$ and $b^i_{j\bar\lambda}$ are the components of α and β, respectively. Then

$$\langle\alpha, \beta\rangle = \sum a^i_{j\bar\lambda}\bar b^i_{j\bar\lambda}\,. \tag{6.16}$$

Since

$$\frac{1}{i}\operatorname{tr}(\alpha\wedge{}^t\bar\beta) = i\sum a^i_{j\bar\lambda}\bar b^i_{j\bar\mu}\theta^\mu\wedge\bar\theta^\lambda\,, \tag{6.17}$$

we may write

$$\langle\alpha, \beta\rangle = \Lambda\left(\frac{1}{i}\operatorname{tr}(\alpha\wedge{}^t\bar\beta)\right) \tag{6.18}$$

where Λ is the adjoint of the operator $L = \Phi\wedge\cdot$. We can also write

$$\frac{n}{i}\operatorname{tr}(\alpha\wedge{}^t\bar\beta)\wedge\Phi^{n-1} = \langle\alpha, \beta\rangle\Phi^n\,. \tag{6.19}$$

Now we define an Hermitian inner product $\hat h$ on Z by

$$\hat h(\alpha, \beta) = \int_M \langle\alpha, \beta\rangle\Phi^n = \int_M \Lambda\left(\frac{1}{i}\operatorname{tr}(\alpha\wedge{}^t\bar\beta)\right)\Phi^n = \int_M \frac{n}{i}\operatorname{tr}(\alpha\wedge{}^t\bar\beta)\wedge\Phi^{n-1}\,. \tag{6.20}$$

The corresponding inner product $\hat g$ on X is given by (see (6.5))

$$\begin{aligned}\hat g(\xi, \eta) &= \hat h(\xi'', \eta'') + \hat h(\eta'', \xi'') \\ &= \int_M \frac{n}{i}\operatorname{tr}(\xi''\wedge{}^t\bar\eta'' + \eta''\wedge{}^t\bar\xi'')\wedge\Phi^{n-1} \\ &= \int_M \frac{n}{i}\operatorname{tr}(\xi'\wedge\eta'' - \xi''\wedge\eta')\wedge\Phi^{n-1}\,.\end{aligned} \tag{6.21}$$

The corresponding 2-form ω on X is given by (see (6.8))

$$\omega(\xi, \eta) = \int_M n\cdot\operatorname{tr}(\xi\wedge\eta)\wedge\Phi^{n-1} \tag{6.22}$$

The affine space $L^2_k(\mathscr{D}(E, h)) \approx L^2_k(\mathscr{D}''(E))$ with $\hat h$ is a flat Kähler manifold with the Kähler form ω. (The tangent bundle of $L^2_k(\mathscr{D}''(E))$ (resp. $L^2_k(\mathscr{D}(E, h))$) is naturally a product bundle with fibre Z (resp. X).)

We can define a natural Hermitian metric on the nonsingular part of the moduli space $\hat{\mathscr{E}}(E, h)/U(E, h)$ of irreducible Einstein-Hermitian connections. Let $[D]$ be a nonsingular point of $\hat{\mathscr{E}}(E, h)/U(E, h)$ represented by $D \in \hat{\mathscr{E}}(E, h)$. The tangent space $T_{[D]}(\hat{\mathscr{E}}(E, h)/U(E, h))$ is identified with $\boldsymbol{H}^{0,1} \approx H^1(M, \operatorname{End}(E^{D''}))$, (see (4.20)). We define an inner product in $\boldsymbol{H}^{0,1}$ by applying the formula (6.20) to harmonic forms $\alpha, \beta \in \boldsymbol{H}^{0,1}$. It is simple to verify that this definition is independent of the choice of D representing $[D]$. Thus we have an Hermitian metric on the nonsingular part of $\hat{\mathscr{E}}(E, h)/U(E, h)$ induced from the flat Kähler metric $\hat{h}$ of $L^2_k(\mathscr{D}''(E))$. This metric is actually Kähler. This fact may be verified as in Itoh [3] by means of the local normal coordinates introduced around $[D]$ by the slice $\mathscr{S}_D$ and the Kuranishi map $k: \mathscr{S}_D \to \boldsymbol{H}^1 \approx \boldsymbol{H}^{0,1}$. However, we shall use instead the symplectic reduction theorem of Marsden-Weinstein proved in the preceding section.

In order to set up the situation to which the reduction theorem can be applied, let V be the nonsingular part of $L^2_k(\mathscr{H}''(E))$ in the following sense. For each $D'' \in \mathscr{H}''(E)$ such that $H^0(M, \operatorname{End}^0(E^{D''}))=0$ and $H^2(M, \operatorname{End}^0(E^{D''}))=0$, consider its $L^2_{k+1}(GL(E))$-orbit. The union V of all these orbits form an open subset of $\mathscr{H}''(E)$ which lie above the nonsingular part of the moduli space $\hat{\mathscr{H}}''(E)/GL(E)$ of simple vector bundles. Let $U \subset \mathscr{S}_{D''}$ be a neighborhood of 0 in the slice $\mathscr{S}_{D''}$ and $N \subset L^2_{k+1}(GL(E))/\boldsymbol{C}^*$ a neighborhood of the identity. From (3.17) and its proof we see that the set $\{(D''+\alpha)^f;\ \alpha \in U, f \in N\}$ is a neighborhood of D'' in $L^2_k(\mathscr{H}''(E))$ homeomorphic to $U \times N$. It is easy to see that V is covered by neighborhoods of this type and hence is a complex submanifold of $L^2_k(\mathscr{D}''(E))$. The Kähler metric $\hat{h}$ of $L^2_k(\mathscr{D}''(E))$ induces a Kähler metric $\hat{h}|_V$ on V, which will be sometimes denoted simply $\hat{h}$. It is convenient to consider V as a submanifold of $L^2_k(\mathscr{D}(E, h))$ as well under the identification $\mathscr{D}''(E) \approx \mathscr{D}(E, h)$. With the Kähler form $\omega|_V$, V may be regarded as a symplectic manifold. We write often ω instead of $\omega|_V$.

Let $G = L^2_{k+1}(U(E, h))/U(1)$, where $U(1)$ is considered as the group of scalar multiplication by complex numbers of absolute value 1. Let $\mathfrak{g} = L^2_{k+1}(\operatorname{End}(E, h))/\mathfrak{u}(1)$ be its Lie algebra.

We define a momentum map $\mu: V \to \mathfrak{g}^*$ by

$$(6.23) \qquad \langle a, \mu(D) \rangle = \int_M i \cdot \operatorname{tr}(a \circ (K(D) - cI))\Phi^n, \qquad a \in \mathfrak{g},$$

where $K(D)$ is the mean curvature of D and the equation $K(D) - cI = 0$ is the Einstein condition as in (1.16).

In order to show that μ satisfies (5.2), using the Kähler metric of M we take the trace of both sides of the equation

$$(6.24) \qquad \partial_t R(D+t\xi)|_{t=0} = \partial_t((D+t\xi)\circ(D+t\xi))|_{t=0} = D\xi$$

to obtain

(6.25) $$\partial_t K(D+t\xi)|_{t=0}=D^*\xi\,.$$

Hence

(6.26) $$\langle a, d\mu_D(\xi)\rangle=\int_M i\cdot\partial_t \operatorname{tr}(a\circ(K(D+t\xi)-cI))\,|_{t=0}\Phi^n$$
$$=\int_M i\cdot\operatorname{tr}(a\circ D^*\xi)\Phi^n\,.$$

On the other hand, the tangent vector a_D of V at D induced by the infinitesimal action of $a\in\mathfrak{g}$ is given by

(6.27) $$a_D=\partial_t(e^{-at}\circ D\circ e^{at})|_{t=0}=Da\,.$$

Hence,

(6.28) $$\omega(a_D,\xi)=\int_M n\cdot\operatorname{tr}(Da\wedge\xi)\wedge\Phi^{n-1}$$
$$=-\int_M n\cdot\operatorname{tr}(a\circ D\xi)\wedge\Phi^{n-1}$$
$$=\int_M i\cdot\operatorname{tr}(a\circ D^*\xi)\Phi^n\,.$$

This establishes

(6.29) $$\langle a, d\mu_D(\xi)\rangle=\omega(a_D,\xi)\,,$$

i.e., μ satisfies (5.2).

In order to verify (5.3) for μ, let $f\in G$. Then

(6.30) $$R(D^f)=f^{-1}\circ R(D)\circ f\,,\qquad K(D^f)=f^{-1}\circ K(D)\circ f\,.$$

Hence,

(6.31) $$\langle a, \mu(D^f)\rangle=\int_M i\cdot\operatorname{tr}(a\circ f^{-1}(K(D)-cI)f)\Phi^n$$
$$=\int_M i\cdot\operatorname{tr}(faf^{-1}\circ(K(D)-cI))\Phi^n$$
$$=\langle faf^{-1}, \mu(D)\rangle\,.$$

This means

(6.32) $$\mu(D^f)=(ad\,f)^*\mu(D)\,,$$

i.e., μ is equivariant with respect to the coadjoint action of G.

Clearly we have

(6.33) $$\mu^{-1}(0)=\{D\in\mathscr{H}(E,h);\ K(D)=cI\}\,,$$

i.e., $\mu^{-1}(0)$ consists of Einstein-Hermitian connections in (E,h).

In order to verify (5.6) and (5.7) for μ, we consider a map

(6.34) $$F\colon V\longrightarrow L^2_{k-1}(A^0(\mathrm{End}(E,h)))/\mathfrak{u}(1)\,,\qquad F(D)\equiv iK(D)\mod \mathfrak{u}(1)\,,$$

where $\mathfrak{u}(1)$ is considered as the space of skew-Hermitian endomorphisms of (E,h) of the form iaI with $a\in\boldsymbol{R}$. (This map F is essentially the momentum map μ.) From (6.25) we obtain

(6.35) $$dF_D(\xi)\equiv iD^*\xi\mod \mathfrak{u}(1)\,.$$

Let $\boldsymbol{H}^0$ be the space of harmonic 0-forms with values in $\mathrm{End}(E,h)$ defined by (2.22). Then $\boldsymbol{H}^0$ consists of endomorphisms of the form iaI_E, $a\in\boldsymbol{R}$, since D is irreducible. On the other hand,

$$L^2_{k-1}(A^0(\mathrm{End}(E,h)))=D^*L^2_k(A^1(\mathrm{End}(E,h)))\oplus\boldsymbol{H}^0\,.$$

Hence, the map $dF\colon T_DV\to L^2_{k-1}(A^0(\mathrm{End}(E,h)))$ is surjective modulo $\boldsymbol{H}^0=\mathfrak{u}(1)$. By the implicit function theorem, $F^{-1}(0)$ is a nonsingular submanifold at D and $T_D(F^{-1}(0))=\mathrm{Ker}(dF_D)$. This shows that (5.6) and (5.7) are satisfied by μ.

We already know that G acts freely (see (1.13)) and properly (see (1.14)) on $\mu^{-1}(0)$. A slice with the desired property is given by (4.7).

Now, by the reduction theorem (5.8) there is a symplectic form ω_W on the reduced phase space $W=\mu^{-1}(0)/G$ such that $\pi^*\omega_W=j^*\omega_V$ on $\mu^{-1}(0)$. From the construction of W it is clear that W is the nonsingular part of the moduli space $\hat{\mathscr{E}}(E,h)/U(E,h)$ and that ω_W is the Kähler form of the Hermitian metric constructed on the nonsingular part of $\hat{\mathscr{E}}(E,h)/U(E,h)$. We have established

(6.36) **Theorem** *The inner product* (6.20) *induces a Kähler metric on the nonsingular part of the moduli space* $\hat{\mathscr{E}}(E,h)/U(E,h)$ *of irreducible Einstein-Hermitian connections in* (E,h).

(6.37) *Remark* The real part of the Kähler metric, i.e., the corresponding Riemannian metric on the moduli space above depends on the Riemannian metric of M but not on its complex structure. For example, if M is a compact Riemannian manifold whose holonomy group is contained in $Sp(m)$, $m=n/2$, so that M is a Ricci-flat Kähler manifold with respect to any one of the complex structures compatible with $Sp(m)$, the Riemannian metric on the moduli space is independent of the complex structure on M we choose although the complex structure and the Kähler structure of the moduli space do depend on the complex structure of M. In fact, the Kähler metric $\hat{h}$ on moduli space was defined by applying (6.20) to harmonic forms $\alpha, \beta \in \boldsymbol{H}^{0,1}$, where $\boldsymbol{H}^{0,1}$ is identified with the holomorphic tangent space of $\hat{\mathscr{E}}(E, h)/U(E, h)$ at $[D]$. The corresponding Riemannian metric is obtained by applying (6.21) to harmonic forms $\xi, \eta \in \boldsymbol{H}^1$, where $\boldsymbol{H}^1$ is identified with the real tangent space of $\hat{\mathscr{E}}(E, h)/U(E, h)$ at $[D]$. Thus,

$$\begin{aligned}\hat{g}(\xi, \eta) &= \hat{h}(\xi'', \eta'') + \overline{\hat{h}(\xi'', \eta'')} \\ &= \int_M (\langle \xi'', \eta'' \rangle + \overline{\langle \xi'', \eta'' \rangle}) \Phi^n \\ &= \int_M \langle \xi, \eta \rangle \Phi^n ,\end{aligned}$$

where $\langle \xi, \eta \rangle$ is defined by the Riemannian metric of M and Φ^n is the Riemannian volume element of M. This shows that $\hat{g}$ does not involve the complex structure of M. This remark will be important in the next section when we consider symplectic Kähler manifolds M.

Theorem (6.36) for $\dim M = 1$ has been obtained by Atiyah-Bott [1]. The same theorem has been proved for moduli spaces of anti-self dual connections on compact Kähler surfaces by Itoh [3]. In order to prove that the Hermitian metric constructed in (6.20) is actually Kähler, instead of using the symplectic reduction theorem Itoh calculated explicitly partial derivatives of the Hermitian metric in local coordinates induced by the Kuranishi map. He calculated also the curvature in these coordinates and obtained the following

(6.38) **Theorem** *If* $\dim M = 1$, *then the holomorphic sectional curvature of the Kähler metric in* (6.36) *is nonnegative.*

This is definitely not true if $\dim M > 1$. We shall derive (6.38) from a geometric theorem on submersions of CR submanifolds.

Let V be a complex manifold (possibly an infinite dimensional complex Banach manifold). Let J be its complex structure, i.e., an endomorphism of the tangent bundle TV such that $J^2 = -I$. Let N be a real submanifold of V. We set

$$T^h N = TN \cap J(TN).$$

If $T^h N$ is a C^∞ complex subbundle of $TV|_N$, then N is called a *CR submanifold* of V. This is the case if $\dim T^h_p N$ is finite and constant.

Assume that V is a Kähler manifold and let $T^v N$ be the orthogonal complement of $T^h N$ in TN; it is a real subbundle of TN. Thus we have a direct sum decomposition:

$$TV|_N = T^h N \oplus T^v N \oplus T^\perp N, \tag{6.39}$$

where $T^\perp N$ is the normal bundle of N in V. Clearly, J leaves $T^h N$ and $T^v N \oplus T^\perp N$ invariant. We *assume that J interchanges $T^v N$ and $T^\perp N$*, i.e., $J(T^v N) = T^\perp N$ and $J(T^\perp N) = T^v N$.

We *assume further that there is a submersion $\pi: N \to W$ of N onto an almost Hermitian manifold W such that* (i) *$T^v N$ is the kernel of π_* and* (ii) *$\pi_*: T^h_p N \to T_{\pi(p)} W$ is a complex isometry for every $p \in N$*. Under these assumptions we have

(6.40) **Theorem** *The almost Hermitian manifold W is Kähler. Let H^V and H^W denote the holomorphic sectional curvature of V and W, respectively. Then, for any unit vector $x \in T^h N$, we have*

$$H^V(x) = H^W(\pi_* x) - 4\,|B(x, x)|^2,$$

where $B: TN \times TN \to T^\perp N$ is the second fundamental form of N in V.

For the proof of (6.40), see Kobayashi [11].

We shall now explain how to derive (6.38) from (6.40). Let

$$V = L^p_k(\mathscr{D}(E, h)), \qquad N = L^p_k(\hat{\mathscr{E}}(E, h)), \qquad W = \hat{\mathscr{E}}(E, h)/U(E, h). \tag{6.41}$$

We may identify W with $L^p_k(\hat{\mathscr{E}}(E, h))/L^p_{k+1}(U(E, h))$. We note also that since $\dim M = 1$, $\mathscr{D}(E, h) = \mathscr{H}(E, h)$ and the moduli space W is nonsingular, (see (4.20)).

The complex structure J of V is given by (6.12). The Kähler metric and the corresponding Riemannian metric on V are given by (6.20) and (6.21). As we have already remarked, these metrics are flat. Now we claim that the decomposition (6.39) of $TV|_N$ at $D \in N$ is given by

$$T^h_D N = \boldsymbol{H}^1, \qquad T^v_D N = D \circ D^* \circ G(L^2_k(B^1)), \tag{6.42}$$

$$T^\perp_D N = D^*_+ \circ D_+ \circ G(L^2_k(B^1)),$$

where $B^1 = A^1(\mathrm{End}(E, h))$, $\boldsymbol{H}^1$ and D_+ are as in Section 2 and G is the Green's operator for the Laplacian $\Delta = D^* \circ D + D_+ \circ D_+^*$. We have to show that J leaves T^hN invariant and interchanges T^vN and $T^\perp N$. Since J preserves the Hilbert space inner product (6.21) of $L_k^2(B^1)$, it suffices to show that (i) $J(T^hN + T^\perp N) \subset T^hN + T^vN$ and (ii) $J(T^hN + T^vN) \subset T^hN + T^\perp N$, i.e., that (i) if $D^*\xi = 0$, then $D_+ J\xi = 0$ and (ii) if $D_+\xi = 0$, then $D^*J\xi = 0$ for $\xi \in L_k^2(B^1)$. But this verification is straightforward, (see (2.16) as well as (III.2.39)). The metric in the moduli space W was constructed in such a way that $\pi_* : T_D^hN \to T_{\pi(D)}W$ is an isometry. Since V is flat, (6.40) implies that the holomorphic sectional curvature of W is nonnegative. This completes the proof of (6.38).

Even when $\dim M > 1$, (6.40) can be applied to $V = L_k^p(\mathscr{H}(E, h))$. However, in this case, $L_k^p(\mathscr{H}(E, h))$ is a complex subvariety of $X = L_k^p(\mathscr{D}(E, h))$ and has nonpositive holomorphic sectional curvature. Thus the holomorphic sectional curvature of W is of the form $b^2 - a^2$, where a^2 is the term determined by the second fundamental form of V in X while b^2 is given, as in (6.40) by the second fundamental form of N in V.

Going back to the 1-dimensional case, we make a few remarks.

(6.43) *Remarks* In general, if W is a Kähler manifold of dimension m, then its scalar curvature σ at $p \in W$ is given as an average of the holomorphic sectional curvature H at p. More precisely (see Berger [2]), we have

$$\sigma(p) = \frac{m(m+1)}{2a_{2m-1}} \int_{X \in S_p^{2m-1}} H(X)\omega \,, \tag{6.44}$$

where S_p^{2m-1} denotes the unit sphere in the tangent space T_pW, ω its volume element and a_{2m-1} its volume. (The scalar curvature σ defined in (I.7.16) is a half of the Riemannian scalar curvature.) From (6.38) and (6.44) it follows that the scalar curvature of the moduli space W is nonnegative.

As we have seen in (2.28), the tangent space H^1 of W at $[D]$ is decomposed as

$$H^1 = \tilde{H}^1 + H^1(M, \boldsymbol{R}) \,.$$

We can deform the holomorphic structure of E by tensoring it with topologically trivial holomorphic line bundles. The subspace $H^1(M, \boldsymbol{R})$ in the decomposition above corresponds to the space of infinitesimal deformations coming from such line bundles. It is not hard to see that the holomorphic sectional curvature vanishes on $H^1(M, \boldsymbol{R})$. The question remains if it is strictly positive on $\tilde{H}^1$. Consider the subspace W_L of the moduli space W consisting of $E^{D''}$ such that its determinant bundle is holomorphically isomorphic to a fixed holomorphic line bundle L so that $\tilde{H}^1$ appears as the tangent space of W_L at $[D]$. It is known that if the rank r of E and the degree d of L are relatively prime and if the genus of M is greater than 1, then W_L is compact, simply connected and unirational, (see Newstead [1] and Atiyah-Bott [1]). Therefore, it would not be outrageous to hope that the holomorphic sectional curvature is positive on $\tilde{H}^1$.

(6.45) *Remark* In order to prove (6.36) we made use of the symplectic reduction theorem. However, it is also possible to derive (6.36) easily from (6.40).

§7 Simple vector bundles over compact symplectic Kähler manifolds

The main purpose of this section is to prove the following (Kobayashi [10])

(7.1) **Theorem** *Let M be a compact Kähler manifold with a holomorphic symplectic structure ω_M. Let E be a C^∞ complex vector bundle over M and let $\hat{\mathscr{H}}''(E)/GL(E)$ be the moduli space of simple holomorphic structures in E. Then ω_M induces, in a natural way, a holomorphic symplectic structure on the nonsingular part*

$$W = \{[D''] \in \hat{\mathscr{H}}''(E)/GL(E);\ H^2(M, \mathrm{End}^0(E^{D''})) = 0\}$$

of $\hat{\mathscr{H}}''(E)/GL(E)$.

As we noted in Section 5 of Chapter II, there are two classes of compact Kähler surfaces which carry holomorphic symplectic structures, namely, (i) complex tori and (ii) $K3$ surfaces. Mukai [1] has shown the theorem above when M is an abelian surface or a $K3$ surface by an algebraic geometric method.

We note that since the canonical line bundle of M is trivial, the Serre duality (III.2.50) implies that if M is a symplectic compact Kähler *surface*, then $\hat{\mathscr{H}}''(E)/GL(E)$ is nonsingular. In fact, $H^2(M, \mathrm{End}^0(E^{D''}))$ is dual to $H^0(M, \mathrm{End}^0(E^{D''}))$, which is zero if $E^{D''}$ is simple, (see the paragraph preceding (3.17)).

Our differential geometric proof relies on the holomorphic version of the symplectic reduction theorem (5.8) of Marsden-Weinstein.

Proof We shall set up the situation to which the holomorphic analogue (5.11) of (5.8) can be applied. Taking $k > \dim M$, we set

$$V = L_k^2(\mathscr{D}''(E)), \qquad G = L_{k+1}^2(GL(E))/\boldsymbol{C}^*, \qquad \mathfrak{g} = L_{k+1}^2(\mathrm{End}(E))/\boldsymbol{C}.$$

Then G acts smoothly and effectively on V.

We define a holomorphic symplectic structure ω_V on V by

$$(7.2) \qquad \omega_V(\alpha, \beta) = \int_M \operatorname{tr}(\alpha \wedge \beta) \wedge \omega_M^m \wedge \bar{\omega}_M^{m-1} \qquad \alpha, \beta \in T_{D''}(V),$$

where α and β are considered as elements of $L_k^2(A^{0,1}(\operatorname{End}(E))) \approx T_{D''}(V)$ and $2m$ denotes the dimension of M.

We define a momentum map $\mu\colon V \to \mathfrak{g}^*$ by

$$(7.3) \qquad \langle a, \mu(D'') \rangle = -\int_M \operatorname{tr}(a \circ N(D'')) \wedge \omega_M^m \wedge \bar{\omega}_M^{m-1}, \qquad a \in \mathfrak{g}, \quad D'' \in V,$$

where

$$(7.4) \qquad N(D'') = D'' \circ D''\colon L_k^2(A^{0,0}(E)) \longrightarrow L_{k-2}^2(A^{0,2}(E)).$$

We may consider $N(D'')$ as an element of $L_{k-2}^2(A^{0,2}(\operatorname{End}(E)))$. We make use of the following formulas to verify (5.2) for μ.

$$(7.5) \qquad \partial_t N(D'' + t\beta)|_{t=0} = D'' \circ \beta + \beta \circ D'' = D''\beta \qquad \text{for} \quad \beta \in L_k^2(A^{0,1}(\operatorname{End}(E))),$$

$$(7.6) \qquad a_{D''} = \partial_t(e^{-at} \circ D'' \circ e^{at})|_{t=0} = -a \circ D'' + D'' \circ a = D''a \qquad \text{for} \quad a \in \mathfrak{g}.$$

The latter means that $D''a$ is the tangent vector $a_{D''} \in T_{D''}(V)$ induced by the infinitesimal action of $a \in \mathfrak{g}$. We verify (5.2) as follows.

$$(7.7) \qquad \begin{aligned} \langle a, d\mu_{D''}(\beta) \rangle &= -\partial_t \int_M \operatorname{tr}(a \circ N(D'' + t\beta)) \wedge \omega_M^m \wedge \bar{\omega}_M^{m-1} \Big|_{t=0} \\ &= -\int_M \operatorname{tr}(a \circ D''\beta) \wedge \omega_M^m \wedge \bar{\omega}_M^{m-1} \\ &= \int_M \operatorname{tr}(D''a \wedge \beta) \wedge \omega_M^m \wedge \bar{\omega}_M^{m-1} \\ &= \omega_V(D''a, \beta) = \omega_V(a_{D''}, \beta). \end{aligned}$$

Since

$$\operatorname{tr}(a \circ N(D''^f)) = \operatorname{tr}(a \circ f^{-1} \circ N(D'') \circ f) = \operatorname{tr}(f \circ a \circ f^{-1} \circ N(D'')) \qquad \text{for} \quad f \in G,$$

we obtain

$$(7.8) \qquad \langle a, \mu(D''^f) \rangle = \langle f \circ a \circ f^{-1}, \mu(D'') \rangle,$$

which verifies (5.3) for μ.

We do not verify (5.6) and (5.7) directly. The proof will proceed in such a way that we will consider only the points of V where (5.6) and (5.7) hold. Let

$$D'' \in \mu^{-1}(0) = \{D'' \in V;\ N(D'') = 0\} = L_k^2(\mathscr{H}''(E)) .$$

If $\langle a, d\mu_{D''}(\beta)\rangle = 0$ for all $\beta \in T_{D''}(V)$, then $D''a = 0$ by (7.7). We consider first the open subset V' of V consisting of D'' such that $a = 0$ is the only solution of $D''a = 0$ in $\mathfrak{g} = L_{k+1}^2(\mathrm{End}(E))/C \approx L_{k+1}^2(\mathrm{End}^0(E))$. Then

$$\text{(7.9)} \qquad \begin{aligned} \mu^{-1}(0) \cap V' &= \{D'' \in V;\ N(D'') = 0 \text{ and } E^{D''} \text{ is simple}\} \\ &= L_k^2(\hat{\mathscr{H}}''(E)) . \end{aligned}$$

To see that G acts freely on $\mu^{-1}(0) \cap V'$, let $f \in L_{k+1}^2(GL(E))$. If $D'' \in \mu^{-1}(0) \cap V'$ and $D''^f = D''$, i.e., $D'' \circ f = f \circ D''$, then $D''f = 0$ and hence $f = aI_E$ with $a \in C^*$. This shows that G acts freely on $\mu^{-1}(0) \cap V'$. We defined a slice $\mathscr{S}_{D''}$ in (3.9). Instead of verifying (5.6) and (5.7) we consider only the nonsingular part W of $\hat{\mathscr{H}}''(E)/GL(E)$ and the portion of $\mu^{-1}(0) \cap V'$ which lies above it. Then there is a unique holomorphic symplectic form ω_W on $W \subset \hat{\mathscr{H}}''(E)/GL(E) = (\mu^{-1}(0) \cap V')/G$ such that $\pi^*\omega_W = j^*\omega_V$ in the notation of (5.11). Q.E.D.

As we remarked in Section 5 of Chapter II, every compact Kähler manifold M with a holomorphic symplectic form ω admits a Kähler metric which makes ω parallel or, equivalently, a Ricci-flat Kähler metric. (It is also part of Yau's theorem that the new Kähler metric can be chosen in the same cohomology class as the given metric.)

Since the moduli space $\hat{\mathscr{E}}(E, h)/U(E, h)$ of irreducible Einstein-Hermitian connections is open in the moduli space $\hat{\mathscr{H}}''(E)/GL(E)$ of simple holomorphic structures, (see (4.21)), the holomorphic symplectic form constructed in (7.1) induces a holomorphic symplectic form on the nonsingular part of $\hat{\mathscr{E}}(E, h)/U(E, h)$. On the other hand, by (6.36) we have also a Kähler metric on the nonsingular part of $\hat{\mathscr{E}}(E, h)/U(E, h)$.

Having made these remakrs, we now state

(7.10) **Theorem** *Let M be a compact Kähler surface with parallel holomorphic symplectic form ω_M. Then there is a natural holomorphic symplectic form on the nonsingular part W of $\mathscr{E}(E, h)/U(E, h)$ which is parallel with respect to the natural Kähler metric of W. In particular, the Kähler metric of W has vanishing Ricci tensor.*

Proof The holonomy group of M is contained in $Sp(1)$, and we have a family of complex structures on M parameterized by a 2-sphere, namely

$$\mathscr{J}=\{J=a_1J_1+a_2J_2+a_3J_3;\ a_1^2+a_2^2+a_3^2=1\}\,,$$

where J_1 may be taken as the given complex structure, (see Section 5 of Chapter II). Every $J\in\mathscr{J}$ is parallel with respect to the given Riemannian metric g of M. In other words, every pair (g, J) with $J\in\mathscr{J}$ defines a Kähler structure on M. In particular, the given Kähler structure is defined by (g, J_1).

Let $V=\boldsymbol{R}^4$ be a typical (real) tangent space of M and g the inner product in V defined by the Riemannian metric of M. We consider $\mathscr{J}$ as a family of complex structures on V compatible with the action of $Sp(1)$. Let

$$\omega_\lambda(X, Y)=g(J_\lambda X, Y) \qquad \text{for} \quad X, Y\in V\,, \quad \text{and} \quad \lambda=1, 2, 3\,.$$

Then each $\omega_\lambda\in\bigwedge^2V^*$ is invariant by $Sp(1)$. Let $(\bigwedge^2V^*)_+$ denote the subspace of $\bigwedge^2V^*$ consisting of elements invariant by $Sp(1)$. Then $\omega_1, \omega_2, \omega_3$ form a basis for $(\bigwedge^2V^*)_+$. Using J_1 we identify $V=\boldsymbol{R}^4$ with $\boldsymbol{C}^2$. Let (z, w) be the natural coordinate system in $\boldsymbol{C}^2$. Let

$$z=x+iy\,, \qquad w=u+iv\,.$$

Then

$$\begin{aligned} &\omega_1=2(x\wedge y+u\wedge v)=i(z\wedge\bar{z}+w\wedge\bar{w})\,,\\ (7.11)\qquad &\omega_2=2(x\wedge u-y\wedge v)=2\cdot\mathrm{Re}(z\wedge w)\,,\\ &\omega_3=2(x\wedge v+y\wedge u)=2\cdot\mathrm{Im}(z\wedge w)\,. \end{aligned}$$

Applying this algebraic fact to tangent spaces of M, we see that the term $B^2_+=A^0(\mathrm{End}(E)\otimes(\bigwedge^2T^*M)_+)$ in the complex (B^*) of Section 2 does not change when we vary the complex structure J within $\mathscr{J}$. Hence, by (2.19), $\mathscr{E}(E, h)$ does not vary with $J\in\mathscr{J}$.

Since the terms B^0 and B^1 in the complex (B^*) are completely independent of the complex structure J, the first cohomology H^1 of the complex (B^*) does not vary with $J\in\mathscr{J}$. We recall that $\boldsymbol{H}^1\approx H^1$ is identified with the (real) tangent space of $\mathscr{E}(E, h)/U(E, h)$ at $[D]$. The inner product in $\boldsymbol{H}^1$ does not depend on J, (see (6.37)).

Let W denote the nonsingular part of the moduli space $\mathscr{E}(E, h)/U(E, h)$. By what we have just stated, the Riemannian metric $\hat{g}$ on W does not depend on $J\in\mathscr{J}$. Each $J\in\mathscr{J}$ induces a complex structure $\hat{J}$ on W. By (6.36), the pair $(\hat{g}, \hat{J})$ defines a Kähler structure on W, i.e., $\hat{J}$ is parallel with respect to $\hat{g}$.

In (6.36), from the Kähler form $\Phi=\omega_1$ on M associated with the pair (g, J_1), we obtained the Kähler form $\hat{\omega}_1$ on W associated with $(\hat{g}, \hat{J}_1)$. Applying (6.36) to the pair (g, J_2), from the Kähler form ω_2 of (g, J_2) we obtain the Kähler form $\hat{\omega}_2$ on W associated with $(\hat{g}, \hat{J}_2)$. But ω_2 is (the real part of) the holomorphic symplectic form ω_M and $\hat{\omega}_2$ is (the real part of) the holomorphic symplectic form ω_W. Being the Kähler form, $\hat{\omega}_2$ is parallel with respect to $\hat{g}$. Hence, ω_W is also parallel. Q.E.D.

§8 Vector bundles over Kähler surfaces

We fix a C^∞ Hermitian vector bundle (E, h) over a compact Kähler surface M. We know that the moduli space $\mathscr{E}(E, h)/U(E, h)$ of Einstein-Hermitian connections in (E, h) is nonsingular at $[D] \in \mathscr{E}(E, h)/U(E, h)$ if $H^0(M, \mathrm{End}^0(E^{D''}))=0$ and $H^2(M, \mathrm{End}^0(E^{D''}))=0$, (see (4.19)). The first condition expresses irreducibility of D, (see (4.14)). By Serre duality (III.2.50), $H^2(M, \mathrm{End}^0(E^{D''}))$ is dual to $H^0(M, \mathrm{End}^0(E^{D''}) \otimes K_M)$, where K_M denotes the canonical line bundle of M. Since $(E^{D''}, h)$ is Einstein-Hermitian, so are $\mathrm{End}(E^{D''})$ and $\mathrm{End}^0(E^{D''})$ with mean curvature 0, i.e., with proportionality constant 0, (see (IV.1.4)). Since K_M is a line bundle, it admits an Einstein-Hermitian structure with mean curvature of the form cI_{K_M}, where the proportionality constant c has the same sign as the degree of K_M:

$$\deg(K_M) = \int_M c_1(K_M) \wedge \Phi = \int_M -c_1(M) \wedge \Phi .$$

The tensor product $F = \mathrm{End}^0(E^{D''}) \otimes K_M$ admits an Einstein-Hermitian with mean curvature cI_F (by (IV.1.4)).

If $c<0$, then F admits no nonzero holomorphic sections (see (III.1.9)). If $c=0$, every holomorphic section of F is parallel by the same theorem (III.1.9). Let f be a nonzero holomorphic section of F. At each point x of M, f defines a traceless endomorphism A_x of E; A_x is unique up to a constant multiplicative factor. Although the eigenvalues of A_x are determined only up to a common multiplicative factor, the eigen subspaces of E_x are well determined. Since f is parallel, it follows that these eigen subspaces of E_x give rise to parallel subbundles of E. But this is impossible if D is irreducible. This shows that if $c=0$ and D is irreducible, then F admits no nonzero holomorphic sections.

We have established (see Kim [1] when $c_1(M)>0$ or $c_1(M)=0$).

(8.1) **Theorem** *Let M be a compact Kähler surfaces with Kähler form Φ. Let (E, h) be a C^∞ Hermitian vector bundle over M. Then the moduli space $\hat{\mathscr{E}}(E, h)/U(E, h)$ of irreducible Einstein-Hermitian connections in (E, h) is a nonsingular Kähler manifold if*

$$\int_M c_1(M) \wedge \Phi \geqq 0 .$$

In order to find the dimension of this moduli space, we calculate $\chi(M, \mathrm{End}(E^{D''}))$ using the Riemann-Roch formula of Herzebruch (II.4.4):

$$\chi(M, \mathrm{End}(E^{D''})) = \int_M \mathrm{td}(M) \cdot \mathrm{ch}(E) \cdot \mathrm{ch}(E^*) . \tag{8.2}$$

From

$$\mathrm{td}(M) = 1 + \frac{1}{2} c_1(M) + \frac{1}{12}(c_1(M)^2 + c_2(M)) ,$$

$$\mathrm{ch}(E) = r + c_1(E) + \frac{1}{2}(c_1(E)^2 - 2c_2(E)) ,$$

$$\mathrm{ch}(E^*) = r - c_1(E) + \frac{1}{2}(c_1(E)^2 - 2c_2(E)) ,$$

we obtain

$$\chi(M, \mathrm{End}(E^{D''})) = \int_M (r-1)c_1(E)^2 - 2rc_2(E) + \frac{r^2}{12}(c_1(M)^2 + c_2(M)) . \tag{8.3}$$

Also by the Riemann-Roch formula, we have

$$1 - h^{0,1} + h^{0,2} = \frac{1}{12} \int_M c_1(M)^2 + c_2(M) . \tag{8.4}$$

For simplicity, write

$$H^p = H^p(M, \mathrm{End}(E^{D''})) , \qquad \tilde{H}^p = H^p(M, \mathrm{End}^0(\mathrm{E}^{D''})) , \qquad \chi = \chi(M, \mathrm{End}(E^{D''})) .$$

Then

$$\dim H^1 = \dim H^0 + \dim H^2 - \chi = 1 + \dim \tilde{H}^0 + h^{0,2} + \dim \tilde{H}^2 - \chi . \tag{8.5}$$

Assume that

$$H^0(M, \mathrm{End}^0(E^{D''})) = 0 \quad \text{and} \quad H^2(M, \mathrm{End}^0(E^{D''})) = 0 .$$

From (8.3), (8.4) and (8.5) we obtain

$$\dim H^1 = 2rc_2(E) - (r-1)c_1(E)^2 + r^2 h^{0,1} - (r^2-1)(1 + h^{0,2}) . \tag{8.6}$$

If $\int c_1(M) \wedge \Phi > 0$, i.e., if the degree of the canonical line bundle K_M is negative, then K_M has no nonzero holomorphic sections, i.e., $h^{0,2} = 0$, (see (III.1.24)).

If $\int c_1(M) \wedge \Phi = 0$, then every holomorphic section of K_M is parallel by the same theorem (III.1.24) so that $h^{0,2} \leqq 1$ and the equality means that K_M is a trivial line bundle.

If $c_1(M) > 0$, then we have $h^{0,1} = 0$ as well as $h^{0,2} = 0$. In fact, by Serre duality and Kodaira's vanishing theorem (III.3.1), we have

$$h^{0,1} = h^{n,n-1} = \dim H^{n-1}(M, K_M) = 0 .$$

In summary,

(8.7) **Theorem** *Let M and (E, h) be as in* (8.1). *Then the dimension of the moduli space $\hat{\mathscr{E}}(E, h)/U(E, h)$ (if nonempty) is given by*

(1) $2rc_2(E) - (r-1)c_1(E)^2 + r^2h^{0,1} + 1 - r^2$, $(r = \mathrm{rank}(E))$, *if* $\int c_1(M) \wedge \Phi \geqq 0$ *and if the canonical line bundle K_M is non-trivial*;

(2) $2rc_2(E) - (r-1)c_1(E)^2 + r^2h^{0,1} + 2 - 2r^2$ *if K_M is trivial*;

(3) $2rc_2(E) - (r-1)c_1(E)^2 + 1 - r^2$ *if* $c_1(M) > 0$.

(8.8) *Remark* If the canonical line bundle K_M is trivial (i.e., M is a torus or a $K3$ surface), then the moduli space above admits a holomorphic symplectic structure. In particular, the dimension appearing in (2) above must be an even integer (provided that the moduli space is nonempty).

(8.9) *Remark* By (IV.4.7), if (E, h) admits an Einstein-Hermitian connection, then

$$2rc_2(E) - (r-1)c_1(E)^2 \geqq 0 .$$

In some special cases, (8.7) gives a better lower bound. For example, for an irreducible Einstein-Hermitian bundle E over a compact Kähler surface M with $c_1(M) > 0$, we have

$$2rc_2(E) - (r-1)c_1^2(E) \geqq r^2 - 1 .$$

We know also (see (IV.4.7) again) that, for an Einstein-Hermitian vector bundle E, the equality $2rc_2(E) - (r-1)c_1(E)^2 = 0$ holds if and only if E is projectively flat. So (8.1), (2) of (8.7) and (7.10) imply that if M is a torus of dimension 2 and if E is a C^∞ complex vector bundle satisfying $2rc_2(E) - (r-1)c_1(E)^2 = 0$, then the moduli space $\hat{\mathscr{E}}(E, h)/U(E, h)$ (if nonempty) is the moduli space of projectively flat Hermitian connections in (E, h) and is a symplectic Kähler manifold of dimension 2.

(8.10) *Remark* We know (see (V.8.3)) that every irreducible Einstein-Hermitian vector bundle over M is Φ-stable. Conversely, if M is a projective algebraic surface and Φ represents the first Chern class $c_1(M)$ of an ample line bundle H, then every H-stable vector bundle E over M admits an irreducible Einstein-Hermitian structure, (see (VI.10.19)). It follows that when M is an algebraic surface with Kähler form Φ representing $c_1(H)$, the moduli space $\hat{\mathscr{E}}(E, h)/U(E, h)$ can be naturally identified with the moduli space of H-stable holomorphic structures in E. Hence, (8.1) and (8.7) can be stated as results on moduli spaces of H-stable holomorphic bundles.

The following result follows from (7.1) and (3.34) in the same way as (8.1) and (8.7), (see Mukai [1]).

(8.11) **Theorem** *Let M be a K3 surface or a complex torus of dimension* 2. *Let E be a C^∞ complex vector bundle over M. Then the moduli space $\hat{\mathscr{H}}''(E)/GL(E)$ of simple holomorphic structures in E is a (possibly non-Hausdorff) complex manifold with a holomorphic symplectic structure. Its dimension is given by*

$$2rc_2(E)-(r-1)c_1(E)^2+r^2h^{0,1}+2-2r^2\,,$$

where $h^{0,1}=0$ (resp. $=2$) if M is a K3 surface (resp. a torus), provided that the moduli space is nonempty.

Bibliography

This bibliography contains some papers which are not directly quoted but are related to the contents of the book. For a more comprehensive bibliography on vanishing theorems, see Shiffman-Sommese [1]. The book by Okonek-Schneider-Spindler [1] contains an extensive bibliography on stable bundles.

Y. Akizuki and S. Nakano [1], Note on Kodaira-Spencer's proof of Lefschetz theorems, Proc. Japan Acad. 30 (1954), 266–272.

A. Andreotti and H. Grauert [1], Théorèmes de finitude pour la cohomologie des espaces complexes, Bull. Soc. Math. France 90 (1962), 193–259.

M. Apte [1], Sur certaines classes caractéristiques des variétés kähleriennes compactes, C. R. Acad. Sci. Paris, 240 (1950), 149–151.

M. F. Atiyah [1], Vector bundles over an elliptic curve, Proc. London Math. Soc. (3) 7 (1957), 414–452.

—— [2], On the Krull-Schmidt theorem with applications to sheaves, Bull. Soc. Math. France 84 (1956), 307–317.

—— [3], The Geometry of Yang-Mills Fields, Lezioni Fermiani, Scuola Normale Pisa, 1978.

M. F. Atiyah and R. Bott [1], The Yang-Mills equations over Riemann surfaces, Phil. Trans. Roy. Soc. London A 308 (1982), 524–615.

M. F. Atiyah, N. J. Hitchin and I. M. Singer [1], Self-duality in four dimensional Riemannian geometry, Proc. Roy. Soc. London A 362 (1978), 425–461.

T. Aubin [1], Équations du type Monge-Ampère sur les variétés kähleriennes compactes, C. R. Acad. Sci. Paris 283 (1976), 119–121.

—— [2], Nonlinear Analysis on Manifolds. Monge-Ampère Equations, Grundlehren Math. Wiss. 252, Springer-Verlag, 1982.

C. Bănică and O. Stănăşilă [1], Méthodes Algébriques dans la Théorie Globale des Espaces Complexes, Gauthier-Villars, 1977.

W. Barth [1], Moduli of vector bundles on the projective plane, Invent. Math. 42 (1977), 63–91.

A. Beauville [1], Variétés kähleriennes dont la première classe de Chern est nulle, J. Diff. Geometry 18 (1983), 755–782.

—— [2], Some remarks on Kähler manifolds with $c_1=0$, "Classification of Algebraic and Analytic Manifolds", Progress in Math. vol. 39 Birkhäuser, 1983, 1–26.

M. Berger [1], Sur les groupes d'holonomie homogène des variétés à connexion affine et des variétés riemanniennes, Bull. Soc. Math. France 83 (1955), 279–330.

—— [2], Sur les variétés d'Einstein compactes, C. R. III Reunion Math. Expression Latine, Namur 1965, pp. 35–55.

F. A. Bogomolov [1], Holomorphic tensors and vector bundles on projective varieties, Izv. Nauk SSSR 42 (1978) (=Math. USSR Izv. 13 (1978), 499–555).

—— [2], Unstable vector bundles and curves on surfaces, Proc. Intern. Congress of Math.

Helsinki, 1978, 517–524.

A. Borel and J.-P. Serre [1], Le théorème de Riemann-Roch (d'après Grothendieck), Bull. Soc. Math. France 86 (1958), 97–136.

R. Bott [1], Homogeneous vector bundles, Ann. Math. 66 (1957), 203–248.

B. Y. Chen and K. Ogiue [1], Some characterizations of complex space forms in terms of Chern classes, Quart. J. Math. Oxford 26 (1975), 459–464.

S. S. Chern [1], Complex Manifolds without Potential Theory, Springer-Verlag, 1979.

M. Deschamps [1], Courbes de genre géométrique borné sur une surface de type général, Sém. Bourbaki 1977/78, Lecture Notes in Math. 710, Springer-Verlag, 1979, 233–247.

S. K. Donaldson [1], A new proof of a theorem of Narasimhan and Seshadri, J. Diff. Geometry 18 (1983), 269–278.

—— [2], Anti self-dual Yang-Mills connections over complex algebraic surfaces and stable vector bundles, Proc. London Math. Soc. 50 (1985), 1–26.

D. S. Freed and K. K. Uhlenbeck [1], Instantons and Four-Manifolds, MSRI Publ. 1, Springer-Verlag, 1984.

A. Friedman [1], Partial Differential Equations of Parabolic Type, Robert E. Krieger Publ. Co., 1983.

P. Gauduchon [1], Fibrés hermitiens à endomorphisme de Ricci non-négatif, Bull. Soc. Math. France 105 (1977), 113–140.

—— [2], La 1-forme de torsion d'une variété hermitienne compacte, Math. Ann. 267 (1984), 495–518.

D. Gieseker [1], On moduli of vector bundles on an algebraic surface, Ann. of Math. 106 (1977), 45–60.

—— [2], On a theorem of Bogomolov on Chern classes of stable bundles, Amer. J. Math. 101 (1979), 77–85.

G. Gigante [1], Vector bundles with semidefinite curvature and cohomology vanishing theorems, Advance in Math. 41 (1981), 40–56.

J. Girbau [1], Sur le théorème de Le Potier d'annulation de la cohomologie, C. R. Acad. Sci. Paris 283 (1976), 355–358.

R. Godement [1], Topologie Algébrique et Théorie des Faisceaux, Hermann, Paris, 1958.

H. Grauert [1], Über Modifikationen und exzeptionelle analytische Mengen, Math. Ann. 146 (1962), 331–368.

—— [2], Analytische Faserungen über holomorph-vollständigen Raume, Math. Ann. 135 (1958), 266–273.

H. Grauert and R. Remmert [1], Coherent Analytic Sheaves, Grundlehren Math. Wiss. 265, Springer-Verlag, 1984.

P. Griffiths and J. Harris [1], Principle of Algebraic Geometry, John Wiley & Sons, 1978.

A. Grothendieck [1], Sur la classification des fibrés holomorphes sur la sphère de Riemann, Amer. J. Math. 79 (1957), 121–138.

—— [2], La théorie des classes de Chern, Bull. Soc. Math. France 86 (1958), 137–154.

R. C. Gunning [1], Lectures on Vector Bundles over Riemann Surfaces, Math. Notes 6, Princeton Univ. Press, 1967.

R. C. Gunning and H. Rossi [1], Analytic Functions of Several Complex Variables, Prentice Hall, 1965.

R. S. Hamilton [1], Harmonic Maps of Manifolds with Boundary, Lecture Notes in Math.

471, Springer-Verlag, 1975.

J.-I. Hano [1], A geometrical characterization of a class of holomorphic vector bundles over a complex torus, Nagoya Math. J. 61 (1976), 197–202.

G. Harder and M. S. Narasimhan [1], On the cohomology groups of moduli spaces of vector bundles on curves, Math. Ann. 212 (1975), 215–248.

R. Hartshorne [1], Ample vector bundles, Publ. Math. IHES 29 (1966), 63–94.

—— [2], Ample Suvarieties of Algebraic Varieties, Lecture Notes in Math. 156, Springer-Verlag, 1970.

—— [3], Stable reflexive sheaves, Math. Ann. 254 (1980), 121–176.

F. Hirzebruch [1], Topological Methods in Algebraic Geometry, Springer-Verlag, 1966.

L. Hörmander [1], Introduction to Complex Analysis in Several Variables, North Holland, 1973.

K. Hulek [1], On the classification of stable rank r vector bundles over the projective plane, Progress in Math. 7, Vector Bundles and Differential Equations, Birkhäuser, 1980, pp. 113–114.

D. Husemoller [1], Fibre Bundles, McGraw-Hill, 1966.

M. Itoh [1], On the moduli space of anti-self dual connections on Kähler surfaces, Publ. Res. Inst. Math. Sci. Kyoto Univ. 19 (1983) 15–32.

—— [2], Geometry of Yang-Mills connections over a Kähler surfaces, Proc. Japan Acad. 59 (1983), 431–433.

—— [3], Geometry of anti-self-dual connections and Kuranishi map, preprint.

L. Kaup [1], Eine Künnethformel für Fréchetgarben, Math. Z. 97 (1967), 158–168.

—— [2], Das topologische Tensorprodukt kohärenter analytischer Garben, Math. Z. 106 (1968), 273–292.

Y. Kawamata [1], A generalization of Kodaira-Ramanujam's vanishing theorem, Math. Ann. 261 (1982), 43–46.

J. L. Kazdan [1], Some Applications of Partial Differential Equations to Problems in Geometry, "Survey in Geometry", Japan, 1983.

G. Kempf and L. Ness [1], On the lengths of vectors in representation spaces, Lecture Notes in Math. 732, Springer-Verlag, pp. 233–244.

H. J. Kim [1], Curvatures and holomorphic vector bundles, Ph. D. thesis, Berkeley, 1985.

—— [2], Moduli of Hermite-Einstein vector bundles, Math. Z., to appear.

F. Kirwan [1], The Cohomology of Quotient Spaces in Algebraic and Symplectic Geometry, Math. Notes 31, Princeton Univ. Press, 1985.

S. Kobayashi [1], On compact Kähler manifolds with positive Ricci tensor, Ann. of Math. 74 (1961) 570–574.

—— [2], Transformation Groups in Differential Geometry, Ergebnisse Math. 70, 1970, Springer-Verlag.

—— [3], First Chern class and holomorphic tensor fields, Nagoya Math. J. 77 (1980), 5–11.

—— [4], Negative vector bundles and complex Finsler structures, Nagoya Math. J. 57 (1975), 153–166.

—— [5], Curvature and stability of vector bundles, Proc. Japan Acad. 58 (1982), 158–162.

—— [6], Differential Geometry of Holomorphic Vector Bundles, Math. Seminar Notes (by I. Enoki) 41 (1982), Univ. of Tokyo (in Japanese).

—— [7], Homogeneous vector bundles and stability, Nagoya Math. J. 101 (1986), 37–54.

—— [8], On two concepts of stability for vector bundles and sheaves, Aspects of Math. & its Applications, Elsevier Sci. Publ. B. V., (1986), 477–484.

—— [9], Einstein-Hermitian vector bundles and stability, Global Riemannian Geometry (Durham Symp. 1982), Ellis Horwood Ltd., 1984, pp. 60–64.

—— [10], Simple vector bundles over symplectic Kähler manifolds, Proc. Japan Acad. 62 (1986), 21–24.

—— [11], Submersions of *CR* submanifolds, Tôhoku Math. J. 39 (1987).

S. Kobayashi and K. Nomizu [1], Foundations of Differential Geometry, Intersci. Tracts Pure & Appl. Math. 15, Vol. 1 (1963), Vol. 2 (1969), John Wiley & Sons.

S. Kobayashi and T. Ochiai [1], On complex manifolds with positive tangent bundles, J. Math. Soc. Japan 22 (1970), 499–525.

S. Kobayashi and H. Wu [1], On holomorphic sections of certain hermitian vector bundles, Math. Ann. 189 (1970), 1–4.

K. Kodaira [1], Complex Manifolds and Deformations of Complex Structures, Grundlehren Math. Wiss. 283, Springer-Verlag, 1985.

N. Koiso [1], Einstein metrics and complex structures, Invent. Math. 73 (1983), 71–106.

M. Kuranishi [1], Deformations of Compact Complex Manifolds, Sem. Math. Sup. Été 1969, Presse Univ. Montreal, 1971.

—— [2], New proof for the existence of locally complete families of complex structure, Proc. Conf. Complex Analysis, Minneapolis, Springer-Verlag 1965, pp. 142–154.

S. Langton [1], Valuative criteria for families of vector bundles on algebraic varieties, Ann. of Math. 101 (1975), 88–110.

A. Lascoux and M. Berger [1], Variétés Kähleriennes Compactes, Lecture Notes in Math. 154, Springer-Verlag, 1970.

J. Le Potier [1], Théorèmes d'annulation en cohomologie, C. R. Acad. Sci. Paris 276 (1973), 535–537.

P. Lelong [1], Fonctions plurisousharmoniques et formes differentielles positives, Gordon and Breach, 1968.

A. Lichnerowicz [1], Variétés kähleriennes à premiere classe de Chern non négative et variétés riemanniennes à courbure de Ricci généralisée non négative, J. Diff. Geometry 6 (1971), 47–94.

M. Lübke [1], Hermite-Einstein-Vektorbündel, Dissertation, Bayreuth, 1982.

—— [2], Chernklassen von Hermite-Einstein-Vektorbündeln, Math. Ann. 260 (1982), 133–141.

—— [3], Stability of Einstein-Hermitian vector bundles, Manuscripta Math. 42 (1983), 245–257.

M. Lübke and C. Okonek [1], Moduli of simple semi-connections and Einstein-Hermitian bundles, preprint.

—— [2], Stable bundles and regular elliptic surfaces, preprint.

J. Marsden and T. Ratiu, Hamiltonian Systems with Symmetry, Fluids and Plasmas, to appear as a book.

J. Marsden and A. D. Weinstein [1], Reduction of symplectic manifolds with symmetry, Reports on Math. Physics 5 (1974), 121–130.

M. Maruyama [1], Stable vector bundles on an algebraic surface, Nagoya Math. J. 58 (1975), 25–68.

—— [2], Openness of a family of torsionfree sheaves, J. Math. Kyoto Univ. 16 (1976), 627–637.

—— [3], On boundedness of families of torsionfree sheaves, J. Math. Kyoto Univ. 21 (1981), 673–701.

—— [4], Moduli of stable sheaves I & II, J. Math. Kyoto Univ. 17 (1977), 91–126 & ibid. 18 (1978), 557–614.

—— [5], On a family of algebraic vector bundles, Number Theory, Algebraic Geometry & Commutative Algebra, in honor of Y. Akizuki, Kinokuniya, Tokyo (1973), 95–146.

—— [6], The theorem of Grauert-Mülich-Spindler, Math. Ann. 255 (1981), 317–333.

H. Matsumura [1], Commutative Algebra, Benjamin/Cummings Publ. Co. 1980.

Y. Matsushima [1], Heisenberg groups and holomorphic vector bundles over a complex torus, Nagoya Math. J. 61 (1976), 161–195.

—— [2], Fibrés holomorphes sur un tore complexe, Nagoya Math. J. 14 (1959), 1–24.

V. B. Mehta [1], On some restriction theorems for semistable bundles, "Invariant Theory", Lecture Notes in Math. 996, Springer-Verlag, 1983, pp. 145–153.

V. B. Mehta and A. Ramanathan [1], Semi-stable sheaves on projective varieties and their restriction to curves, Math. Ann. 258 (1982), 213–224.

—— [2], Restriction of stable sheaves and representations of the fundamental group, Invent. Math. 77 (1984), 163–172.

J. W. Milnor and J. D. Stasheff [1], Characteristic Classes, Ann. of Math. Studies 76, Princeton Univ. Press, 1974.

P. K. Mitter and C. M. Viallet [1], On the bundle of connections and the gauge orbit manifold in Yang-Mills theory, Comm. Math. Physics 79 (1981), 457–472.

Y. Miyaoka [1], On the Chern numbers of surfaces of general type, Invent. Math. 32 (1977), 225–237.

M. Miyanishi [1], Some remarks on algebraic homogeneous vector bundles, Number Theory, Algebraic Geometry & Commutative Algebra, in honor of Y. Akizuki, Kinokuniya, Tokyo (1973), 71–93.

H. Morikawa [1], A note on holomorphic vector bundles over complex tori, Nagoya Math. J. 41 (1971), 101–106.

A. Morimoto [1], Sur la classification des espaces fibrés vectoriels holomorphes sur un tore complexe admettant des connexions holomorphes, Nagoya Math. J. 15 (1959), 83–154.

J. Morrow and K. Kodaira [1], Complex Manifolds, Holt, Reinehart & Winston, 1971.

S. Mukai [1], Symplectic structure of the moduli space of sheaves on an abelian or $K3$ surface, Invent. Math. 77 (1984), 101–116.

—— [2], Semi-homogeneous vector bundles on an abelian variety, J. Math. Kyoto Univ. 18 (1978), 239–272.

D. Mumford and J. Fogarty [1], Geometric Invariant Theory, 2nd ed., Springer-Verlag, 1982.

S. Nakano [1], Vanishing theorems for weakly 1-complete manifolds, Number Theory, Algebraic Geometry & Commutative Algebra, in honor of Y. Akizuki, Kinokuniya, 1973, 169–179; II, Publ. Res. Inst. Math. Sci., Kyoto Univ. 10 (1974), 101–110.

M. S. Narasimhan and C. S. Seshadri [1], Stable and unitary vector bundles on compact

Riemann surfaces, Ann. of Math. 82 (1965), 540–567.

A. Newlander and L. Nirenberg [1], Complex analytic coordinates in almost complex manifolds, Ann. of Math. 65 (1957), 391–404.

P. E. Newstead [1], Rationality of moduli spaces of stable bundles, Math. Ann. 215 (1975), 251–268; correction: 249 (1980), 281–282.

—— [2], Introduction to Moduli Problems and Orbit Spaces, Tata Inst. Lecture Notes 51, Springer-Verlag, 1978.

S. Nishikawa and T. Ochiai [1], Existence of Harmonic Maps and Applications, Reports on Global Analysis (in Japanese), 1980.

V. A. Norton [1], Analytic moduli of complex vector bundles, Indiana Univ. Math. J. 28 (1978), 365–387.

—— [2], Nonseparation in the moduli of complex vector bundles, Math. Ann. 235 (1978), 1–16.

N. R. O'Brian, D. Toledo and Y. L. L. Tong [1], The trace map and characteristic classes for coherent sheaves, Amer. J. Math. 103 (1981), 225–252.

—— [2], Hirzebruch-Riemann-Roch for coherent sheaves, Amer. J. Math. 103 (1981), 253–271.

T. Oda [1], Vector bundles on abelian surfaces, Invent. Math. 13 (1971), 247–260.

T. Ohsawa [1], Isomorphism theorems for cohomology groups of weakly 1-complete manifolds, Publ. Res. Inst. Math. Sci. Kyoto Univ. 18 (1982), 191–232.

C. Okonek, M. Schneider and H. Spindler [1], Vector Bundles on Complex Projective Spaces, Progress in Math. 3 (1980), Birkhäuser.

S. Ramanan [1], Holomorphic vector bundles on homogeneous spaces, Topology 5 (1966), 159–177.

A. Ramanathan [1], Stable principal bundles on a compact Riemann surface, Math. Ann. 213 (1975), 129–152.

C. P. Ramanujam [1], Remarks on the Kodaira vanishing theorem, J. Indian Math. Soc. 36 (1972), 41–51.

M. Raynaud [1], Fibrés vectoriels instables — applications aux surfaces, (d'après Bogomolov), "Surfaces Algébriques", Lecture Notes in Math. 868, Springer-Verlag, 1981, pp. 293–314.

M. Reid [1], Bogomolov's theorem $c_1^2 < 4c_2$, Proc. Intl. Symp. on Alg. Geometry, Kyoto, Kinokuniya Book Co. 1977, pp. 623–642.

G. Rousseau [1], Instabilité dans les espaces vectoriels, "Surfaces Algébriques", Lecture Notes in Math. 868, Springer-Verlag, 1981, pp. 277–292.

G. Scheja [1], Fortsetzungssätze der komplex-analytischen Cohomologie und ihre algebraische Charakterisierung, Math. Ann. 157 (1964), 75–94.

M. Schneider [1], Ein einfacher Beweis des Verschwindungssatzes für positive holomorphe Vektorbündel, Manuscripta Math. 11 (1974), 95–101.

—— [2], Chernklassen semi-stabiler Vektorraumbündel vom Rang 3 auf dem komplexen projektiven Raum, Crelles J. 315 (1980), 211–220.

G. Schumacher [1], On the geometry of moduli spaces, Manuscripta Math. 50 (1985), 229–267.

J.-P. Serre [1], Faisceaux algébriques cohérents, Ann. of Math. 61 (1955), 197–278.

—— [2], Un théorème de dualité, Comment. Math. Helv. 29 (1955), 9–26.

C. S. Seshadri [1], Fibrés Vectoriels sur les Courbes Algébriques, Astérisque 96, Soc. Math. France, 1982.

—— [2], Theory of moduli, Algebraic Geometry - Arcata 1974, Proc. Symp. Pure Math. 29, Amer. Math. Soc., 1975, 263–304.

—— [3], Space of unitary vector bundles on a compact Riemann surfaces, Ann. of Math. 85 (1967), 303–336.

S. Shatz [1], The decomposition and specialization of algebraic families of vector bundles, Compositio Math. 35 (1977), 163–187.

B. Shiffman and A. J. Sommese [1], Vanishing Theorems on Complex Manifolds, Progress in Math. 56, Birkhäuser, 1985.

Y. T. Siu and G. Trautmann [1], Gap-Sheaves and Extension of Coherent Analytic Subsheaves, Lecture Notes in Math. 172, Springer-Verlag, 1971.

S. Soberon-Chavez [1], Rank 2 vector bundles over a complex quadric surfaces, Quart. J. Math. 36 (1985), 159–172.

H. Spindler [1], Die Modulräume stabiler 3-Bündel auf P_3 mit den Chern-klassen $c_1=0$, $c_3=c_2^2-c_2$, Math. Ann. 256 (1981), 133–143.

K. Takegoshi and T. Ohsawa [1], A vanishing theorem for $H^p(X, \Omega^q(B))$ on weakly 1-complete manifolds, Publ. Res. Inst. Math. Sci. Kyoto Univ. 17 (1981), 723–733.

F. Takemoto [1], Stable vector bundles on algebraic surfaces, Nagoya Math. J. 47 (1973), 29–48; II, ibid. 52 (1973), 173–195.

C. H. Taubes [1], Stability in Yang-Mills Theories, Commum. Math. Phys. 91 (1983), 235–263.

A. N. Tjurin [1], The geometry of moduli of vector bundles, Uspehi Math. Nauk. 29 (1974), 59–88 (=Russian Math. Survey 29 (1974) 57–88).

G. Trautmann [1], Moduli for vector bundles on $P_n(C)$, Math. Ann. 237 (1978), 167–186.

—— [2], Zur Berechnung von Yang-Mills Potentialen durch holomorphe Vektorbündel, (Proc. Conf., Nice, 1979), pp. 183–249, Progress in Math., 7 Birkhäuser, 1980.

K. K. Uhlenbeck [1], Connections with L^p bounds on curvature, Comm. Math. Phys. 83 (1982), 31–42.

—— [2], Removable singularities in Yang-Mills fields, Comm. Math. Phys. 83 (1982), 11–30.

H. Umemura [1], Some results in the theory of vector bundles, Nagoya Math. J. 52 (1973), 97–128.

—— [2], A theorem of Matsushima, Nagoya Math. J. 54 (1974), 123–134.

—— [3], Stable vector bundles with numerically trivial Chern classes over a hyperelliptic surface, Nagoya Math. J. 59 (1975), 107–134.

—— [4], On a certain type of vector bundles over an abelian variety, Nagoya Math. J. 64 (1976), 31–45.

—— [5], On a theorem of Ramanan, Nagoya Math. J. 69 (1978), 131–138.

—— [6], Moduli spaces of the stable vector bundles over abelian surfaces, Nagoya Math. J. 77 (1980), 47–60.

J. L. Verdier [1], Le théorème de le Potier, Astérisque 17 (1974), 68–78.

E. Vesentini [1], Osservazioni sulle strutture fibrate analitiche sopra una varieta kähleriana compatta, Nota I e II, Atti Accad. Naz. Lincei Rend. Cl. Sci. Mat. Natur. (8) XXIV (1958), 505–512; ibid. XXIV (1957), 232–241.

E. Viehweg [1], Vanishing theorems, Crelles J. 335 (1982), 1–8.

H. Wakakuwa [1], On Riemannian manifolds with homogeneous holonomy group *Sp*(*n*), Tôhoku Math. J. 10 (1958), 274–303.

A. Weil [1], Introduction à l'Étude des Variétés Kähleriennes, Hermann, 1958.

H. Wu [1], The Bochner technique, Proc. 1980 Beijin Symp. Diff. Geom. & Diff. Eq's, Science Press, Beijin (Gordon & Breach), 1984, pp. 929–1071.

J. H. Yang [1], Einstein-Hermitian vector bundles, Ph. D. thesis, Berkeley, 1984.

K. Yano and S. Bochner [1], Curvature and Betti Numbers, Annals of Math. Studies 32, Princeton Univ. Press, 1953.

S. T. Yau [1], On the Ricci curvature of a compact Kähler manifold and the complex Monge-Ampère equation, I. Comm. Pure & Appl. Math. 31 (1978), 339–411.

A. Zandi [1], Quaternionic Kähler manifolds and their twistor spaces, Ph. D. thesis, Berkeley, 1984.

Index

Notations

Throughout the book:

M	Usually a complex manifold, sometimes a real manifold, except in Sections 3 & 4 of Chapter V where it is a module.
g	an Hermitian metric on M, usually a Kähler metric.
Φ	the Kähler form corresponding to g, see p. 27.
E	a complex (often holomorphic) vector bundle over M.
h	an Hermitian structure in a vector bundle E.
A^r	the space of complex r-forms on M.
$A^r(E)$	the space of complex r-forms on M with values in E.
$A^{p,q}$	the space of (p, q)-forms on M.
$A^{p,q}(E)$	the space of (p, q)-forms on M with values in E.
D	a connection in a vector bundle E and also the exterior covariant differentiation $A^r(E) \to A^{r+1}(E)$ defined by the connection, see p. 1, p. 3.
D', D''	the $(1,0)$ and $(0,1)$ components of D so that $D = D' + D''$, see p. 8.
$\omega = (\omega^i_j)$	the connection form corresponding to D, see p. 1, p. 9.
$\delta' = D'^*$	the adjoint operator of D'. δ' is used in Chapter III while D'^* is used in Chapter VII, see p. 62.
$\delta''_h = D''^*$	the adjoint operator of D''. δ''_h is used in Chapter III while D''^* is used in Chapter VII, see p. 64.
$R = D \circ D$	the curvature of a connection D, see p. 3, p. 8.
$\Omega = (\Omega^i_j)$	the curvature form corresponding to R, p. 3, p. 8.
K	the mean curvature transformation of E, (the trace of R with respect to g), see p. 26, p. 51.
$\hat{K}$	the mean curvature form on E corresponding to K, see p. 26, p. 51.
$c_i(E, h)$	the i-th Chern form of an Hermitian vector bundle (E, h), see p. 41.
$\det E$	the determinant line bundle of a vector bundle E, see p. 18.
$\det \mathscr{S}$	the determinant line bundle of a coherent sheaf $\mathscr{S}$, see p. 162.
$c_1(\mathscr{S})$	$= c_1(\det \mathscr{S})$.
$\deg(E)$	$= \int_M c_1(E) \wedge \Phi^{n-1}$, see p. 55, p. 133.
$\deg(\mathscr{S})$	$= \int_M c_1(\mathscr{S}) \wedge \Phi^{n-1}$, see p. 168.
$\mu(E)$	$= \deg(E)/\mathrm{rank}(E)$, see p. 134.
$\mu(\mathscr{S})$	$= \deg(\mathscr{S})/\mathrm{rank}(\mathscr{S})$, see p. 168.

In Chapter V:

$\mathrm{dh}(M)$	homological dimension of a module M, see p. 144.
$\mathrm{codh}(M)$	homological codimension of a module M, see p. 145.
$S_m(\mathscr{S})$	m-th singularity set of a coherent sheaf $\mathscr{S}$, see p. 155.

In Chapter VI:

$\mathrm{Herm}(r)$	the space of $r\times r$ Hermitian matrices, see p. 193.
$\mathrm{Herm}^+(r)$	the space of $r\times r$ positive definite Hermitian matrices, see p. 193.
$\mathrm{Herm}(E)$	the space of Hermitian forms in a vector bundle E, see p. 196.
$\mathrm{Herm}^+(E)$	the space of Hermitian structures in E, see p. 196.
$L(h, k)$	Donaldson's Lagrangian, see p. 198.

In Chapter VII:

$GL(E)$	the group of C^∞ bundle automorphisms of E, see p. 238.
$U(E, h)$	the subgroup of $GL(E)$ preserving h, see p. 239.
$\mathrm{End}^0(E)$	the subbundle of the endomorphism bundle $\mathrm{End}(E)$ consisting of tracefree endomorphisms of E, see p. 250.
$\mathrm{End}(E, h)$	the bundle of skew-Hermitian endomorphisms of (E, h), see p. 240.
$\mathrm{End}^0(E, h)$	the bundle of tracefree skew-Hermitian endomorphisms of (E, h), see p. 250.
$E^{D''}$	a complex vector bundle E with the holomorphic structure given by D'', see p. 244.
$\mathscr{D}''(E)$	the set of $D''\colon A^0(E)\to A^{0,1}(E)$ satisfying (VII.1.1), see p. 238.
$\mathscr{H}''(E)$	the set of holomorphic structures on E, i.e., the subset of $\mathscr{D}''(E)$ consisting of D'' such that $D''\circ D''=0$, see p. 239.
$\mathscr{D}(E, h)$	the set of connections D in E preserving h, see p. 239.
$\mathscr{H}(E, h)$	$=\{D\in\mathscr{D}(E, h);\ D''\circ D''=0\}$, see p. 240.
$\mathscr{E}(E, h)$	the subset of $\mathscr{H}(E, h)$ consisting of Einstein-Hermitian connections, see p. 242.
$\hat{\mathscr{H}}''(E)$	the set of simple holomorphic structures in E, see p. 261.
$\hat{\mathscr{E}}(E, h)$	the set of irreducible Einstein-Hermitian connections in (E, h), see p. 266.

Library of Congress Cataloging-in-Publication Data

Kobayashi, Shoshichi, 1932-
Differential geometry of complex vector bundles.

(Publications of the Mathematical Society of Japan; 15. Kāno memorial lectures ; 5)
Bibliography: p.
Includes index.
1. Vector bundles. 2. Vanishing theorems. I. Title.
II. Series: Publications (Nihon Sūgakkai) ; 15.
III. Series: Publications (Nihon Sūgakkai). Kanō memorial lectures ; 5.
QA612.63.K63 1987 516.3′62 87-45526
ISBN 0-691-08467-X (Princeton)